MW00787895

Firearms in American History

Firearms in American History

A Guide for Writers, Curators, and General Readers

Charles G. Worman

WESTHOLME
Yardley

Frontispiece: "Charles G. Alexander, King George County, Virginia. Taken on his way to California. April 1849." He holds a Model 1841 Mississippi rifle and has what appears to be a pepperbox thrust in his belt. (*Courtesy: Herb Peck, Jr.*)

Westholme Publishing, LLC
Eight Harvey Avenue
Yardley, Pennsylvania 19067
Visit our Web site at www.westholmepublishing.com

First Edition: June 2007
10 9 8 7 6 5 4 3 2 1

ISBN: 978-1-59416-044-8
ISBN 10: 1-59416-044-9

Printed in United States of America

For Michael, Steven, Heidi, Tyler, Madison, Max, and Davi. May they always be blessed and have more friends than their homes can hold.

Contents

Preface

My primary intent in writing this book is to offer a very general history of firearms in the United States, the years from colonization to the end of the nineteenth century—a survey that as a college course perhaps could be titled "Firearms in American History 101." The gun, like the surveyor's compass, the axe, and the plow, was a vital element in the exploration, settlement, and preservation of the United States. It was a necessary tool, for it defended life and property against two- or four-legged marauders and also put food on the table. It settled disputes between individuals and armies which unfortunately could not be resolved in a peaceful manner and even provided recreational opportunities. Thus it was present in many aspects of the first four centuries of American life.

This book also is intended to fulfill a second purpose as an aid to writers. Because of the gun's close ties with early United States history, the use of firearms is rather frequently mentioned in both fiction and non-fiction writing, but sometimes by authors who are not sufficiently familiar with firearms history to avoid mistakes. One I recall was a writer's statement that the typical trapper or mountain man's rifle in the early 1800s was between .28 and .40 caliber. Yet a ball under about .40 caliber would have little more effect on a bison or grizzly bear that a nuisance bite from an insect. Another misconception is a rather widespread belief that America's Revolutionary War was won in large measure by rifle-armed backwoodsmen firing on their red-coated enemy from behind trees and rocks. But it just wasn't so.

While a writer can't be an expert in all fields, an attempt at achieving accuracy whenever possible with a little background research is certainly desirable and advantageous. I hope that this book will help a novelist or writer of short stories to make sure a character picks up a gun that is appropriate to the setting and that the character handles it correctly. For the writer of nonfiction, I hope this book will provide a guide to the role of firearms in whichever period of pre-1900 American history is under discussion.

If as an author you rely on this book and your setting is the South during the War Between the States, Civil War, or War of Northern Aggression—whatever you choose to call it—you'll feel confident arming your Rhett Butler with an English revolver such as a Tranter imported by either Hyde & Goodrich or A. B. Griswold & Co., both of New Orleans, and allowing your Scarlett O'Hara to slip a compact four-barrel Sharps pepperbox pistol into her handbag. You'd show your knowledge of arms of the west if you mention a trapper or mountain man of the 1830s carrying a spare wooden ramrod in the barrel of his rifle, or of the slaughter of the bison by your reference to a Kansas hunter in the early 1870s purchasing his 12-pound Sharps rifle from F. C. Zimmerman of Dodge City. Throughout this book, I've included such assists to aid writers in giving credibility to their work.

This book doesn't dwell on those many details to be found in available books on specialized topics but should help place the evolution of firearms development and use in this country in perspective. Many of the titles listed in the bibliography offer far more information on various elements of the story. But even though this volume is intended more for the individual with only a casual interest in or modest knowledge of antique guns and their role in pre-1900 American life, I hope the more knowledgeable individuals can also find useful information.

I *From Matchlock to Flintlock*

The Era of Colonization (1500–1775)

Arms of the Early Colonists

Beginning in the early 1500s, Spain undertook to explore and establish colonies in what today is the United States, establishing outposts in Virginia, North Carolina, Florida, along the Gulf Coast, and in the southwest. France followed in the 1560s with posts in South Carolina and Florida and then in Canada. The Swedes, Dutch, and English weren't far behind. For all of these early European arrivals, life was a battle against an uncompromising environment, sometimes the American Indians, and occasionally the inhabitants of competing nations' colonies. Regardless of the colonists' country of origin, guns became crucial to survival whether in securing food or providing protection from two- or four-legged antagonists.

Even before colonization of our country began, the first portable firearm ("hand cannon") had already appeared—a simple hollow tube of iron or bronze sealed at one end and loaded from the muzzle with powder and a projectile. To ignite the powder charge, a glowing wire or smoldering bit of tinder was inserted through a vent (touch hole) passing into the barrel at the closed end. When these crude weapons first arrived on the scene is debatable, but they were present in Europe by 1340 or perhaps even a few decades earlier. Christopher Columbus in his second voyage to the New World (1493) is known to have brought at least 100 guns, probably in the form of hand cannon.[1]

The initial hand cannon were simple, although some later were fitted with crude sights and mounted on the end of a short pole or a form of wooden stock, but they had features in common with other guns which still were being loaded from the muzzle as late as the mid-1800s. Three necessary ingredients were propellant, a projectile, and a means of igniting the former. Black powder (its name derived from the charcoal it contained) was composed of varying proportions of saltpeter, charcoal, and sulfur. It was a rather unstable mixture, hygroscopic (it took up water), gave off a cloud of smoke revealing the shooter's position and sometimes obscuring one's vision on a windless day, and left a nasty residue which could foul a barrel or clog the vent. Black powder with its faults was not replaced by so-called smokeless powder until the late 1800s.

Unless one cleaned the barrel occasionally when shooting, powder fouling as it built up made it increasingly difficult to ram the projectile down the barrel to seat against the powder charge. Firing a gun with the ball positioned only partway down the barrel was a dangerous proposition since the excessive pressure thus created could cause the barrel to rupture. Any buildup of powder residue in

1. M. L. Brown, *Firearms in Colonial America: The Impact on History and Technology, 1492-1792* (Washington, D.C., 1980), 36.

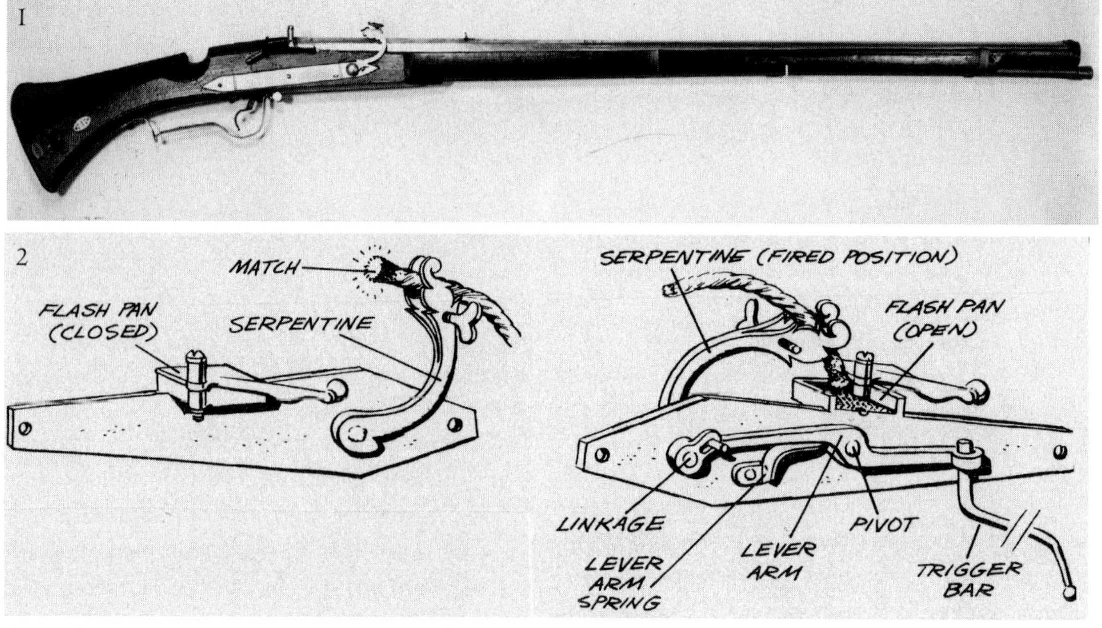

1. Matchlock arquebus from about 1600. (*Courtesy: National Park Service, Fuller Gun Collection, Chickamauga and Chattanooga National Military Park*) 2. Matchlock mechanism. (*Courtesy: George Neumann from Battle Weapons of the American Revolution*)

the touch hole could be cleared with a short wire or vent pick, which soon became a common shooting accessory used for centuries. Although projectiles of various materials were tried initially, a round lead ball eventually proved best. Lead is soft and has a relatively low melting point making it easy to mold into shape, even in a rustic cabin's fireplace.[2]

When gunpowder failed to ignite and the gun misfired or if in the field one carelessly rammed the ball down the barrel before the powder, an obvious question is how the load was removed from any firearm which loaded from the muzzle. If the barrel had a screw-in plug at the rear end and one had access to a bench vise, the remedy was not particularly complicated. Otherwise the answer was another simple device used throughout the centuries of muzzle loaders, a corkscrew-like worm or a ball screw. Either was threaded to screw onto the end of the iron or wooden ramrod. After thrusting it down the barrel and screwing it into the offending soft lead ball or wadding over a charge of shot, the load hopefully could be pulled out with minimal effort. Similarly, a "scourer" to scrape the inside of the barrel or to which a piece of cloth or tow (a tuft of flax or hemp) could be affixed to remove powder residue and a little later a brush to clean residue from the priming pan were other early accessories.

The Matchlock

Aside from the crude hand cannon, projectile-firing arms brought to North America with the early explorers and colonists were the crossbow and the matchlock gun. Distance sometimes was recorded in terms of "crossbow shots," and such Spanish explorers in the south and west as De Vaca, Coronado, and De Soto in the 1500s incorporated

2. Both gunpowder and lead were scarce in the seventeenth-century English colonies and had to be imported from Europe. Bullets were sometimes used as barter, and in Boston about 1630 each musket ball was worth about a farthing.

substantial numbers of crossbowmen in their ranks. Crossbows were more accurate and projected an arrow (more properly called a "bolt" or "quarrel") with greater force than a hand-drawn bow. The most powerful examples, however, required time to reload and prepare for firing, allowing perhaps no more than one shot a minute, insufficient when the target was fast moving. A necessary crossbow accessory was a crank or windlass-like affair to draw back the bow string or a separate "goat's foot" lever to cock less powerful, faster firing examples (Figures 1, 2).

By the mid-1500s, the matchlock gun was gaining predominance over the crossbow in Europe and in the New World. Its mechanism was rather simple although it was refined as time passed. It incorporated a cord or match, soaked in a solution of saltpeter and dried, and which in the simplest form of matchlock was held in a "serpentine" shaped somewhat like a backward "C" or an "S" and secured to the side of the stock. Pressure on the other end of the serpentine lowered the smoldering match into contact with the priming powder contained in the small pan or trough at the side of the barrel. When the priming ignited, a flame passed through the barrel's touch hole or vent to ignite the main powder charge in the barrel. The cord usually was made of twisted strands of cotton or hemp and burned at a variable rate, often about four or five inches an hour. Various contemporary terms referred to different forms of the matchlock, but "musket" usually meant a heavy military piece, often 16 to 20 pounds in weight requiring a separate forked rest to support the barrel when firing. More convenient were the lighter arquebus and caliver or the shorter carbine-like petronel (Figure 3).[3]

The process of preparing to fire the awkward matchlock was a rather tedious one. To minimize the risk of igniting loose powder, the match, kept lit at each end, was removed from the serpentine before a charge of powder was poured into the muzzle, often from one of a number of wooden cylinders suspended from a bandolier. A round ball, a little smaller than the diameter of the bore, then was taken from a pouch and followed (a soldier or hunter might hold several in his mouth in readiness), and finally a wad of paper or tow was pushed down the barrel with the ramrod to keep the ball in place. From a separate flask, the shooter poured a charge of finer grain priming powder into the pan, then closed the cover over the pan before replacing the cord in the serpentine. The shooter then blew on the glowing end and uncovered the pan before bringing the match into contact with the priming powder. The undersize ball literally bounced from side to side as it passed down the barrel, losing velocity as some of the gas blew past it.

3. Drawings from the 1640s of a French musketeer with matchlock gun, cord or match, bandoleer of wooden cylinders containing powder charges, and a forked rest. (*Courtesy: S. James Gooding, The Canadian Gunsmiths 1608-1900*)

In addition to its slow rate of fire, the matchlock had several drawbacks. The smoldering match gave one's position away in darkness, wet weather could render the gun inoperable, and the danger of accidental ignition of powder was always present. Captain John Smith of the Jamestown

3. One of the firearms in the Pilgrim Hall museum collection at Plymouth, Massachusetts, is an Italian matchlock musket of late sixteenth-century origin which is thought to have been used in that English colony.

colony in Virginia was seriously burned when powder he was carrying was inadvertently ignited by a lit match. In 1609, heavy rain rendered the matchlocks useless when Indians threatened a Dutch party under Henry Hudson. Also, the musketeer had to adjust the match in the serpentine as it slowly burned and there needed to be a source of fire available in case the match went out. In warfare, the Indian's bow and arrow was a more convenient weapon, but the impact of the musket ball was substantial when it struck, and the smoke and noise of a matchlock initially had a profound psychological impact on Indian foes unfamiliar with it.

Despite its disadvantages, the matchlock musket was an essential element in maintaining the security of the colonies in the face of intermittent conflict with American Indians, conflict that sometimes resulted from the colonists' own treachery. In 1638, for example, the Rhode Island colony enacted militia laws which required each adult male to provide himself with a musket, one pound of powder, 20 bullets, about 12 feet of match cord, sword, bandoliers, and a rest for his musket. In England by the mid-1600s, the caliber of the military matchlock musket was standardized at 10 gauge or three-quarters of an inch (.75 caliber), a bore size later carried over in the famed English "Brown Bess" flintlock musket. Some more sophisticated matchlock muskets remained in continued although diminished use in the colonies as late as the 1670s; others were updated with the substitution of more efficient lock mechanisms. A 1676 shipment of arms to Virginia included 300 matchlock and 200 flintlock muskets. The use of the sturdy and inexpensive matchlock in Spanish territory in the south and southwest persisted even longer, into the early 1700s even after Spain officially adopted a form of flintlock in 1703.[4]

The Wheel Lock, a Choice for the Wealthy

A more reliable and faster firing gun than the matchlock appeared in Europe by the 1520s, the wheel lock. It operated somewhat on the principle of a cigarette lighter. A heavy mainspring was wound or compressed with a separate iron key or "spanner," sometimes quite artistically chiseled and often with one end shaped to serve as a screwdriver. When the trigger was pulled, a "dog head" or vise holding a piece of iron pyrites in its jaws came into contact with the spring-activated rotating serrated wheel to create sparks which ignited the priming powder in the pan. Due to the wheel lock mechanism's greater complexity and need for closer production tolerances, difficulty in repairing one, and expense, it never completely replaced the matchlock as a military arm. However, it often was applied to the sporting guns of the wealthy and made the pistol a practical weapon for the first time. Some of the most elaborate firearms found in museums today are wheel locks with exquisite ivory, pearl, and bone inlays and other embellishments. These were forerunners of ornate American Winchester, Marlin, Smith & Wesson, Colt, and other long and short guns of the mid- to late 1800s with gold inlays or intricate scenes engraved on their metal or wood surfaces (Figures 4, 5).[5]

Some wheel locks arrived in America, both in the form of the newly developed hand held pistol or as a shoulder arm. It is recorded that at least a few wheel locks were present in 1586 at the ill-fated English settlement on Roanoke Island. In 1598 Luis de Velasco led a Spanish expedition into New Mexico which established the town of Santa Fe and which carried 19 matchlocks and 19 wheel locks. Archaeologists also have uncovered several

4. Brown, *Firearms in Colonial America,* 227. Well into the nineteenth century, bore size sometimes was described as the number of round balls that could be cast from a pound of lead. For example 18 bore or 18 to the pound was about .64 caliber, 100 to the pound equated to .36 caliber, etc.

5. Before the growing popularity of gunflints, gunspalls made of chert produced the required sparks. Flint was less brittle and didn't have to be replaced after about 20 strikes as did the gunspall. Brown, *Firearms in Colonial America,* 79.

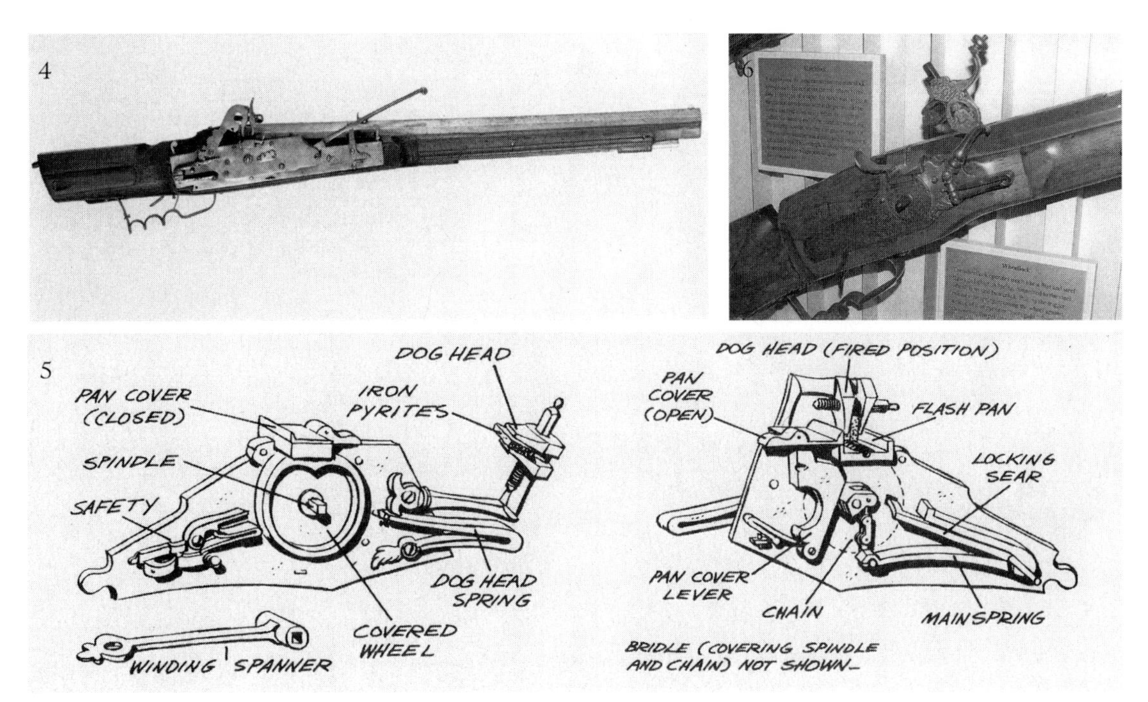

4. German wheel lock rifle dated 1591 with 1 1/2-inch bore. (*Courtesy: Smithsonian Institution, Ralph G. Packard Collection*) **5.** Interior and exterior views of a wheel lock mechanism. (*Courtesy: George Neumann from Battle Weapons of the American Revolution*) **6.** Closeup of the lock mechanism of a European wheel lock on exhibition at the Springfield Armory National Historic Site.

wheel lock mechanisms plus a few spanners at the site of the Jamestown, Virginia, colony founded in 1607. Among the firearms exhibited at the National Rifle Association's museum in Fairfax, Virginia, is a wheel lock carbine thought to have been owned by John Alden of Plymouth Colony, Massachusetts, and found when his family home was being restored (Figure 6).[6]

One of the first gunsmiths to arrive in the English colonies was Peter Keffer, who landed at Jamestown, Virginia, in 1608. He probably fulfilled a dual role as blacksmith. Others in his trade who arrived in the seventeenth-century generally concentrated their efforts on repairing existing firearms rather than attempting the more difficult task of producing them. It was not until the early 1700s that manufacturing efforts began more in earnest, following increased production of iron from ore found in the colonies and the establishment of foundries, forges, and blast furnaces. Even then such colonial-made arms often incorporated some components imported from Europe, particularly England, such as locks and barrels, making them more properly termed as American assembled rather than American made.

The Flintlock and Its Variations

Less complicated that the wheel lock was a mechanism which appeared in Europe by the mid-1500s and was applied to both civilian and military hand and shoulder guns and eventually evolved in France in the early 1600s into what we know today as the true flintlock. Differences in design resulted from regional preferences and evolutionary changes as the snaphaunce (Dutch), snaplock (Scandinavian), dog lock (English), and miquelet lock with its external mainspring. The latter was particularly

6. Ibid., 112.

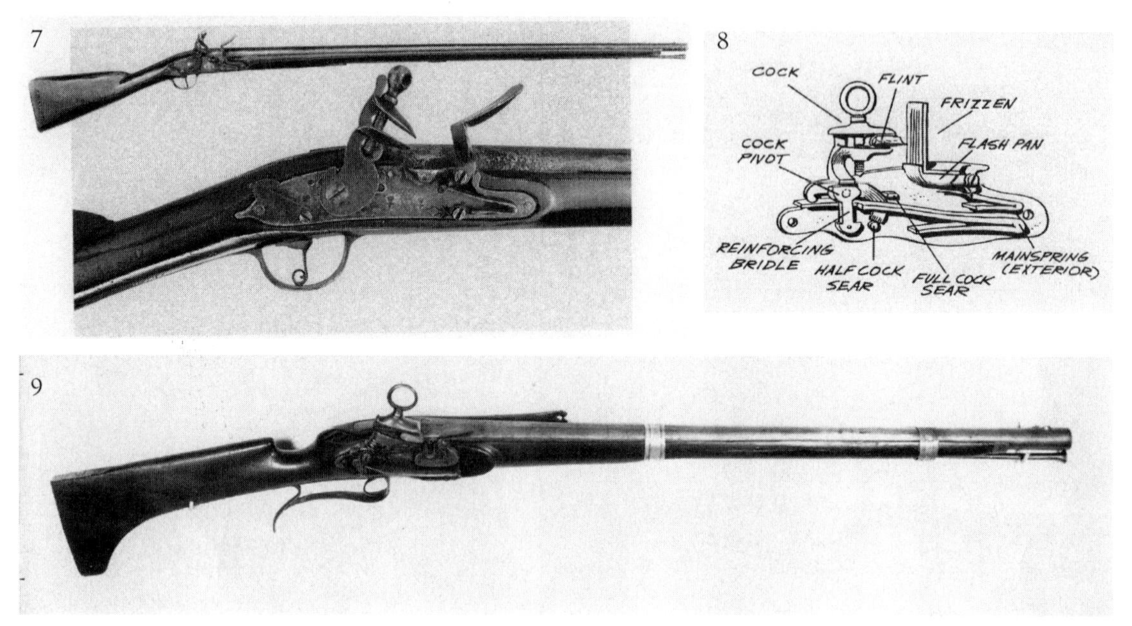

7. English "dog lock" version of a flintlock musket. Most common between the 1620s and 1670s, the catch behind the hammer engages an external hammer notch and acts as a primary or secondary safety to hold the hammer in position before being released. This one, with a 1707 lock, is branded "J. DANFORTH" on the left side of the stock, the name of a prominent Connecticut family of pewterers who came to America in the 1600s. The stock originally extended to the muzzle and used a plug bayonet but it was cut back about four inches in the early 1700s and a stud welded to the barrel to accept a socket bayonet which slipped over the muzzle. (*Courtesy: C. W. Slagle, photo by Ron Paxton*) **8.** Miquelet lock with exterior mainspring. (*Courtesy: George Neumann from Battle Weapons of the American Revolution*) **9.** A Spanish smoothbore escopeta with a miquelet lock with external mainspring. Such guns were common among soldiers and civilians alike in Spain's colonies in North America from the mid-1600s to the early 1800s. (*Courtesy: Arizona Historical Society*)

popular in the Spanish possessions in North America, often in the form of the light smoothbore *escopeta* used for hunting or for military use by militia (Figures 7, 8, 9, 10).

The basic principle of these various systems was the same. A piece of chert or later flint held in the jaws of the cock (today often called the hammer) created ignition sparks when a spring drove it in a downward arc against a vertical bar of steel. Thus the hazard and unreliability of the lighted match and the expensive complexity of the wheel lock were overcome. An early improvement probably made in England was the combining of the cover over the pan and the vertical "steel" into an L-shaped "battery" (later often termed "frizzen") rather than a separate sliding pan cover. As the cock struck the battery, tilting it forward, the priming powder in the pan was exposed to the sparks. It was this flintlock form which became most common in the English sea board colonies of North America (Figure 11).[7]

The flintlock generally was reliable except in inclement weather. As long as the vent was clear, the flint sharp and properly positioned, the frizzen dry and not oil covered, and the priming powder dry and of good quality, the system worked well. The modern expression "flash in the pan," referring to a dazzling but ineffective occurrence, was

7. Often a piece of leather (or a thin sheet lead cap on a heavy military flintlock) between the jaws of the cock helped hold the gunflint tightly in place.

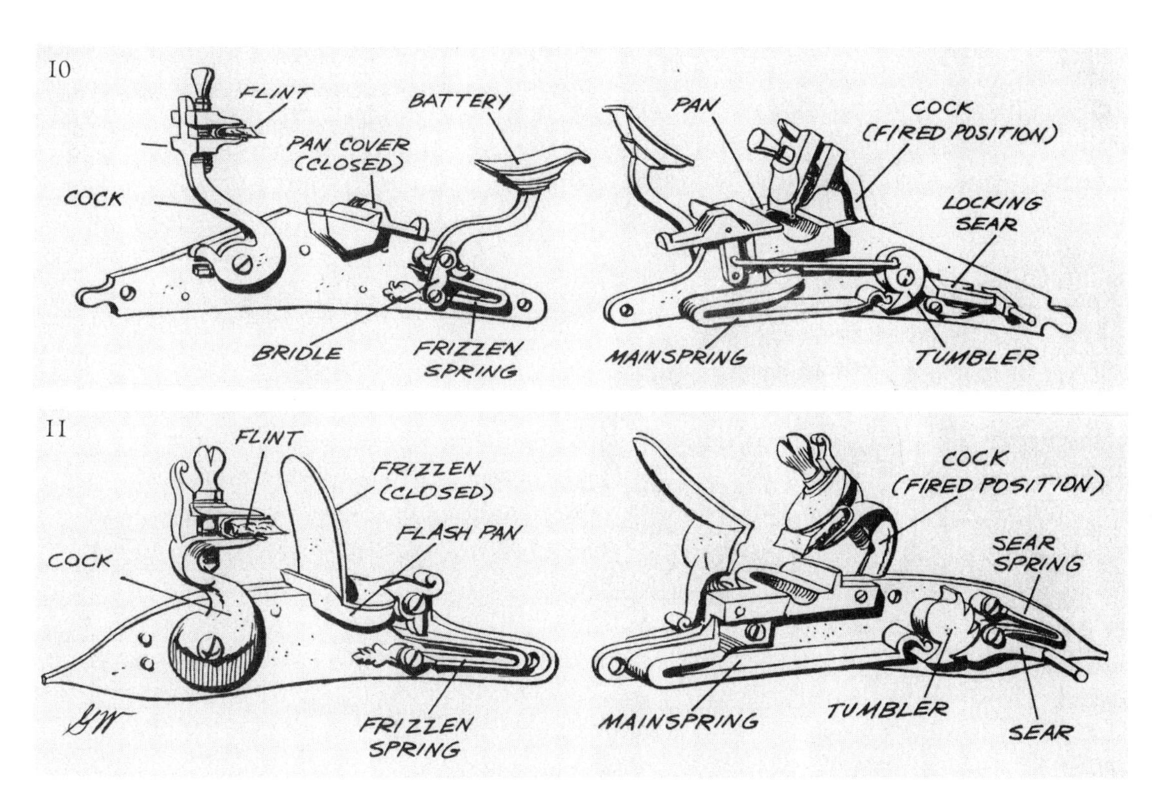

10. Snaphaunce lock with separate pan cover and frizzen or battery. **11.** True flintlock with combined frizzen and pan cover. (*Courtesy: George Neumann from Battle Weapons of the American Revolution*)

derived from a situation in which the priming powder ignited but failed to set off the main powder charge. A few innovative mid-eighteenth-century European makers even offered flintlocks with the lock on the underside or inverted at the side, removing the distraction of the flash near the shooter's face. Although the powder began to fall out of the pan as the hammer knocked the frizzen aside, sparks ignited the falling powder and a flame still passed through the vent. Modern tests have demonstrated the practicality of the design, although it never was widely accepted.[8]

In the era of the flintlock, an essential element was the flint itself. There were about nine standard sizes ranging upward from those for use in small pocket pistols to the largest size appropriate for a cannon lock. Flints were chipped into proper shape by workers known as "knappers" using knapping hammers. Knapping never became a significant colonial American industry and most flints were imported, those of a grayish color from continental Europe or black flints from England. A skilled knapper could chip out a good flint in about a minute; a good one might last for perhaps 50 or so firings, a poor one substantially fewer. On a long hunt, a civilian hunter had to carry spare flints; soldiers were often issued one per 20 cartridges, although this ratio varied. Flintlock muskets, sometimes referred to as "firelocks," remained in military use from about the mid-1600s for almost 200 years

8. Other expression derived from firearms include "lock, stock and barrel" (to indicate completeness), to go off "half cocked" (act impulsively), and "hair trigger temper" (referring to the ease with which a gun with an adjustable hair trigger can be discharged).

12. English military style brass barrel blunderbuss on a swivel, appropriate for mounting on a wall or the railing of a ship or small boat. It weighs 28 pounds and the bore diameter at the muzzle is 2 1/4 inches. **13.** English brass barrel blunderbuss made by a London gun maker named Green and owned by Ethan Allen, who in 1775 reportedly loaned it to Colonel Benedict Arnold when the two with a patriot band captured Fort Ticonderoga from the British. The stock is inscribed with Allen's name. (Exhibited at Fort Ticonderoga, New York)

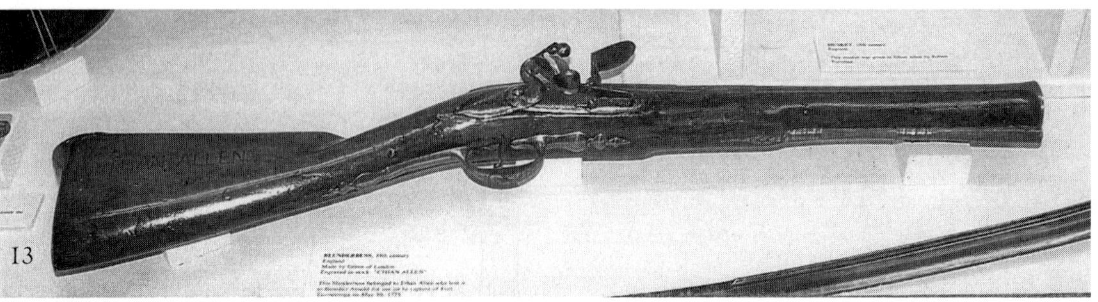

in Europe and from the mid- to late 1600s forward as well in English and French colonies in the New World. In civilian and military service the flintlock could eventually be found in all forms of firearms from the pocket pistol to the musket to the short, large-bore blunderbuss with its flared muzzle.

The blunderbuss is one of the better known types of firearms of the flintlock and into the percussion era. Some believe with its flared muzzle it could be loaded with irregular bits of scrap iron, glass, or other debris. But even though small pebbles probably could have been used as projectiles in an emergency, the intended load for this early version of a "riot shotgun" was a measured charge of round lead shot. Any projectiles other than those of soft lead would have worn the bore excessively, particularly if the barrel was of soft brass, as was often the case. There has been some controversy as to how effective the bell-shaped or oval muzzle was in spreading the pattern of shot, but firing tests have shown that the closer to the rear of the barrel the expansion of the bore began the more the pattern of the shot expanded. If there was no increase in bore size except near the muzzle, there was little spread, although the wide muzzle did facilitate loading as in the swaying rigging of a ship or on a bouncing mail coach and the psychological advantage over the highwayman or other at whom a blunderbuss was pointed can't be ignored (Figure 12).

The blunderbuss, occasionally fitted with a folding bayonet mounted above the barrel, saw little use in colonial America before the 1700s. Thus those November calendar scenes portraying a Pilgrim father in the 1620s carrying a bell-mouthed blunderbuss over his shoulder are in error. A noteworthy brass barrel flintlock blunderbuss made about 1785 by H. W. Mortimer of London has a spring bayonet and the appropriate words inscribed around the flared muzzle "HAPPY HE THAT ESCAPES ME." The blunderbuss style muzzle occasionally was applied to pocket or larger holster-size pistols as well, these too sometimes equipped with a folding bayonet or dagger (Figure 13).

A news item in the November 11, 1768, edition of England's *Bath Journal* illustrated the blunderbuss's usefulness in defeating a highwayman's threat.

"This morning about one of the clock, between Hounslow and Staines, the Bath to Bristol coach was attacked by a lone highwayman. The guard, having observed him coming up and thinking him suspicious, made ready his blunderbuss with shot, and fired. A slug lodged in the body of the villain who fell from his horse and expired."[9]

Military Arms in the English Colonies

In America from 1689 until the end of the French and Indian War in 1763, English colonists were drawn into a series of wars between France and England, each nation with its Indian allies joining in the conflicts. The competing armies came to rely largely on the flintlock smoothbore musket, quick to load but of little use at long range. European military tactics throughout the eighteenth century, and during the Revolutionary War, relied on a heavy volume of firepower delivered by infantrymen standing shoulder to shoulder in two or three ranks against a similarly massed enemy. Another rank in the rear served as file closers, taking the place of those wounded or killed. Britain's infantrymen were trained to load and fire four shots a minute, executing each step of the process on command. A typical engagement involved a formation marching to within perhaps 50 to 75 yards of an enemy force, firing volleys until the enemy force collapsed or firing a volley or two, then charging with the bayonet.

The earliest bayonets are thought to have been developed in the cutlery center of Bayonne, France, in about the 1640s. These were daggers with a wooden handle tapered so one could be inserted into the muzzle of the musket. This obviously prevented the loading or firing of the gun and converted the musket into a pike. If the bayonet fit too loosely it might fall out, or if it stuck in place, the gun could not be fired until it was removed. The plug bayonet gradually was replaced by about 1700 with an all metal socket type with a base which slipped over the outside of the muzzle eliminating these hindrances. One unusual musket of Revolutionary War vintage made by Ambrose Peck of Massachusetts had a cavity in the butt stock in which to store a bayonet.

In the excitement and confusion of battle, not all men actually were discharging their muskets, despite the high rate of fire sought in training. Misfires, inadvertently loading more than one charge on top of another, and the eventual need to clear powder fouling from the pan and touch hole diminished the rate of fire, as did the occasional need to tighten, readjust, or replace one's flint. Furthermore, if an excited soldier forgot to remove his ramrod before firing, it went whistling toward the enemy rendering him unable to reload his musket until he retrieved a replacement from a dead or wounded comrade.

Except for men serving as rangers or skirmishers on the edge of a formation or in advance of the main battle lines, there was little need for aimed fire—one merely leveled his musket in the direction of the massed enemy line before him and loaded and fired as he was commanded. However, such tactics were impractical in heavily wooded areas, as General Edward Braddock discovered when his army was ambushed and routed by a force of French and Indians south of Fort Duquesne (later Pittsburgh) in 1755.

When firing a musket from the shoulder and loading with a single ball, modern tests show it is possible to keep one's shots all on a man-size silhouette target at 50 yards or so, but much beyond that point accuracy drops off sharply. As one British officer pointed out in the 1770s, the musket might strike a man at up to 100 yards (the length of a football field), but one would be very unfortunate to be hit at 150 yards by someone aiming at him. At the 1775 battle of Bunker Hill (actually fought on Breed's Hill outside Boston), Colonel William Prescott's instructions to his men

9. Cyril Bracegirdle, "The Stage-Coach Driver's Friend," *The Gun Report*, 29, no. 3, August 1983, 34.

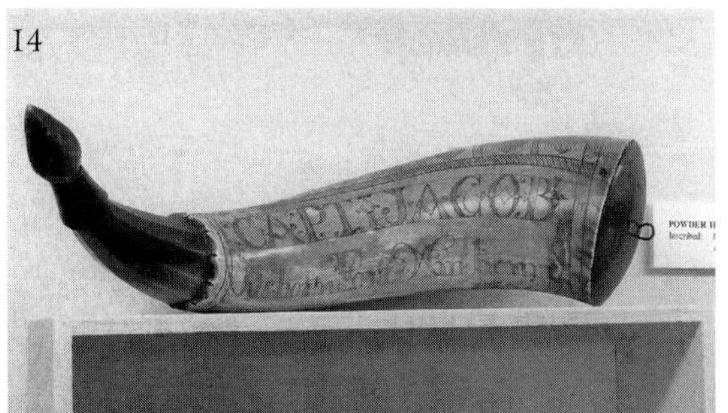

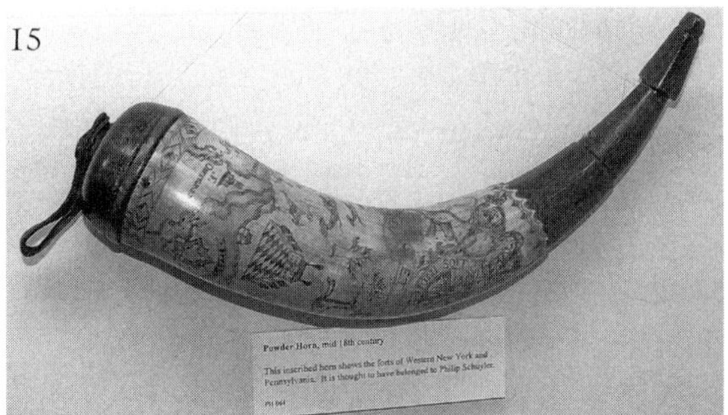

14. Powder horn from the French and Indian War, inscribed "CAPT. JACOB/his horn Fort Wm Henry 1756." The fort was built at the south end of Lake George by Sir William Johnson as a staging point for attacks on France's Fort Carillion, later renamed Fort Ticonderoga. Ticonderoga is located at a strategic point between Lakes George and Champlain. (*Exhibited at Fort Ticonderoga, New York*) **15.** Inscribed with the outline of forts in Pennsylvania and western New York, this powder horn on exhibit at Fort Ticonderoga is thought to have belonged to Philip Schuyler, who fought in the French and Indian War and served as a major general in America's Continental Army during the Revolutionary War.

to hold their fire until they could see the whites of their British enemies' eyes wasn't such an unreasonable order in view of the colonists' shortage of gunpowder and the inaccuracy of their smoothbores.[10]

This inherent inaccuracy of a musket as its undersize ball literally bounced from side to side down the barrel was compensated for by the frequent use of a multi-ball load, a musket ball plus three or more smaller buck shot on top ("buck and ball"). The popularity of the load was such that some bullet molds had multiple cavities (gang molds), one to cast a musket ball and others buck shot. Captain Henry Dearborn, later to rise to general officer rank under General George Washington, participated in the failed American attack on Quebec on New Year's Eve of 1775. He recalled that his smoothbore "was Charged with a ball and Ten Buck shott." However, his gun misfired in the blowing snow as he sought to shoot a British sentry.

Much earlier, in 1609, an encounter between France's Samuel de Champlain and Iroquois Indians at the south end of the lake later named for him resulted in tragedy. As Champlain later explained, "When [at a distance of 30 yards] I saw them make a move to draw their bows upon us, I took aim with my arquebus and shot straight at one of the three chiefs, and with this shot two fell to the ground and one of their companions was wounded who died thereof a little later. I had put four bullets into my arquebus." It is not known whether his gun was a matchlock or wheel lock or even an early form of flintlock, but unfortunately for French fortunes in the region, that encounter fostered a lengthy hatred of the French by many within the Iroquois confederacy.[11]

By the later 1600s, paper cartridges containing ball and powder in a single unit were coming into greater use with European armies and in the colonies. Soldiers now could carry these loose in leather cartridge boxes (cartouche boxes) or tin canisters more conveniently than the bandolier with its individual powder containers. Greater protec-

10. Recent firing tests of a British Brown Bess musket kept all shots in a 15-inch circle at 50 yards and all in a four-foot circle at 100 yards but these results were achieved using modern black powder of better than eighteenth-century quality and firing from a rest rather than unsupported from the shoulder. Joe D. Huddleston, *Colonial Riflemen in the American Revolution* (York, Pa., 1978), 21.

11. S. James Gooding, *The Canadian Gunsmiths 1608 to 1900* (West Hill, Ontario, 1962), 13.

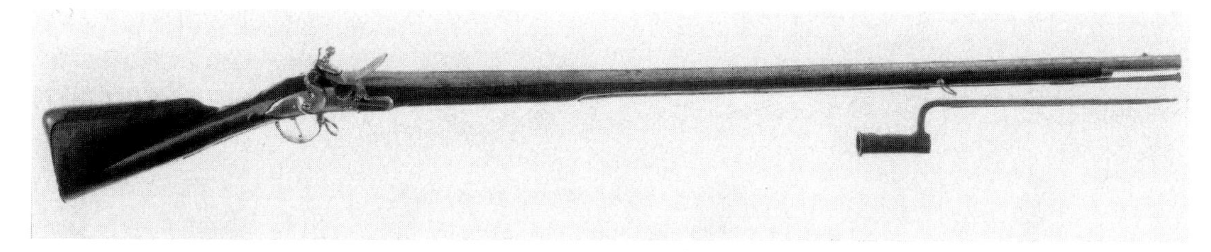

16. A rare example of an early British musket with 46-inch barrel and its socket bayonet. In the early 1700s British regiments were known by the commanding colonel's name. This barrel is stamped "POCOCK C3 No 17" dating its use in 1720-21 by Colonel Pocock's Regiment of Foot which eventually became the 36th Foot. (*Courtesy: Jerry A. Barrows, James D. Julia, Inc.*)

tion for the fragile cartridges came when wooden blocks were added to the interior of the cartridge boxes, drilled with 20 to 30 holes, each to hold one paper cartridge. An inner flap was preferred as well as an outer flap to help protect the cartridges from rain. An infantryman normally carried his cartridge box on a shoulder belt; mounted troops often used a waist belt. The cartridges could be tied with string at each end or tied at one end and twisted at the other and the torn cartridge paper rammed down the barrel on top of the ball to keep the load in place.

At the time of the Revolutionary War, the British method of making cartridges involved placing the ball in the concave end of a forming stick the same diameter as the ball. A pre-cut piece of cartridge paper slightly longer than the stick was wrapped around it, and the excess protruding beyond the ball was tied. The stick was withdrawn to leave a tube, the powder charge with that needed for priming was poured in, and the open end of the cartridge was twisted closed.

Among civilians, both for hunting and militia duty, powder was carried separate from balls or smaller shot either in flasks of wood, leather, tin, or other material or in containers made from hollowed cow horns. A raw horn could be obtained anywhere cattle were slaughtered, and if one were far enough west, the black horn from a bison was a suitable substitute. The horn's interior was removed after boiling and the exterior surface was then scraped smooth. Cow's horns sometimes were scraped thin enough to be translucent so when held to the light the powder level could be seen. The small end was cut off and a stopper fitted while a wooden plug was made for the large end. Although many powder horns were plain, engraved horns are a true form of American folk art whether the decoration was the work of the owner or a professional horn engraver. Such horns provided a means of self-expression and although some might carry only the owner's name and a date crudely scratched into the horn's surface, others might display the outline of a fort, detailed maps of a region, fully rigged ships, floral designs, a mermaid holding a comb and mirror, a bird's eye view of a city, game animals, or even a bit of erotica among the many forms of decorative art (Figures 14, 15).

Many horns decorated with military designs originated during the French and Indian War, often handcrafted by colonial soldiers to relieve boredom. "I, powder, With My Brother Ball/Am hero like to Conquor all" is the inscription on Addison Fitch's horn dated 1766. One elaborately decorated horn of the Revolutionary War era bears an accurate copy of Paul Revere's engraving of the Boston Massacre. Unfortunately, too many engraved specimens found today are modern fakes, aged artificially by the unscrupulous in an attempt to fool the unwary.

About 1728 Britain began production of a musket which, with modest modifications, would serve that nation for the next 100 years. It eventually came to be known unofficially as the "Brown

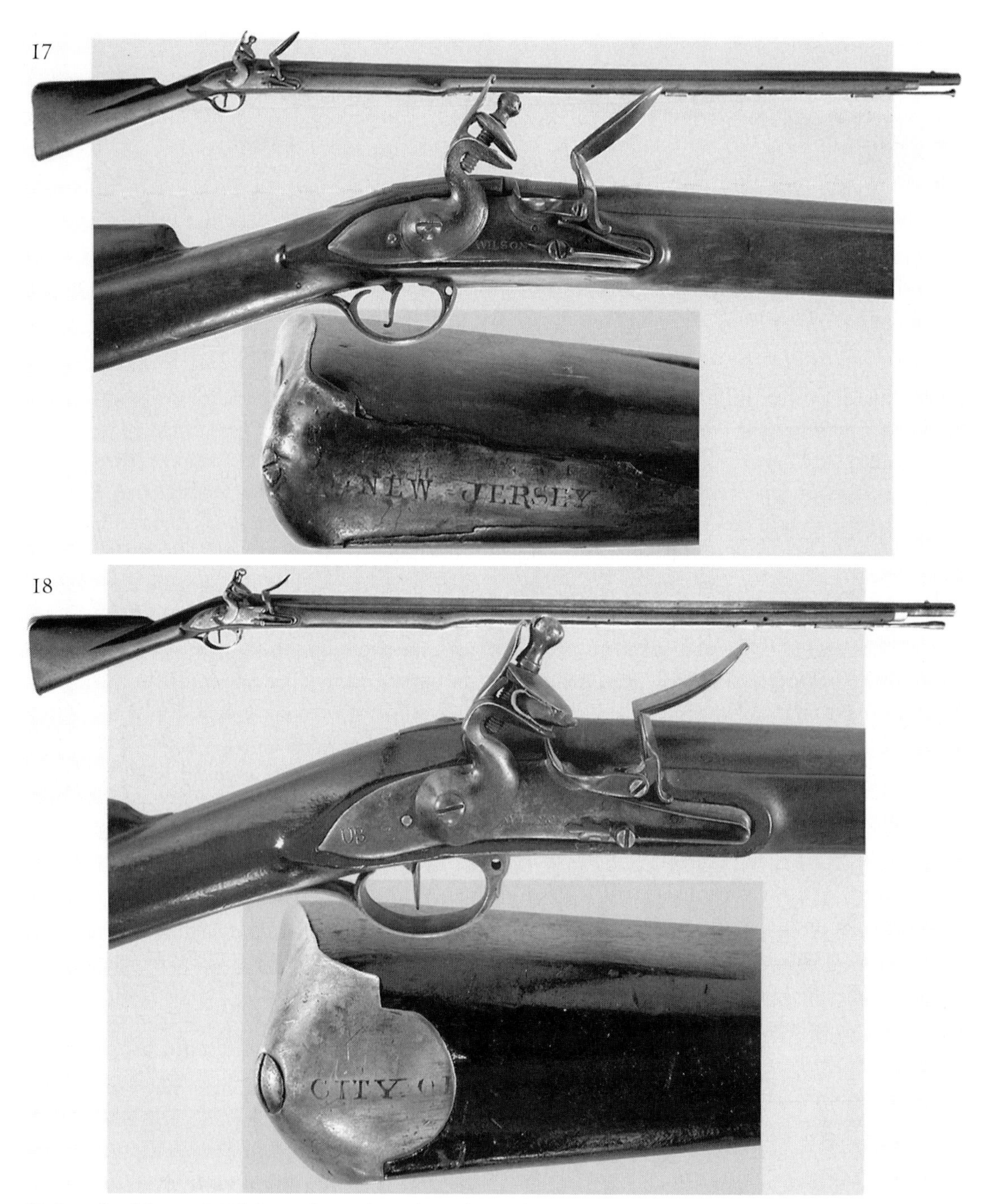

17. Long Land Pattern British Brown Bess style musket with 45 1/4-inch barrel made by Wilson for commercial sale. "NEW JERSEY" engraved on the tang of the butt plate identifies it as one of those purchased by that colony to equip troops raised for service during the French and Indian or Seven Years War. The front sight engages the bayonet. (*Courtesy: C. W. Slagle, photo by Ron Paxton*) **18.** A historic City of New York flintlock musket purchased by the city in 1755. It may have been one of those which armed a portion of a regiment in General James Abercrombie's army in their 1758 expedition against France's Fort Carillion (Fort Ticonderoga). It may also have been carried by one of the New York regiments participating in the 1775 American attack on Montreal. The rear of the lock plate is surcharged "US" and the butt stock was shortened along with the butt plate tang which removed a portion of the city marking. (*Courtesy: C. W. Slagle, photo by Ron Paxton*)

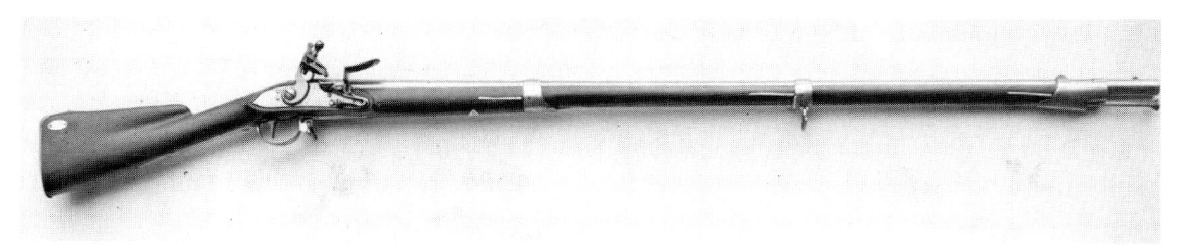

19. French Model 1763 .69 caliber musket, commonly known as the Charleville. (*Courtesy: National Park Service, Fuller Gun Collection, Chickamauga and Chattanooga National Military Park*)

Bess"; the earliest known printed use of that term was in the early 1770s. Its nickname came from the color of the walnut stock, which in the early 1700s replaced beech wood which had been painted black. The other suggested source of the term, rust brown staining of barrels, didn't become common until the late 1700s. "Bess" probably was an anglicized version of the German "busche" for gun. With this arm, Britain replaced its mixed array of army muskets, which included matchlocks converted into flintlocks. The Brown Bess was .75 caliber but fired a .71 caliber ball, originally from a 46-inch barrel in what was known as the "Long Land Pattern" but which was reduced to 42 inches about 1750 in a model designated as the "Short Land Pattern." The latter was the musket most often used by British forces in America during the Revolutionary War. Finally, the barrel length was reduced to 39 inches by 1800. Early specimens used a wooden ramrod, generally replaced by the 1760s with one of iron (Figure 16).[12]

Famed English poet Rudyard Kipling (1865–1936) wrote a lengthy tribute to the "old girl," which included the following final lines.

If you go to Museums—there's one in Whitehall—
Where old weapons are shown with their names writ beneath,
You will find her, upstanding, her back to the wall,
As stiff as a ramrod, the flint in her teeth.
And if ever we English had reason to bless
Any arm save our mothers', that arm is Brown Bess!

The Bess was a sturdy weapon with brass fittings. Its barrel was held to the stock by pins and a tang screw at the rear or breech end. Like infantry muskets throughout Europe, it usually was fitted with two sling swivels so it could be carried over the shoulder suspended by a leather or cloth sling. Weighing between 10 and 11 pounds and with an overall length of about six feet with the bayonet attached, it was somewhat unwieldy for the average soldier who stood only about five feet seven inches tall. The 17-inch bayonet was an essential accessory, carried in a scabbard suspended from a waist or shoulder belt, usually the former. Thousands of these muskets were sent to Britain's colonies in North America. Some Besses were made for commercial sale to colonels equipping their own regular British army regiments and some for sale to English volunteer militia units (Figures 17, 18).[13]

Commissioned and non-commissioned officers and some special units after about 1750 often carried a fusil (or fuzee or fusee), a version of the musket but reduced in weight and dimensions and .65 to .70 rather than .75 caliber. Officers' fusils were privately purchased and more ornate than the standard musket. Artillerymen, dragoons, and fast-moving light infantry units sometimes were issued

12. Bill Ahearn, *Flintlock Muskets in the American Revolution and Other Colonial Wars* (Lincoln, R.I., 2005), 24. Brown, *Firearms in Colonial America*, 229.

13. George D. Moller, *American Military Shoulder Arms, Vol. 1* (Niwot, Colo., 1993), 231.

what were known as carbines, .65 rather than .75 caliber, and with a barrel of 36 to 42 inches.[14]

Despite the preponderance of muskets in British army units in North America, a limited number of rifles were present. Swiss-born Lieutenant Colonel Henry Bouquet in 1756 was appointed commander of the First Battalion, 60th Foot (Royal Americans). An exceptionally capable leader, he sought to prepare his men to fight in the woods found in the American wilderness and obtained rifles for some of his marksmen. His success in defeating American Indians at the battle of Bushy Run in western Pennsylvania in 1763 earned him hero status in the eyes of many.[15]

French muskets were shipped to that nation's forces in North America—Canada and France's holdings in the Ohio valley and on both sides of the Mississippi River southward to Spanish controlled Louisiana. About 1700 France abandoned its remaining matchlocks replacing them with flintlocks and in 1717 chose a standardized flintlock musket which through various models persisted in its basic design until 1777. This musket was lighter than the Bess and .69 caliber with iron rather than brass butt plate and other mountings, and after the late 1720s bands rather than pins secured the barrel to the stock. This use of barrel bands facilitated disassembly and allowed use of a lighter stock, a definite relief to a foot soldier who welcomed the reduction of the weight of the gear he carried by even a pound or so. "Charleville" is a name often heard today when referring to these French muskets, particularly the lightened version of the 1763 model adopted in 1766, even though they were made at several other royal arsenals as well. As in the British army, lighter and better finished officers' muskets were available as were carbines (or musketoons) for mounted troops who often fought on foot (Figure 19).[16]

The French Model 1763 is often cited as the arm supplied in large numbers to the Americans during the Revolutionary War and as that on which the first muskets made after the war at the new U.S. arsenals were based. To be correct, however, the French musket adopted in 1763 was replaced by a version about a pound lighter three years later, although many bore the 1763 date on the barrel. It was this 1766 variation, some with the later addition of a spring to retain the rearmost barrel band, which figured prominently in the later history of firearms development in America.[17]

The Civilian Fowler

Civilian fowlers of the colonial era were intended for hunting birds and small game with shot but also could be used to hunt larger animals with a single ball. Typically they were more graceful, longer, lighter in weight and less sturdy than a military musket. A barrel length of 50 inches or more was not uncommon, and most used transverse pins to secure barrel to stock, a stock that sometimes had a massive "club butt." They had no rear sight, although there might be a groove at the breech to help guide one's eye toward a simple front sight. Among the French colonists, such a civilian hunting gun was known as a *fusil de chasse* and it served with irregular French militia forces as well.

These civilian fowlers probably represent the first guns to be made in what would become the

14. During the Revolutionary War, Washington discouraged the carrying of firearms by officers thinking that it distracted them from their primary duties. The terms "fusee" and "fusil" are thought to have come from the Italian word for flint, *fucile.*

15. Ian M. McCulloch and Timothy J. Todish, eds., *Through So Many Dangers: The Memoirs and Adventures of Robert Kirk, Late of the Royal Highland Regiment* (Fleischmanns, N.Y., 2004), 133.

16. In this period, a carbine was distinguished from a musket by a smaller caliber rather than necessarily a shorter barrel.

17. Rene Chartrand, *Uniforms and Equipment of the United States Forces in the War of 1812* (Youngstown, N.Y., 1992), 86-87. George C. Neumann, *Battle Weapons of the American Revolution* (Texarkana, Tex., 1998), 93-94.

20. Colonial Hudson Valley long fowler with a curly maple stock and a barrel 59 3/8 inches long. The overall length is a couple of inches less than six and one-half feet. The gun almost certainly was assembled in America. (*Courtesy: C. W. Slagle, photo by Ron Paxton*)

United States. However, many fowlers were imported from England and France, and even after fabrication in the colonies began in the early 1700s, the use of a European barrel or lock was common. Gunsmith James Geddy in 1737 advertised in the *Virginia Gazette* "Fowling pieces and large Guns fit for killing wild fowl in Rivers at a reasonable rate." The only existing long gun definitely identified as being owned by George Washington at the time of his death is a plain half-stocked English fowler made about 1750 by Richard Wilson of London, an artifact displayed for many years at Mount Vernon (Figure 20).[18]

Well before 1700, militia laws in most of the English colonies required each able-bodied man to own a firearm suitable for use in the event he should be called to military duty. In 1687, for example, Connecticut's militia act required that every man physically able to serve as a foot soldier be provided with "a well fixed musket, the barrel not under three foot in length, and the bore for a bullet of twelve to the pound [.73 caliber], a collar of bandoliers or cartouche box with twelve charges of powder and bullets at the least, and a sword, or, if the officer so appoint, with a good pike and sword." Annually each militia captain was to inspect each man's gun and ammunition.[19]

By the mid-1700s, except in Pennsylvania and those limited surrounding areas where rifles gradually became more common, most colonists met this militia obligation with what often was the only gun in the house. That generally would have been the smoothbore flintlock fowling piece—today we would call it a shotgun—often of about .75 caliber, acceptable on muster or military training day but more often used for hunting. These dual-purpose long fowlers rarely were fitted with sling swivels or a bayonet, but militia ordnances usually permitted a sword or perhaps a hatchet as a substitute for the latter.[20]

18. Ahearn, *Flintlock Muskets in the American Revolution,* 105, 138.

19. Ross K. Harper, "Historical Archaeology on the Eighteenth Century Connecticut Frontier: The Ways and Means of Captain Ephraim Sprague," *Museum of the Fur Trade Quarterly,* 41, no. 2, Summer 2005, 12.

20. Muster days during peacetime often provided a convenient excuse for drinking and carousing rather than for actual military training, it seems. The *New England Farmer* published in Boston in 1827 carried the observation: "We never before the present season witnessed so general and decided symptoms of the disgust and contempt that are entertained of military trainings. The inequality of the duty required has made it intolerable; while the farcial style in which it is performed awakens nothing but ridicule and derision."

21. Flintlock Pennsylvania rifle by Samuel Grove, Jr., of York County, Pennsylvania. A .44 caliber rifle with a 43 1/2–inch barrel, it features an ornate brass patch box, silver eagle inlay in the left side of the curly maple stock, and relief and incised carving with checkering at the wrist and forward of the lock at the balance point and dates from the early 1800s. (*Courtesy: C. W. Slagle, photo by Ron Paxton*)

When the use of small lead pellets or shot in smoothbore fowling pieces was inaugurated for hunting small game isn't known, but it was in practice at least by the early 1500s. Although shot initially was produced in molds or by tumbling small pellets to round their edges, by the 1600s some were doing it by pouring molten lead through a sieve and letting the drops cool and solidify as they fell, landing in a container of water as a cushion. William Watts of England in 1769 made it possible to produce spherical pellets in commercial quantities by erecting a tall wooden tower from which the molten pellets fell. By the early 1800s wooden towers were being joined by masonry towers, all using the natural tendency of a falling drop of liquid to take on a spherical shape as it falls. The size of the shot was determined by the size of the holes in the sieve, from a few hundredths of an inch in diameter to what is known today as "triple ought" (000) buckshot of about .36 caliber. In 1828 in Baltimore the Merchants Shot Tower Company built what then was one of the world's largest such structures, 250 feet tall, 50 feet wide at the base, and 25 feet in diameter at the top.

Two of the characteristics of those fowlers made or assembled in the colonies were the usual lack of a patchbox and the frequent use of American maple or walnut or some fruit wood such as cherry for the stock. In the northeast, particularly in the Hudson River valley, some of these hunting guns were deserving of the term "long fowlers," with an overall length of seven feet or more! By 1800 as gun powder improved in quality, the barrel length diminished, since a long barrel no longer was thought necessary to prevent the waste of unburned powder. But the graceful single-shot fowlers retained their popularity well into the nineteenth century, some as original flintlocks and others made in percussion form or converted to that new ignition system even though double-barrel fowlers were becoming available by the 1790s.

Early in the nineteenth century, from 1810 or so into the 1830s, New England makers again were producing light dual-purpose flintlock militia muskets, many ordered for use in the state of Massachusetts. They usually were rather plainly finished .69 caliber arms with a pin-fastened stock and about a 39-inch round barrel, somewhat shorter than the earlier long fowler. The base of the front sight doubled as a lug on which to secure a bayonet, and the guns easily could serve for hunting in peacetime.

The Pennsylvania-Kentucky Rifle Appears

By the mid-1700s, the first truly American arm was evolving, a civilian hunter's weapon that eventually would be frequently termed the Kentucky rifle. Rifling is a series of parallel spiral grooves cut into the barrel to give the projectile a spinning motion, stabilizing it in flight and increasing range and accuracy. No one knows precisely where or when the concept of rifling a gun barrel originated; it may have been Austria or Germany, probably about 1500. The word "rifle" may be a derivation of the German word *riffeln,* meaning "to groove." It is known that in the mid-1500s shooting matches were held in Germany with matchlock rifles. There were a few modest attempts by the mid-1600s by several European nations (including Denmark) to introduce rifled arms into their armies on a very small scale, but for many years it would remain a feature found occasionally in sporting and rarely in military arms. A few rifled arms were brought to North America by Russians, Spaniards, and other Europeans before the early 1700s, but their use at that time was uncommon.

It's been said that rifling at first consisted of straight grooves intended to reduce the problem of loading a gun with a bore fouled by powder residue from previous firings. Straight rifling is definitely known and this explanation of its purpose may have some merit. However, citing it as the source of spiral rifling is questionable since the theory of rotational stabilization was already known and had been applied to some crossbows.

22. Fancy edged patch box cover with what may be an engraving of a long-haired Indian. The rifle on which it appears was probably made in Bedford County, Pennsylvania, in the 1820s. Elaborate patch box cover designs generally didn't come into vogue until the 1790s and into the next century but came to vary widely in style. (*Courtesy: Norm Flayderman*)

German immigrants to the southeast region of William Penn's colony were familiar with rifled guns of German and Swiss make, typically the short hunting rifle with a barrel length of about 30 inches, caliber between .60 and .70, and with a thick buttstock. In their homeland, such guns had been useful when hunting wild boars and deer in the forested countryside and shooting contests with rifles had been a popular diversion. A distinctive form of new rifle combined characteristics of the graceful long fowler already present in America and the German *jaeger* rifle. Traditionally these rifles emerged first from the gun shops in the Lancaster County area of Pennsylvania, but before long similar rifles were being crafted in Maryland, the Shenandoah Valley of Virginia, and the piedmont region of North Carolina as well.[21]

21. Martin Meylin of Lancaster appears to have been one of the first Pennsylvania rifle makers, perhaps active as a gunsmith as early as 1719, although precise verification is lacking. Other early Lancaster rifle smiths included J. Metzger (1728) and Philip Lefever (1731).

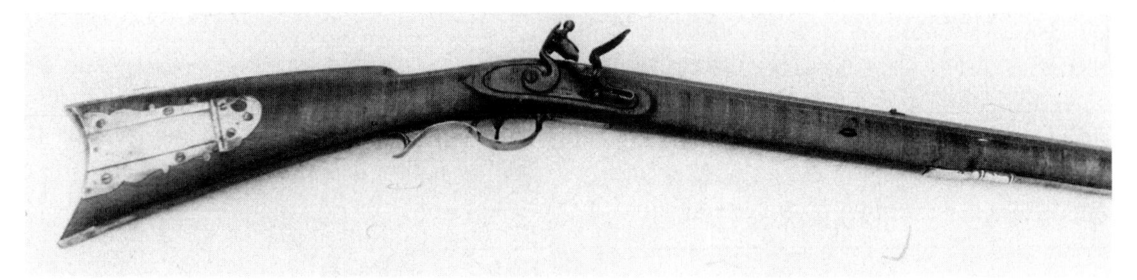

23. Flintlock Pennsylvania rifle by Henry Deringer, Jr., of Philadelphia and so marked on both lock plate and barrel.

These rifles (sometimes known among colonists as rifle guns) didn't incorporate any mechanical innovations but rather had new stylistic features such as an octagonal barrel heavier than that of a smoothbore fowling piece. Curly maple wood was a preferred choice for the graceful stock. Plain maple wood sometimes was given an artificial grain, probably by using stain or dye, although some say it was accomplished by wrapping a tarred string around the stock and burning it to give a slightly charred effect. Trigger guard, butt plate, thimbles beneath the barrel to hold the hickory ramrod, and other fittings usually were brass (Figure 21).

A typical rifle by the early 1770s had an octagonal barrel of around 42 inches, often about .45 to .50 caliber, and a slower twist of the rifling than a *jaeger* allowing use of a smaller powder charge. The smaller caliber economized on lead, for a pound produced only 11 round .75 caliber balls, but yielded about 50 balls of .45 caliber. With use, rifling in the soft iron barrel became worn, and the barrel might be bored smooth, then recut, resulting in a larger caliber than when first made. If the rifleman was a skilled marksman, his rifle was accurate at distances of 200 yards or sometimes more, a far cry from a smoothbore's accurate range of no more than perhaps 60 yards.

As large game animals such as elk, bison, and bear became fewer later and smaller game predominated, particularly in the more settled areas of the east, calibers of rifles intended for use in that part of the country by the 1830s were diminishing and "squirrel rifles" of .30 to .36 caliber or so became common. Not until well into the 1800s were there any carefully developed standards as to caliber or science of the depth or width of rifling grooves or the speed of their twist. Each maker often had his own preferences, and a customer needed a bullet mold of the proper size for his new rifle. Makers might keep a few ready-made rifles on hand, but often one was made to the customer's specifications as to caliber, weight, length, and ornamentation.

One feature often found on these rifles was a covered rectangular cavity in the butt stock in which one could carry such items as precut patches to enclose the lead ball, a vent pick, and spare flints. This butt trap became known as a patch box in America. A sliding wood cover was found on some *jaegers,* but in the Pennsylvania rifles this gradually was replaced by a hinged sheet brass patch box cover not subject to loss as was the sliding cover. More rifles in the early 1800s were featuring decorative stock carving and a patch box cover of elaborate design (Figure 22).

The final result of this evolutionary process was an aesthetically pleasing and distinctly American arm. Through the centuries it has been known as the American rifle or "long rifle," and its frequent designation as the Kentucky rifle probably originated with a ballad popular after the War of 1812, "The Hunters of Kentucky," containing the phrase "Kentucky rifles." The reference was to the dramatic 1815 victory at New Orleans by a smaller heterogeneous American force of infantry and artillery under General Andrew Jackson, aided

24, 25, 26. Three views of a classic Pennsylvania rifle of the 1780s or 1790s. Although it is unsigned, various features indicate it probably was made by Wolfgang Haga of Reading (Berks County), Pennsylvania. Tax records as early as 1767 list him as a gunsmith, but he may have been engaged in the profession earlier. He died in 1796. The handsome rifle has a 42-inch octagonal barrel, curly maple stock with tiger-striped grain, and incised carving on the left side of the stock. The star inlay is silver and the rest of the fittings are brass. The bore is worn almost smooth and now is about .62 caliber. (*Courtesy: Norm Flayderman*)

by a number of rifle-armed Kentucky volunteers. But Pennsylvania or Pennsylvania-Kentucky rifle is perhaps the most correct term in view of its location of origin (Figures 23, 24, 25, 26).[22]

It's sometimes stated that one innovation appearing in the Pennsylvania rifle was the use of a thin leather or cloth patch around the ball, providing a gas seal and gripping the rifling to give the ball the spin necessary for its greater range and accuracy. This greased or spit-moistened patch also helped keep powder residue soft and reduced its buildup. However, the use of a patched ball had been known in Germany earlier.

22. The pertinent stanza of Samuel Woodworth's poem, later put to music by Noah Ludlow, is "But Jackson he was wide awake, and wasn't scar'd at trifles;/For well he knew what aim we take with our Kentucky rifles./So he led us down to Cypress swamp, the ground was low and mucky;/There stood John Bull in martial pomp but here was old Kentucky."

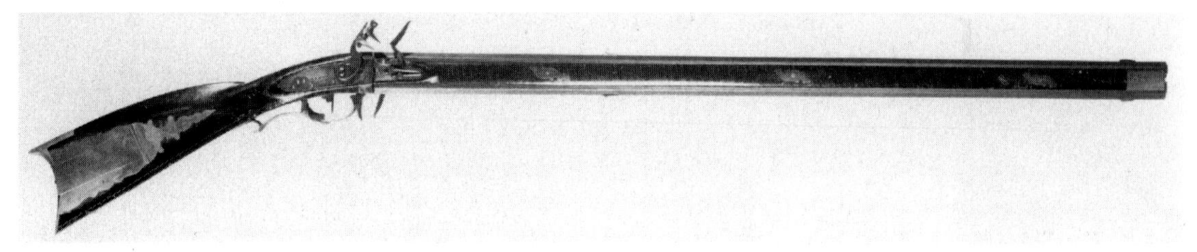

27. Swivel barrel Pennsylvania rifle by Daniel Boyer of Orwigsburg, Pennsylvania. (*Courtesy: Smithsonian Institution*)

The skilled rifleman experimented until he found the proper powder charge that gave optimum accuracy in his rifle, then prepared a measure, perhaps made from the hollow tip of a cow's horn. One common method of estimating an approximate charge was to place a rifle ball in the palm of the hand held flat, then from the powder horn pour enough powder in a mound over the ball to just cover it. Another method was to place a piece of white cloth or paper beneath the muzzle and fire the gun, seeing if there was evidence of unburned powder being blown out. If so, the charge was reduced until no powder appeared while retaining accuracy and range. One could also use the earlier European ratio of three grains of powder for every seven grains of ball weight.

Sights on a Pennsylvania rifle were simple and not adjustable once secured in place—a small blade front sight and a fixed rear with a "V"-sighting notch. With fixed sights and when firing on a windy day, the shooter had to estimate how much effect the wind might have on the ball in its flight and adjust his aim accordingly. "Kentucky windage" is a common term for this form of wind correction. The accuracy of which these rifles were capable was not phenomenal compared with later standards, but it was a substantial improvement over the smoothbore. Early makers may have thought a long barrel gave added range, an incorrect assumption, but the longer sight radius (distance between front and rear sights) did improve the shooter's accuracy.

The muzzle-loading rifle's inherent slowness in loading prompted a few makers to offer double-barrel rifles to provide a ready second shot. In some the barrels were fixed in position side-by-side with two locks while with "over and under" double rifles the barrels swiveled on a center pin, an arrangement which required only a single lock. These guns offered the advantage of a second shot without reloading but at the expense of substantial extra weight, added cost, and in the case of the swivel gun a certain degree of fragility. The concept of swivel barrels was applied to a few pistols during the colonial era as well. Side-by-side double barrel fowling pieces did not possess the same weight disadvantage as did a double rifle and were more practical and common than their two-barrel rifled cousins (Figure 27).

Some rifle makers also produced smoothbore fowling pieces (shotguns), many plain but functional and others handsomely finished. These fowlers usually were single shots with a long barrel often octagonal at the breech but becoming round a third of the way or so toward the muzzle, as had been common with some of the earliest Pennsylvania rifles. Some riflesmiths didn't confine their work to the civilian trade but also received state or federal contracts to produce military rifles and muskets.

The term "smoothbore rifle" sounds like a contradiction, but it describes some Pennsylvania rifles with an octagonal barrel intentionally bored smooth or with straight rather than spiral rifling grooves. These could serve a dual purpose for firing either shot when hunting small game or a single or even two round balls. Even without the spin imparted by rifling, a smoothbore rifle with a snug

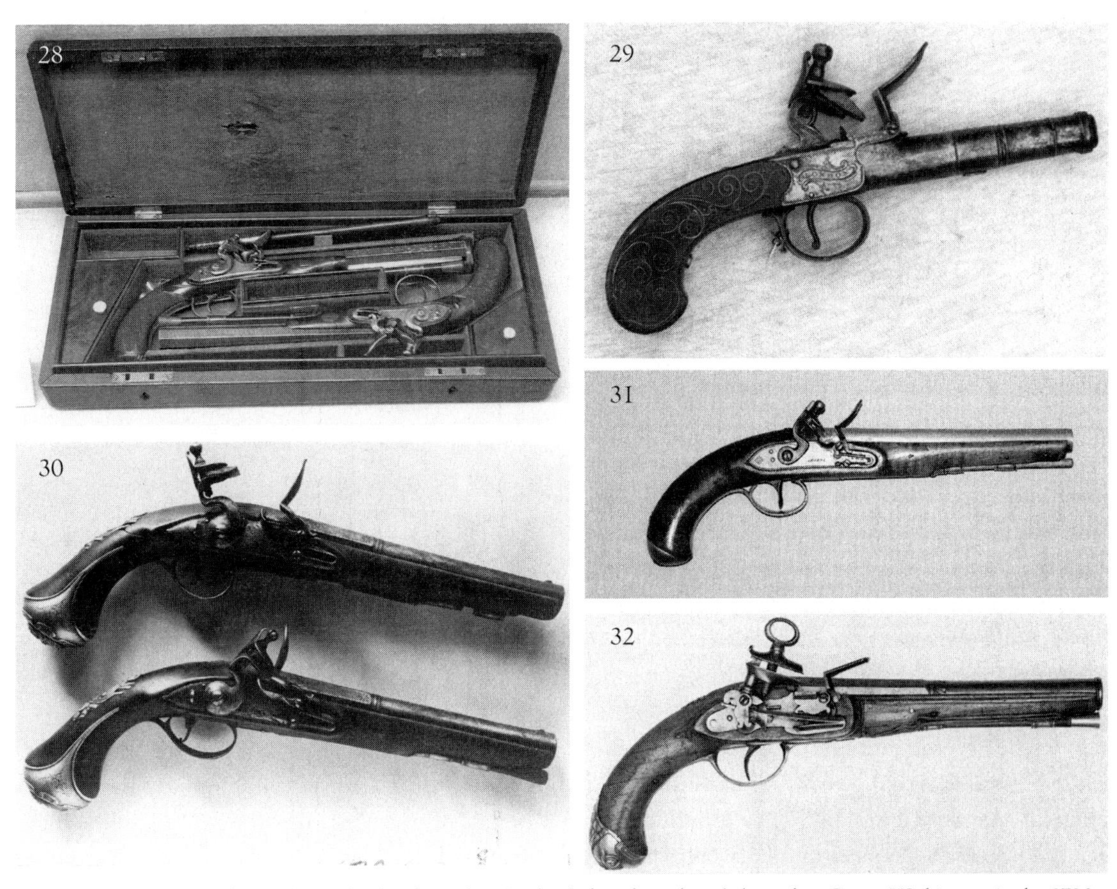

28. Cased flintlock pistols by John Richards of London, England, thought to have belonged to George Washington in the 1790s. (*Exhibited at Fort Ticonderoga, New York*) **29.** Screw barrel .45 flintlock pocket pistol by Archer of London with a 2 1/8-inch barrel. The small lug on the underside of the barrel facilitated unscrewing it to load, using a wrench that often was incorporated into one handle of the bullet mold. **30.** Pair of smoothbore .60 caliber English officer's style holster pistols made by Thomas Bates of London probably about the 1790s. These were owned by trader and Indian agent Charles M. de Langlade, active in the Wisconsin region. (*Courtesy: State Historical Society of Wisconsin*) **31.** Brass barrel English .54 pistol of the late 1700s or early 1800s marked on the lock "SHARPE" and on the nine inch barrel "SHARPE" and "EXTRA PROOF." It could have served as a moderately priced officer's pistol or been intended for the general trade. Similar pistols by Ketland, Sharpe, and several other English makers are relatively often found at gun shows in the United States today which probably indicates their popularity. **32.** Spanish-type miquelet pistol of the early nineteenth century made by Joseph Bustindui[?], perhaps in Mexico City. It's representative of such used in California and the southwestern U.S. while under Spanish and later Mexican rule. (*Courtesy: Los Angeles County Museum of Natural History*)

fitting patched ball could achieve acceptable accuracy at moderate ranges as when hunting deer in a forested region. Firing tests were conducted between a spiral groove rifle, straight-groove rifle, and smoothbore rifle in the early 1900s from a bench rest using a greased patch with each ball against a man-size silhouette target. Each produced 10 hits at 100 yards; at 200 yards the respective hits were 10, 5, and 4; and at 300 yards accuracy fell off to 5, 2, and 1 hit.[23]

23. Captain John G. W. Dillin, *The Kentucky Rifle* (York, Pa., 1959), 71.

A backwoodsman on a long hunt away from home with his rifle had to carry a number of accessories, many of which were contained in a leather shoulder bag or hunting pouch. These usually included a powder horn slung over his shoulder, plus a second smaller one for finer grain priming powder, pre-cast balls, linen or other material for patches, a patch cutting knife, spare flints, a hank of tow to facilitate cleaning, and perhaps even a bar of lead, ladle in which to melt lead, and hinged bullet mold of iron or brass, although bullet molds made of soft soapstone were not unknown in the seventeenth and eighteenth centuries. Also vital was a small wire pick (or occasionally a feather) to clear any fouling from the touch hole. This pick could be carried in the hunting bag or sometimes was held in two small staples set into the butt stock. Another possible accessory was a "cow's knee," a conveniently shaped piece of leather from that part of bovine anatomy which could be slipped over the lock in rain or snow to keep it dry.

In 1773–74, J. F. D. Smyth of England traveled in the more remote areas of Virginia and the southern colonies and left a description of the arms and dress of Shenandoah Valley backwoodsmen as the militia assembled during what was known as Lord Dunmore's War.

> [Their tomahawks were] more useful than anything except the rifle-barreled firelocks, both of which all the male inhabitants . . . constantly carry along with them every where. Their whole dress is also very singular, and not very materially different from that of the Indians, being a hunting shirt, somewhat resembling a waggoner's frock, ornamented with a great many fringes, tied around the middle with a broad belt, much decorated also, in which is fastened a tomahawk, an instrument that serves every purpose of defence and convenience, being a hammer at one side and a sharp hatchet at the other, the shot bag and powder-horn, carved with a variety of whimsical figures and devices, hang from their necks over one shoulder, and on their heads is a flapped hat. . . . Sometimes they wear leather breeches, made of Indian dressed elk, or deer skins, but more frequently thin trousers. On their legs they have Indian boots or leggings, made of coarse woolen cloth, that either are wrapped round loosely and tied with garters, or are laced upon the outside, and always come better than half way up the thigh; these are a great defence and preservative, not only against the bite of serpents and poisonous insects, but likewise against the scratches of thorns, briars, scrubby bushes, and under wood, with which this whole country is infested and overspread. On their feet they sometimes wear pumps of their own manufacture, but generally Indian moccasins, of their own construction also, which are made of strong elk's, or buck's skin, dress thence round to the forepart of the middle of the ankle, without a seam in them, yet fitting close to the feet, and are indeed perfectly easy and pliant. Thus habited and accoutered, with his rifle upon his shoulder, or in his hand, a backwoods's man is completely equipped for visiting, courtship, travel, hunting, or war.[24]

Civilian Pistols in the Colonies

Civilians in eighteenth- and early nineteenth-century America often found a pistol to be a necessity or at least a comfort, whether in the city or traveling on isolated roads. A sea captain might rely on a pistol or two as a means of enforcing his absolute authority on shipboard. When traveling, it wasn't unusual for a gentleman of means to carry among his pieces of luggage a pair of sizeable .50 or .60 caliber flintlock pistols in a handsome wooden case of walnut or oak trimmed in brass, the interior lined with felt or velvet and partitioned into compartments for the guns and such accessories for their care and use as a cleaning rod, screwdriver, bullet mould, powder flask, cast bullets, and oil can. Whether of small pocket size or a larger belt or holster gun, pistols in the colonies usually orig-

24. Peter F. Copeland and Brooke Nihart, "Valley of Virginia Militia in Dunmore's War, 1774," *Military Collector and Historian*, 57, no. 4, Winter 2005, 191.

inated in Britain or on the European continent. Among the better known English makers of such pistols were Robert Wogdon, Durs Egg, and H. W. Mortimer, each of whom made guns in the late 1700s and into the next century (Figures 28, 29, 30, 31, 32).[25]

Highly prized by today's collectors are those flintlock and later percussion pistols from the makers of Pennsylvania rifles, graceful and similar in style to their long guns. Commonly known today as Kentucky pistols, specimens of pre-Revolutionary War origin frequently were European in style, but they later took on the characteristics of the Pennsylvania rifle. They often were stocked in curly maple or sometimes cherry or walnut, with an iron or brass round, octagonal, or part round/part octagonal barrel, some rifled and others smoothbore, and normally with brass but sometimes silver fittings. Kentucky pistols were made in far fewer numbers than rifles and unlike the rifle which could be used for hunting or defense, the pistol was intended primarily as a weapon. Most were a larger belt or holster size rather than a compact pistol for the pocket. Some were chosen as personal sidearms by Revolutionary War and War of 1812 officers. Perhaps one could argue that the term "Pennsylvania pistol" is more appropriate, but it has not caught on with today's collectors (Figure 33).[26]

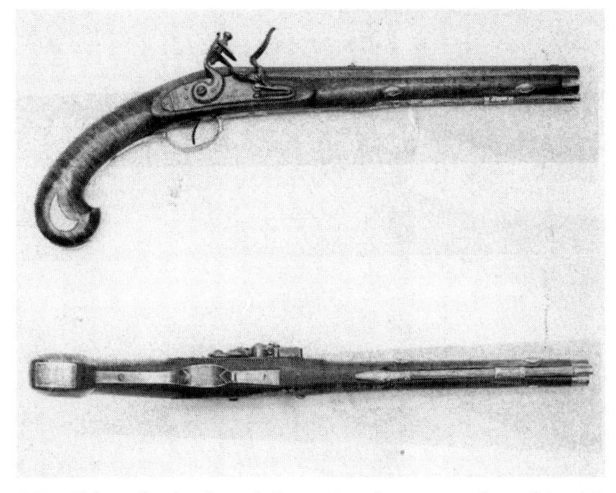

33. Although the barrel is unsigned as to maker, this .44 Kentucky flintlock pistol features coin silver mounts including fore end cap, ramrod thimbles, and a stylized half moon set in the right side of the butt. The lock, perhaps imported from England, is stamped "J. Holland."

25. Collectors today recognize two different styles of casing for civilian pistols—English and French. In the former, thin wood partitions divide the interior into compartments for gun(s) and accessories. French style casing provides cavities shaped to fit the exact contour of gun or guns and accessories.

26. A brass barrel, generally a copper-tin alloy, was easier to fabricate that one of ferrous metal, more resistant to corrosion, and strong enough for the black powder loads of the eighteenth and early nineteenth century.

2 Military Arms from the Revolutionary War to the 1820s

Guns of the War of Independence

Great Britain had prohibited the large-scale manufacture of arms in its North American colonies, relying on them for raw materials rather than manufactured goods. Several of the colonies, including Massachusetts, in the year or so before actual war with Great Britain broke out had purchased some arms overseas, many of which were shipped clandestinely from France. But during the chaos at the outbreak of hostilities with the mother country in 1775 and well into the war years and the open shipment of large numbers of French muskets, the American army was forced to rely on a potpourri of firearms obtained by any means, including seizure from some of the many citizens who were neutral or Tories who favored the British cause. Civilian hunting guns, European military arms already present in the colonies or purchased abroad, Brown Bess muskets seized from British stores, and some rifles—examples of all of these found use in American hands. One British officer present at the battles at Lexington and Concord in April 1775 commented on the colonials who harassed the retreating British force: "These fellows were generally good marksmen, and many of them used long guns made for duck shooting."[1]

In addition, early in the war some American gunsmiths made a very limited number of muskets under contracts with the Continental Congress, individual colonies, or Committees of Safety, muskets which frequently were patterned after the Brown Bess. It is often written that makers of Committee of Safety muskets purposely didn't place their name on them for fear of reprisal in case of ultimate British victory. However, contracts for these muskets sometimes demanded that the arms be marked with the maker's name or initials, even though gunsmiths did not always do so (Figure 34).

William Whetcroft, a goldsmith, contracted with a Maryland Council of Safety for military arms. He agreed to produce "good substantial proved Musquets 3 1/2 feet long in the Barrell and of three quarters an Inch in the Bore [.75 caliber], with good double Bridle Locks, black Walnut or maple Stocks, and plain strong Brass mounting, Bayonets with Steel Blades 17 inches long, steel Ramrods, double Screws, priming Wires and Brushes fitted thereto, with a pair of brass Moulds for every 80 Musquets, to cast 12 Bullets on one Side, and on the other Side to cast Shot of such Size as the Musquet will chamber three of them."[2]

Blacksmiths sometimes were encouraged to learn the gunsmithing trade to help alleviate the arms shortage with the frequently added incentive of exemption from military service. Those men skilled in wood working, such as cabinet makers

1. Ahearn, *Flintlock Muskets in the American Revolution,* 105.

2. Daniel D. Hartzler, *Arms Makers of Maryland* (York, Pa., 1977), 27.

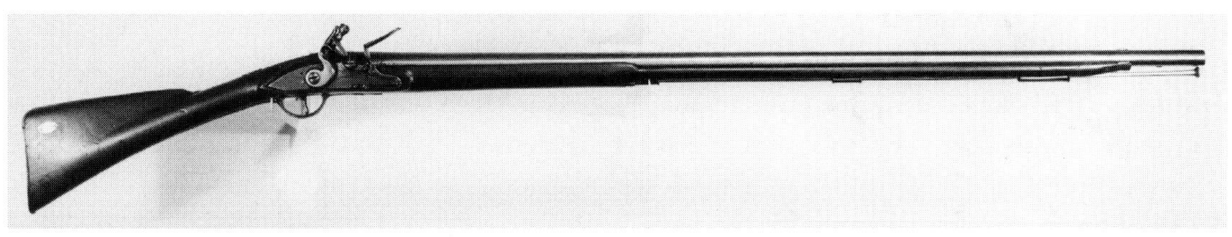

34. Committee of Safety musket by Samuel Barrett of Concord, Massachusetts. It doesn't follow the lines of a British Brown Bess musket as closely as some Committee of Safety muskets did and except for its iron rather than wooden ramrod resembles a civilian fowler. (*Courtesy: National Park Service, Fuller Gun Collection, Chickamauga and Chattanooga National Military Park*)

and carpenters, sometimes worked with gunsmiths to stock muskets. Some craftsmen concentrated on making gunlocks, a specialty which had been carried on in England and continental Europe for generations. Locks required steel for springs and the frizzen, and steel was not produced in quantity in the United States until well into the 1800s, another reason locks had often been imported rather than made locally.

Americans gained independence with a true conglomeration of arms, most of them of foreign origin.[3] Of course during the war in addition to those muskets obtained from France, arms were captured from British and from German mercenary troops and others were recovered from the battlefield. Adding to the mix and confusion were those assembled into serviceable weapons by gunsmiths in the colonies using a variety of parts salvaged from damaged or obsolescent weapons, cannibalized for spare parts or restocked thus creating hybrid arms that can be difficult to identify today. Colonial makers also provided a number of bayonets under contract, for the ultimate outcome of a Revolutionary War engagement often was decided by that edged weapon.[4]

A Lack of Arms and Training

The colonies' arms shortage as war continued was of such concern that Benjamin Franklin, apparently in complete seriousness, in 1776 suggested that bows and arrows be considered. Although it would take longer to train a bowman than a musketeer, such a weapon was as accurate and as effective at a modest range, one could launch a dozen or so arrows in a minute, and the weapon certainly was easier and cheaper to produce in quantity. But the war in the colonies provided France with an opportunity to yank the British lion's tail and later to dispose of muskets which had been superseded by an improved model in 1777. Clandestine shipments of French arms to the rebels began in February 1776. Franklin and Silas Deane ultimately arranged for loans of almost $10,000,000 from France and the purchase of large numbers of French arms. The former reported in April 1777 that: "We have purchased 80,000 fusils [muskets], a number of pistols, etc. . . . They were King's arms and secondhand, but so many . . . are unused and

3. The Brigade of the American Revolution is a modern living history association encompassing almost 3,000 men and women who re-create the life of the common soldier in the Revolutionary War. With careful attention to authenticity, members adopt the uniforms, arms and accoutrements, tents, and other equipage of an American or foreign unit which fought in the war, including infantry, marines, sailors, cavalry, riflemen, engineers, artillery, musicians, and others. More than 130 units are represented today.

4. For authors' possible benefit, a few makers thought to have made muskets for Committees of Safety are Abijah Thompson of Woburn, Mass.; Philadelphia's Lewis Prahl and Benjamin Town; Samuel Boone of Berks County, Pa.; Andrus Waters of Sutton, Mass.; Hugh Orr, Bridgewater, Conn.; and Timothy Bloodworth of North Carolina.

exceptionally good that we esteem it a great bargain if only half of them should arrive." Other American agents were successful in purchasing Dutch, Belgian, German, and Spanish arms in lesser numbers.[5]

The infusion of a large number of French muskets of the same caliber (.69) also eased the problem caused by the need to supply ammunition for a variety of guns of different bore sizes. When in 1795 the United States authorized the production of its first official musket at the federal armories at Springfield, Massachusetts, and later at Harpers Ferry, (West) Virginia, it used the French light variation of the 1763 "Charleville" musket as the pattern. After that time, the French caliber of .69 would remain the standard for U.S. muskets until 1855. (Because of the rapid buildup of powder fouling in the barrel, a .69 musket generally fired a .64 caliber ball.)

Persistent shortages of gunpowder and lead also hampered the American war effort. At the beginning of 1775 there apparently were no operating powder mills in the colonies, although several had existed earlier. But within the next two years numerous mills began operating, including five near Philadelphia alone, although the cumulative output of all the mills couldn't fulfill the entire wartime need. In New York City in July 1776 when an equestrian statue of King George III was torn down it yielded an estimated 4,000 pounds of lead for bullets. Paper suitable for cartridge making also was a scarce commodity. There even exist today rare cartridges of Revolutionary War vintage in which paper currency was used. In camp at Valley Forge in April 1778 Washington directed that any noncommissioned officer or soldier who discharged his musket or otherwise wasted ammunition was subject to harsh discipline. However, there was some authorized firing when in camp as an easy means of unloading muskets when sentries or a detachment returned from outpost duty or scouting might fire into a bank of earth. Later the lead could be dug out and remolded into balls.

To discourage the theft of muskets, those which were not privately owned sometimes were stamped or branded in the wood "US," "UNITED STATES," or "U STATES" in accordance with a February 1777 resolution by the Continental Congress that all government owned weapons were to be so identified. Jeremiah Greenman in his diary in the spring of 1777 while near Morristown, New Jersey, mentioned an order "for all our arms to be carried to town to have them stamped US."[6]

Initially the Continental Army was an untrained "rabble in arms," as one British general called it. A few officers and men had served in the French and Indian War, but generally there was little military experience within this new army. Enlistments were short term and whoever could raise a company of fifty men became a captain and 500 men a colonel, regardless of qualification for command. The army was held together largely through Washington's leadership and personality.

But in early 1778 the situation improved with the arrival of a Prussian officer, Baron Frederick Wilhelm Augustus von Steuben, who with encouragement from Benjamin Franklin and Silas Deane offered his services, at first without pay. Washington made him a major general and at Valley Forge thrust upon him the job of training the undisciplined and ill-equipped army. Von Steuben arrived with his giant Russian wolfhound named Azor and spoke very little English but instructed through personal example and interpreters. His hearty swearing in a mixture of German, French, and English on the field undoubtedly produced numerous clandestine grins as he gradually turned the army into an effective and confident fighting force. The training manual he developed remained in use for several decades after

5. Brown, *Firearms in Colonial America,* 319. Throughout the first half of the eighteenth century, Spain had an enviable reputation of producing some of the highest quality European gun barrels.

6. Ahearn, *Flintlock Muskets in the American Revolution,* 47-48.

the war. In addition, he established standards for camp sanitation and layout and also urged officers to treat their men with respect and kindness.

It appears that the baron's military career was exaggerated by Franklin and Deane; he had never been a general in the Prussian army and in fact was never ranked higher than a captain. However, there is no question about the invaluable service he rendered to Washington's disorganized army. As the baron later recalled: "The arms at Valley Forge were in a horrible condition, covered with rust, half of them without bayonets, many from which a single shot could not be fired. The [cartridge] pouches were quite as bad as the arms. A great many of the men had tin boxes instead of pouches, others had cow horns; and muskets, carbines, fowling pieces, and rifles were to be seen in the same company."[7]

Efficient handling of one's musket was but one skill the baron demanded, emphasizing rapidity of loading and firing. In simple terms, the soldier brought the hammer to the half cock position, bit off the end of the paper cartridge, poured a little powder in the pan as priming and tipped the frizzen back to cover it, then poured the remaining powder down the barrel, rammed the ball with the cartridge paper on top of it to keep the ball in place, brought the hammer to full cock and was ready to fire. The cartridges often were made by contractors, sometimes by children as young as 12. The government in 1781 was paying one contractor five cents for each cartridge.

Some iron or brass bullet molds were provided to soldiers in the field, gang molds with multiple cavities enabling them to cast round balls and buckshot, and ladles in which to melt lead. War by its nature is not a civilized affair, and at least early in the war both sides are known sometimes to have been guilty of firing musket balls altered to increase the severity of wounds. Most vicious was the use a musket ball cut in half and affixed to the ends of a nail, similar on a small scale to bar shot used in naval cannon to destroy an enemy ship's rigging. A hollowed out bullet also offered a courier or spy a means of concealing a message. In October 1777 a captured British officer was seen to swallow a bullet. Before he was hanged, he was forced to swallow "emetic tartar" and the recovered bullet was found to contain a message from General Clinton to General Burgoyne.[8]

Riflemen usually carried their own mold of a size appropriate for the caliber of their particular rifle, typically a single cavity mold. Other accessories common to musket and rifle alike included a vent pick to keep the touchhole clear, a brush to clean the pan of powder residue, a screwdriver to tighten the jaws of the hammer around the flint, and perhaps a leather cover to wrap around the breech to keep the lock and powder dry in inclement weather.

The Military Role of the Pennsylvania Rifle

Despite the popularly held view, the Revolutionary War was not decided by rifle-armed Americans fighting from behind rocks and trees. The Pennsylvania rifle was not well suited for military use, and its employment in that struggle was limited. The Continental army relied largely on smoothbore muskets and standard mass formation European tactics of that period. Note that the silver flintlock musket and not the rifle became the centerpiece on the army's combat infantryman's badge in the twentieth century.

The Pennsylvania rifle—call it the Kentucky rifle if you prefer—unquestionably was far more accurate than the smoothbore musket of any nation, and it had seen modest use during the French and Indian War (1754-63) in the hands of frontiersmen. British Major George Hanger, taken in 1777

7. Rudolf Cronau, *The Army of the American Revolution and Its Organization* (New York, 1923), 15.

8. Ahearn, *Flintlock Muskets in the American Revolution*, 44-45, 47. During the American Revolution British soldiers with grim humor sometimes referred to buckshot as "Yankee peas."

as a prisoner of war by the Americans, later wrote that he often had been told by backwoodsmen that skilled riflemen consistently could hit a target the size of a man's head at 200 yards. Hanger, himself an excellent marksman, felt if he was standing still a rifleman could hit him at 300 yards.

Among the first volunteers to reach the Continental Army encamped outside of Boston in the summer of 1775 were riflemen from Pennsylvania, Maryland, and Virginia clad in long hunting shirts and organized into 12 companies to act as light infantry, serving as scouts and skirmishers. One was commanded by fiery Daniel Morgan, who would rise to the rank of brigadier general. In camp before admiring onlookers, these men readily demonstrated their surprising marksmanship with their long rifles. On the way to Boston in August 1775, one rifle company was reported to have conducted such a demonstration. A rifleman took a board about five inches wide and seven inches long and held it between his knees while his brother from a distance of 60 feet or so without a rest put eight shots through it without injury to his kin. When spectators expressed surprise at the feat, they were told that many men in the company could do the same. Not all riflemen were marksmen but, being used to the challenges of life on the frontier, these men might endure hardships better than their more civilized contemporaries from the seaboard farms and towns. They often were hard drinking, hard fighting, and reluctant to accept rigid military discipline, but effective when used appropriately and with competent leadership.

Writing of those riflemen who marched to Cambridge, Massachusetts, that summer, one observer described them as: "remarkable for the accuracy of their aim, striking a mark with great certainty at two hundred yards' distance. At a review, a company of them, while on quick advance, fired their balls into objects of seven inches in diameter at the distance of two hundred and fifty yards. They are now stationed on our lines, and their shot have frequently fatal to British officers and men who expose themselves to view, even at more than double the distance of common musket shot."[9]

A claim of such accuracy at 250 yards was unrealistic and may have been a planned exaggeration for propaganda purposes. However, the rifle's greater accuracy over that of a smoothbore musket was achieved at the expense of speed in loading. First the rifleman poured a measured charge of powder from his horn down the barrel, then placed a greased or spit-moistened cloth or buckskin patch on the muzzle, removed a round lead ball from his leather hunting bag and seated the ball in the muzzle with his thumb, then pushed the patched ball down the barrel with the wooden ramrod. He moved the hammer to the half cock position and from a smaller horn poured a little finer grain, faster burning priming powder into the pan, cocked the piece, tilted the frizzen back into place to cover the pan, and was ready to fire. A musketeer, using a paper cartridge containing both powder and an undersize ball, could load in roughly half the time. In an emergency to speed loading a rifle or in the event the barrel became heavily fouled with powder residue, the patch could be left off the ball but with some loss of accuracy. Loading a rifle also could be speeded a little by using a flat piece of wood with holes bored in it, each hole of a size to hold an already patched ball. Positioning this loading block with a hole aligned over the muzzle allowed the rifleman to load with his ramrod without handling either patch or ball.

Another disadvantage of a Pennsylvania rifle was that its octagonal barrel could not accept a bayonet without alteration. Furthermore, the rifle was fabricated as a civilian hunting gun, with a stock that might not be sturdy enough to withstand the added strain of using it with a bayonet in place or as a club in hand-to-hand combat. When in the fall of 1776 Maryland offered to add a rifle company to the Continental army's strength, it was rejected by the Secretary of the Board of War. The

9. Hartzler, *Arms Makers of Maryland,* 41.

army would be pleased to have the men, but they should be armed with muskets "as they are more easily kept in order, can be fired oftener and have the advantage of Bayonetts."[10]

In May 1777, General Washington created a "corps of rifle troops" made up of marksmen chosen from the Continental Army at large to serve on detached duty as needed. At the end of a campaign their publicly owned rifles would be placed in storage and the men returned to their respective regiments. At his direction, a form of folding spear was issued to some of these riflemen to counter their lack of bayonets since riflemen, one British officer noted, often could be routed with the bayonet as they reloaded after they had fired a single shot. The Continental Army's General "Mad Anthony" Wayne was not the only officer who had a decided preference for the musket and bayonet.[11]

Limited though the use of the rifle was, riflemen could and sometimes did have a profound effect on an enemy when concentrating their fire on officers. One of Daniel Morgan's riflemen, Timothy Murphy, had a reputation as being a particularly skilled marksman. At the battle of Saratoga, New York, in 1777 from his position in a tree, he fatally wounded Britain's General John Frazier, a loss which helped demoralize the redcoats that day. Murphy's rifle sometimes has been described as an over-and-under double-barrel Pennsylvania rifle. However such guns were quite rare until early in the nineteenth century, and contemporary accounts of the battle and Murphy's exploit make no reference to such a rifle. It's far more likely his rifle was a typical single shot.

After that same engagement, an American doctor recalled: "A number of our soldiers [riflemen] placed themselves in the boughs of high trees, in the rear and flanks, and took every opportunity of destroying the British officers by single shot; in one instance, General Burgoyne was the object, but the aide-de-camp . . . received the ball through his arm, while delivering a message to Burgoyne; the mistake, it is said, was occasioned by having his saddle furnished with rich lace, and was supposed by the marksman to be the British commander."[12]

Frontiersmen experienced in fighting Indians at home on the frontier with their long rifles also were valuable in campaigns in which American Indian adversaries were expected to be encountered. Use of the rifle perhaps was more common in the Carolinas than in the north since in the south much of the fighting was done by frontier militiamen rather than units of the Continental Army. Washington at one point advocated the adoption of the rifleman's dress, including the knee-length rifle shirt or frock. "It is a dress which is justly supposed to carry no small terror to the enemy, who think every such person a complete marksman."[13]

However riflemen, unless supported by quicker loading muskets and bayonets, were at a decided disadvantage if they attempted to employ the common open field massed formation infantry tactics of the day. As one German officer after the 1776 battle of Long Island put it, with some exaggeration and apparent arrogance: "The greater part of the riflemen were pierced with the bayonet to the trees. These dreadful people ought to be pitied rather than feared; they always require a quarter of an hour to load a rifle, and in the meantime they feel the effect of our balls and the bayonet." Among the estimated 160,000 militiamen and 230,000 men of the Continental Army who fought in the war, the rifle's presence was uncommon. Yet Britain too made modest use of rifle guns during the war.

10. In the face of a shortage of rifles as well as other arms, the Continental Congress in June 1776 directed General Washington to purchase those privately owned "riffles" from riflemen who refused to reenlist. (Moller, *American Military Shoulder Arms, Vol. 1*, 179).

11. James L. Kochan, "Continental Army Rifle Corps, 1778-1779," *Military Collector and Historian*, 52, no. 4, Winter 2000, 185.

12. Huddleston, *Colonial Riflemen in the American Revolution*, 44.

13. Ibid., 16.

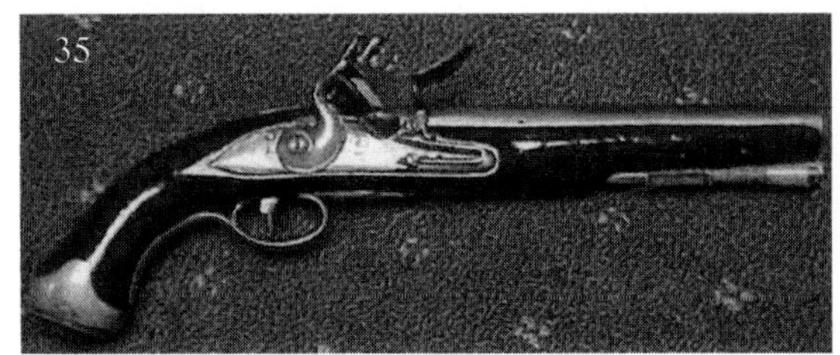

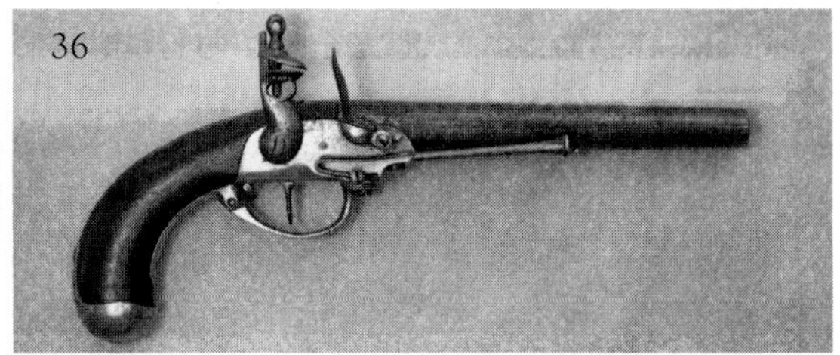

35. British Light Dragoon pistol of 1759 pattern, the standard enlisted dragoon's handgun of the Revolutionary War. (*Courtesy: Neil and Julia Gutterman*) **36.** North & Cheney, first official model pistol adopted by the U.S. government.

As early as 1746, Britain had acquired a few rifled carbines for the expedition to relieve the captured fortress of Louisbourg in Nova Scotia. A dozen were issued to engineers accompanying General Braddock's ill-fated march on Fort Duquesne at today's Pittsburgh in 1755, and there were other instances of rifled guns being issued during the French and Indian War. But in 1775, Britain had only a limited number of rifle-armed companies, although there were some rifles dispersed among light infantry of regular regiments. Britain in 1776 did produce 800 .65 caliber rifles, brass mounted with a 29-inch octagonal barrel and an iron ramrod initially secured at the muzzle with an innovative swivel. Two hundred more such rifles were made in Hanover, Germany, all of which later were shipped to America. In addition, there were perhaps as many as 4,000 short *jaeger* rifles (generally without a bayonet) in the hands of some of the Hessian and other German troops hired by King George III to fight in this country. Thus it is possible that there were more rifles in regular use in British and German hands than within the American army.[14]

Military Pistols, Carbines, Musketoons, and Wall Guns

Military pistols during the Revolutionary War in both armies were of lesser importance than shoulder arms. They were intended for use at close range and usually were carried only by mounted officers, seamen, and such horsemen as dragoons, often issued in pairs and generally carried in holsters crossed over the horse's withers at the front of the saddle. Britain's mounted troopers during the war generally carried the light dragoon pattern adopted in 1759 with a .65 caliber nine-inch barrel. British seamen's pistols were fitted with a belt hook on the left side. Enlisted men's handguns made in this country often were patterned after British dragoon models. Officers' pistols, sometimes rifled, occasionally silver mounted, and sometimes with a brass rather than iron barrel, generally were privately owned and were better balanced and better finished than those of standard military issue. Those of British manufacture predominated among American officers, particularly during the first years of the war, and British influence on the limited number of pistols made in the colonies whether for civilian or military use was strong. But as the war progressed, imported French military arms included pistols as well as muskets. When in 1799 the U.S. government contracted for its first official pistols for the army, 2,000 horse or dragoon pistols to be made by Simeon North of Berlin, Connecticut, with Elisha Cheney's backing, they were patterned very closely after the French model of 1777 (Figures 35, 36).[15]

14. De Witt Bailey, *British Military Flintlock Rifles, 1740-1840* (Lincoln, R.I., 2002), 11-12, 28, 199-200.

15. Neumann, *Battle Weapons of the American Revolution*, 236.

General George Washington is known to have owned many guns during his life, including a pair of silver-mounted brass barrel holster pistols made by John Hawkins of London. An audacious thief apparently stole one pair of his pistols, for the general's expense account entry for September 1, 1775, at Cambridge, Massachusetts, noted he paid 1 pound 10 shillings for their recovery and subsequent repair. In March 1776 while inspecting American defenses at Dorchester Heights overlooking Boston, he lost another, one with a "turn-off" barrel which he described as "a screw barreled pistol mounted with silver and a head resembling that of a pug dog at the butt." Again on July 2, 1777, a letter described yet another of his pistols which was lost or stolen: "You will know it by being a large brass barrel and the lock is also brass, with the name of Gabbitas, the Spanish armorer, thereon. It also has a heavy brass butt. His Excellency is much exercised over the loss of this pistol, it being given him by Gen. Braddock, and having since been with him through several campaigns."[16]

A ledger kept by Colonel Josiah Harmar noted that in November 1782 in South Carolina he purchased from a British deserter "an elegant case of pistols, silver mounted, for which I paid him six guineas." These probably are the same silver-mounted pistols which Harmar wrote he had in 1787 when as a brigadier general he was seeking to rid the Ohio country of hostile Indians.[17]

One distinctive style of military pistol used during the war but not copied by American makers was that which armed officers and men in those Scottish infantry regiments which served with the British army in North America during the French and Indian and Revolutionary wars. These were the only infantrymen in the British army to be issued handguns, and although in 1776 they were ordered to discard broadswords and pistols, some such pistols remained in continued use. These pistols were made in both Scotland and England and were distinctive in design having a stock of iron or brass rather than wood, no trigger guard, and a trigger often ending in a ball shape at its lower end. Major John Pitcairn of the Royal Marines had such a pair when present at the Battle of Lexington in April 1775. His pistols were captured and exist in a museum today near the battle site (Figure 37).[18]

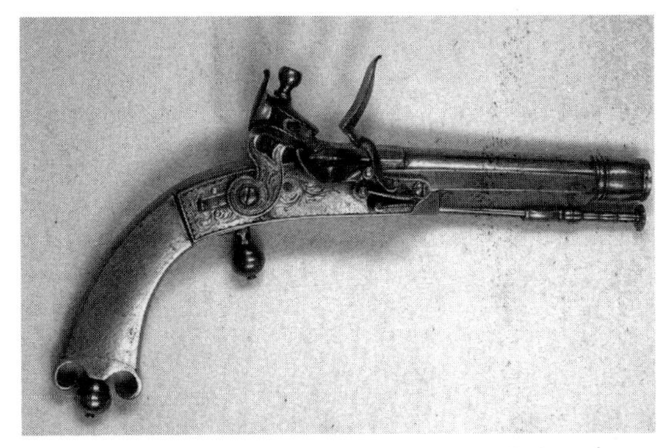

37. Scottish pistol of post-Revolutionary War vintage, somewhat more elaborately finished than the usual military style but illustrating the general configuration of those used by such Scotch regiments as the famed Royal Highlanders (Black Watch). The ram's horn butt shown here was common as was the round head screw-out vent pick located between the "horns." A belt hook on the left side of such pistols used by infantrymen and junior officers allowed the pistol to be suspended from a strap hanging over the left shoulder. (*Courtesy: Michael F. Carrick*)

Cavalry or dragoon units in the Continental army were limited in number and often fought on foot as well as on horseback. Their weapons were swords and pistols. When shoulder arms were available and carried by mounted troops, these were carbines or musketoons, terms usually used interchangeably at that time and applicable to shoulder guns that had a smaller bore than a musket and sometimes were a little shorter than the standard

16. Brown, *Firearms in Colonial America*, 322.

17. William H. Guthman, *U.S. Army Weapons, 1784-1791* (American Society of Arms Collectors Publication, 1975), 20-21.

18. Neumann, *Battle Weapons of the American Revolution*, 244, 246.

infantry musket. It's possible that no such shoulder arms were made in the colonies and when long guns were needed and British or French carbines weren't available, infantry muskets might be pressed into service. Washington even suggested the use of blunderbusses by dragoons, but there is no evidence this was done.[19]

One other unusual form of firearm used during the war was then called the amusette, a wall or rampart gun. An oversize weapon, it was too heavy to be fired from the shoulder and instead was fired from a wall or palisade or on a swivel in a small boat and could be used to defend a frontier fort or even fire against vessels on a river or close to shore. Although today the terms "wall gun" and "rampart gun" are often used interchangeably, they were listed separately in eighteenth-century returns, and the latter may be the lighter and more readily transported of the two styles. Several examples of rifled wall guns about five feet long and weighing 50 pounds exist today which were made at the Rappahannock Forge across the river from Fredericksburg, Virginia. There is a report of one rifled wall gun firing a two-ounce ball and striking a half sheet of paper at 500 yards. The wall gun concept didn't die at war's end, for in 1807-8 seven more were produced at Harpers Ferry. Then in 1847 four of .75 caliber were built at Harpers Ferry and at least one at Springfield, with about a 30-pound barrel, firing about a three-ounce conical ball, sighted for use up to 800 yards, and mounted on a tripod "adapted to be carried on a pack horse, and to be managed by one man."

Naval Small Arms

American naval forces during the war consisted of three elements—the Continental navy, state navies such as those of Massachusetts and Virginia, and hundreds of privateers, privately owned armed vessels issued letters of marque authorizing them to prey on an enemy nation's shipping. As was true of the Continental army, these American sea-going forces had to rely on foreign arms for many of their pistols and muskets, predominantly ones of British and French origin plus some obtained from American contractors. These locally produced ship's muskets are difficult to identify for little or no documentation exists describing their characteristics. They often were a few inches shorter than the standard infantry musket and sometimes were coated with japan blacking to resist corrosion.[20]

Muskets often were used by marines and seamen in a vessel's tops firing down on the enemy's decks, sometimes successfully picking off officers, as later happened to Britain's famed Lord Horatio Nelson at the battle of Trafalgar in 1805. (Ironically, Nelson on his flagship *Victory* refused to allow the use of small arms in the rigging for fear of setting sails or lines on fire.) A few rifles too saw naval use for the same purpose both onboard navy and privateer ships, but the number was small and the guns undoubtedly were Pennsylvania-style rifles. The blunderbuss also proved useful on shipboard, ranging in size from compact shoulder-fired pieces to larger guns mounted on swivels, all fired as a form of shotgun whether from tops or deck rails or the gunwale of a small boat. This presence of blunderbusses on privateers, naval vessels, and merchantmen persisted at least through the War of 1812.

19. American forces apparently made some use of blunderbusses during the late 1700s since there was a total of 68 listed in 1793 returns from Philadelphia, Fort Pitt, and Fort Washington. An 1810 military dictionary published in Philadelphia described a blunderbuss as: "A well-known firearm, consisting of a wide, short, but very large bore, capable of holding a number of musket or pistol balls; very fit for doing great execution in a crowd, making good a narrow passage, defending the door of a house, staircase, etc., or repelling an attempt to board a ship." Guthman, *U.S. Army Weapons, 1784-1791*, 21.

20. Britain had its own sea service musket, patterned closely after the Brown Bess but with a shorter barrel, usually about 37 inches. Moller, *American Military Shoulder Arms, Vol. 1*, 236.

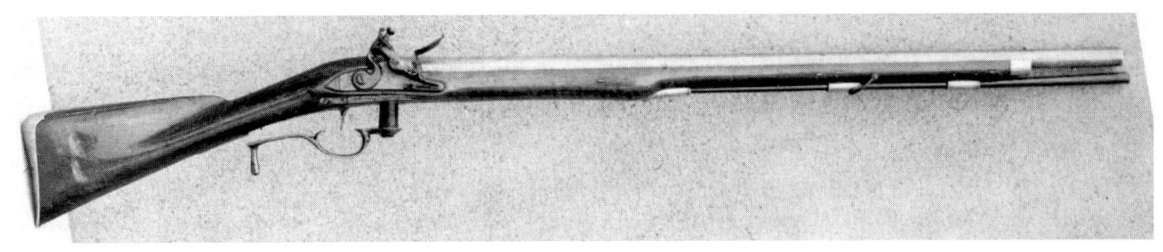

38. Ferguson breech-loading rifle, shown with the trigger guard and breech screw in the down position to expose the breech for loading. (*Courtesy: National Park Service, Morristown National Historical Park, Morristown, New Jersey*)

Handgun use extended to common seamen as well as officers since these, along with axes, cutlasses, and pikes, were useful for hand-to-hand boarding operations and for landing parties ashore. Officers' pistols often were privately owned, and it wasn't until 1797 that the American navy issued its first contracts for handguns, at least 600 of which were delivered by a half dozen makers. In 1808, Simeon North received a contract for 3,000 .64 caliber single shot flintlock handguns (the caliber was later increased to .69). One feature of the Model 1808 and found on most of this nation's naval flintlock pistols throughout much of the flintlock era was an iron belt hook on the left side opposite the lock enabling a sailor to carry the weapon suspended from a belt or sash. The navy normally allotted pistols to ships on the ratio of two per gun (cannon), but it seems this number often was exceeded.

Early Attempts at Military Repeaters and Breechloaders

By the early 1500s there had been attempts at making repeating firearms as well as ones which loaded at the breech rather than from the muzzle. Few approached practicality. In the mid-1600s Michele Lorenzoni of Florence, Italy, was making a lever-operated repeating magazine flintlock utilizing two magazines in the butt, one for powder and the other for balls, plus a smaller magazine in the lock for priming powder. Other repeaters drawing on this design were made in England in the later 1600s by Abraham Hill and John Cookson and in the 1750s in Massachusetts by another John Cookson. Such guns were neither common nor inexpensive and were not practical for military application. In 1777 Joseph Belton of Philadelphia offered Congress an eight-shot repeater and, although that body agreed to test 100 muskets converted to his system, there is no evidence that all these guns were ever produced. Belton's repeating musket used an old concept, that of multiple charges of powder and ball loaded one on top of the other. Each ball was pierced and contained a fuse mixture which burned through to ignite the powder behind it, so when the first charge was fired in the usual manner, the remaining loads behind it ignited one after the other a few seconds apart.

Britain's army in 1762 had experimented with screw plug breech-loading rifles and 15 years later such a design, although not a repeater, did see very limited service during the Revolutionary War in British hands. The inventor was Captain Patrick Ferguson and his rifle incorporated a screw plug at the rear of the barrel which was lowered by one complete revolution of the trigger guard, exposing the chamber into which a ball and then loose powder were loaded. Rotating the guard in the opposite direction closed the breech and the gun was ready to prime and fire. It was simple to operate, and in 1776 Ferguson put on an impressive demonstration for Britain's Board of Ordnance which, despite rain and high winds, included firing a surprising six shots in one minute and four shots in the same length of time while advancing at a walk. Only about 100 of these .62 caliber rifles were sent to America to equip his green-clad corps of light infantry. They are known to have been used at the 1777 battle of Brandywine where the inventor was wounded seriously in his right arm, but soon after the battle the corps was dissolved. Later in 1780 Ferguson was killed by American rifles at the critical battle of King's Mountain in South Carolina by which time interest in his breechloader had largely disappeared. As advanced in design as his rifle was, a primary weakness was in the stock at the breech area where much wood had to be removed making it difficult to maintain under the rigors of field service (Figure 38).[21]

At Brandywine near Philadelphia, Ferguson may have held the fate of the new nation in his hands. Armed with one of his rifles, he encountered a mounted American officer of obvious prominence who despite Ferguson's order to halt slowly continued on his way. "I . . . could have lodged a half a dozen balls in or about him before he was out of my reach . . .. but it was not pleasant to fire at the back of an unoffending individual who was acquitting himself very coolly of his duty so I let him alone." Evidence exists that the man in his sights probably was General George Washington. (Despite Ferguson's account of the incident, some have questioned whether it ever occurred.)[22]

Post-Revolutionary War Military Arms

Independence brought with it the need for the infant government to develop new sources for arms rather than to continue reliance on foreign sources. However, to prepare for any national emergency, the emphasis was on the militia for defense rather than a large standing army. In 1792 Congress established a uniform state militia system and directed that except for exemptions all free white males between eighteen and forty-five years of age could be enrolled. Members were to arm themselves with a musket, bayonet, accouterments, and at least twenty-four cartridges or a rifle. Private contractors who had received orders from Committees of Safety and others during the Revolutionary War rarely had been able to provide even a portion of the numbers for which they had contracted. Congress in April 1794 took a major step forward. It authorized the construction of government armories at Springfield, Massachusetts, already the site of an arsenal established early in the Revolutionary War at General Washington's direction, and at the picturesque confluence of the Potomac and Shenandoah Rivers at Harpers Ferry, Virginia (now West Virginia) for the production of small arms. Musket manufacture began first at Springfield in 1795 and five years later at the new armory in Virginia. By 1813 there would be 231 employees at the latter facility. Much effort at the armories would be devoted to repairing small arms as well as their manufacture.[23]

Some foreign arms remained in use into the early 1800s as production at the armories gradually evolved. After 1782, many of these foreign muskets not already marked were stamped to indicate government ownership. Generally this was done in a more workmanlike manner than during the war and the marking "US" was applied to barrel and lock plate. Between 1782 and 1785, armorers on contract with the Continental Congress stamped more than 33,000 muskets to identify them as gov-

21. Bailey, *British Military Flintlock Rifles*, 19, 37, 55.

22. Ibid., 48.

23. The Springfield Armory was immortalized in 1844 by Henry Wadsworth Longfellow after he composed the poem "The Arsenal at Springfield" following a visit there when his new bride commented how much the storage racks of arms resembled an organ. It begins: "This is the Arsenal, from floor to ceiling,/Like a huge organ, rise the burnished arms;/But from their silent pipes no anthem pealing/Startles the villages with strange alarms./Ah! What a sound will rise, how wild and dreary,/When the death-angel touches those swift keys!/What loud lament and dismal Miserere/Will mingle with their awful symphonies!"

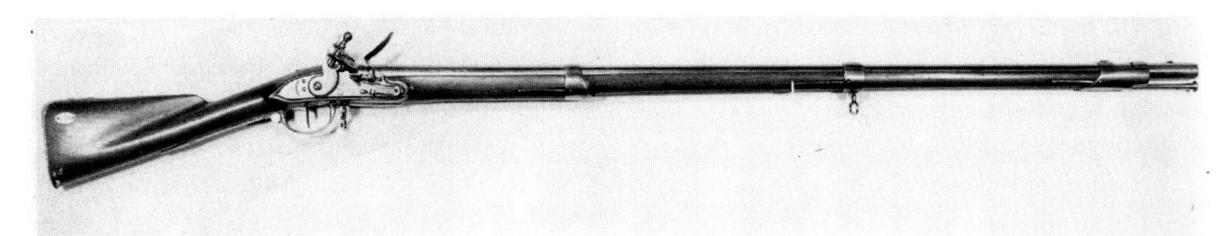

39. Springfield Model 1795 .69 musket. (*Courtesy: National Park Service, Fuller Gun Collection, Chickamauga and Chattanooga National Military Park*)

ernment property. There was no official designation as the Model 1795, which is the somewhat erroneous but common modern terminology for those early U.S. muskets made at Springfield and Harpers Ferry. In government records they were designated by such terms as "Charleville Pattern" or "New Muskets, Charleville Pattern." There were distinct differences between the guns made at the two armories and numerous evolutionary changes among the muskets produced at the same armory. Interchangeability of parts was not a consideration and it was not until about 1816 that an effort was made to standardize the muskets made at the two sites (Figure 39).[24]

During the Revolutionary War, the shortage of arms had forced sometimes rather crude repairs of damaged muskets in the field, such as keeping muskets with broken stocks in service. In the mid-1780s immediately after the war at national armories more professional efforts were made to clean and repair damaged muskets. Stocks broken at the wrist sometimes were mended by gluing the pieces together and adding two screws or wrapping them with wire or heavy cord, on occasion adding two neat brass plates before applying the wrapping. On some occasions stocks were repaired by splicing undamaged wood salvaged from others. Some of the Revolutionary War period flintlock muskets found today with professional appearing stock repairs may be examples of such post-war salvage operations.[25]

In 1798 tensions between the United States and its former ally France reached a critical stage and war seemed a definite possibility. Despite large quantities of arms remaining after the Revolutionary War, many of these had been subsequently poorly stored, stolen, dispersed to frontier militia, or otherwise lost. Now Congress was forced to seek bids from private contractors to supplement those limited numbers of available arms. Twenty-seven makers replied, but not all actually produced weapons under these contracts. The usual contract cost to the government for a musket with bayonet, ramrod, wiper, ball screw, and touchhole pick and brush was about $11. Previously in 1797 both Virginia and Pennsylvania had issued contracts for Charleville pattern muskets for their own state militia organizations. Some contractors also produced muskets for private sale to such as seagoing privateers during the Quasi War with France (1797-1800) and the War of 1812. The government was encouraging the development of a domestic arms industry, but the U.S. government also purchased some 11,000 Brown Bess muskets from England between 1798 and 1800. The lock was the most complex and difficult part to manufacture, particularly in most contractors' modest facilities. In the late 1790s the government ordered some musket and rifle locks from the firm of Thomas and John Ketland of England through their Philadelphia sales office.[26]

24. Brown, *Firearms in Colonial America*, 389 (n. 113). Muller, *American Military Shoulder Arms, Vol. 1*, 160.

25. Guthman, *U.S. Army Weapons, 1784-1791*, 13.

26. Chartrand, *Uniforms and Equipment of the United States Forces in the War of 1812*, 85.

Eli Whitney, Sr., of New Haven, Connecticut, had had no prior gun making experience yet he emerged as a leader among those contractors who received federal and state orders in the 1790s and later. He had earned fame but little initial financial gain for his invention in 1793 of the cotton gin, soon after graduation from Yale College. It took Whitney a full decade to deliver the 10,000 guns ordered under a 1798 contract, but in doing so he introduced significantly advanced manufacturing procedures to facilitate mass production. Later in 1818 Whitney would be recognized and designated by the government as a "private armory" along with Henry Deringer, Simeon North, Asa Waters, Nathan Starr, and Lemuel Pomeroy. Except for Deringer of Philadelphia, each was located in Connecticut or Massachusetts.

Whitney often is credited with developing the system of interchangeability of parts as applied to firearms manufacture and of making musket parts by machine. Such is inaccurate, but he did achieve a major advance in industrial development in the United States by designing tools, jigs, and machinery that could be operated by unskilled workers to turn out large quantities of parts, previously a tedious procedure involving much hand work. True interchangeability of parts from one gun to another in American military arms would not be achieved until manufacture of the Hall breech-loading rifle two decades later.

Thomas Jefferson in 1798 while Minister to France had observed the use of interchangeable parts made by machines when he visited Honoré Blanc's musket factory in Paris. Despite Jefferson's enthusiastic promotion of the concept, Congress failed to grasp the significance of this technological advance and the idea remained dormant in this country for another twenty years. Jefferson took a particular interest in firearms and between 1773 and 1800 purchased at least three long guns and eight pistols. One of the former was a double barrel fowling piece purchased in France in 1797 for 60 francs. Only two of Jefferson's guns exist as properly authenticated, a pair of .44 English brass barrel pistols that the future president purchased in 1786.[27]

In 1808 despite the presence of two national armories, the government still had to rely on private arms makers in a crisis situation. Just such a critical incident occurred under Jefferson's administration in that year when the British ship *Leopard* stopped the American warship *Chesapeake* at sea and seized some American sailors claiming they were British subjects. Faced with an increasingly uncertain world situation, Congress authorized the War Department to contract for some 85,000 muskets, only about half of which eventually were delivered. At this time there were no detailed blueprints from which contractors could work. Instead, they were provided with a gun to be used as a pattern which was to be followed closely.

This also was the year in which Simeon North received a contract for navy pistols, each fitted with a belt hook. Until 1836 North was the major producer of pistols for the army, navy, and militia, making thousands of smoothbore military handguns in various models. Except for those navy pistols with a belt hook attached, flintlock and later percussion single-shot martial pistols were too large to carry conveniently on one's person. Instead the usual method of carrying these bulky handguns was in single or double saddle holsters. The army's pistols by North were .69 "musket" caliber until .54 "rifle" size bore was adopted in 1816, reducing the heavy recoil somewhat. They generally measured about 15 inches in total length and their bulk and weight were such that they easily could serve as a "skull cracker" when empty. The wooden ramrod

27. Brown, *Firearms in Colonial America,* 381. These Jefferson pistols also represent a rather common style of breechloader available throughout much of the eighteenth century with a screw off or "turn off" barrel which was unscrewed and removed so the powder charge and ball could be loaded in the chamber without the need for a ramrod, offering improved range and accuracy with its tight-fitting ball.

carried in a channel in the stock beneath the barrel was subject to loss or breakage until North's model of 1819 in which the metal rod was secured to the barrel by means of a swivel (Figure 40).

One exception to the frequently graceless and rather awkward appearance of the flintlock era U.S. army pistols was a .54 caliber smoothbore produced at Harpers Ferry Armory in the 1806-8 period. These apparently were intended primarily for the militia since no regular cavalry regiment existed between 1802 and 1808. Issued in pairs as "horsemen's pistols," the brass butt cap and flowing lines gave it an attractive styling and it has the distinction of being the first U.S. handgun made at a national armory. Captain Philip St. George Cooke of the 1st Dragoons was still carrying one in 1845, although perhaps by then converted to percussion, for in describing an impromptu bison hunt on horseback in that year he mentioned his success in downing a female with "my old Harper's Ferry 'buffalo slayer.'" In 1922 a pair of crossed Model 1805 pistols was approved as the insigne for the army's military police (Figure 41).

Far fewer in number than the muskets made under government contracts of 1808 were military style pistols ordered in 1807-8 from private arms makers. There was little commonality to their design but they usually were somewhat more graceful in appearance than many martial pistols, typically around .54 caliber, with a full length stock. A few of the names of makers known to have provided small numbers of such pistols, contracted for at about $10 per pair, include Joseph Henry, Jacob Cooke, William Calderwood, and John Shuler, all of Pennsylvania. Militia units of the post-Revolutionary War period and even later often were very parochial, showing little uniformity among the various groups in terms of weapons or dress. Wealthy commanders of militia companies, as in some major cities, sometimes garbed their men in gaudy and distinctive rather than practical uniforms, making a splendid showing at peacetime parades, shooting contests, and other public gatherings.

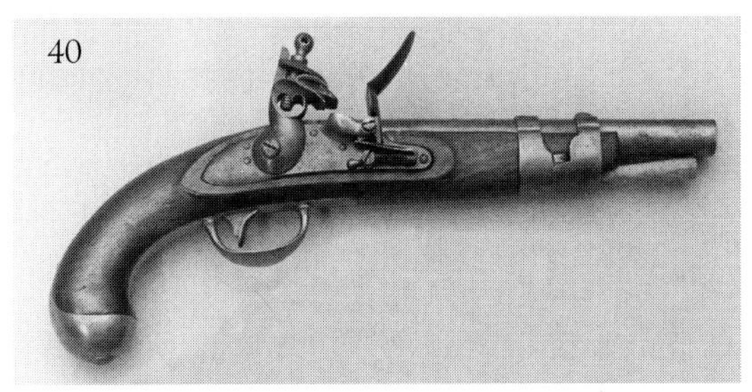

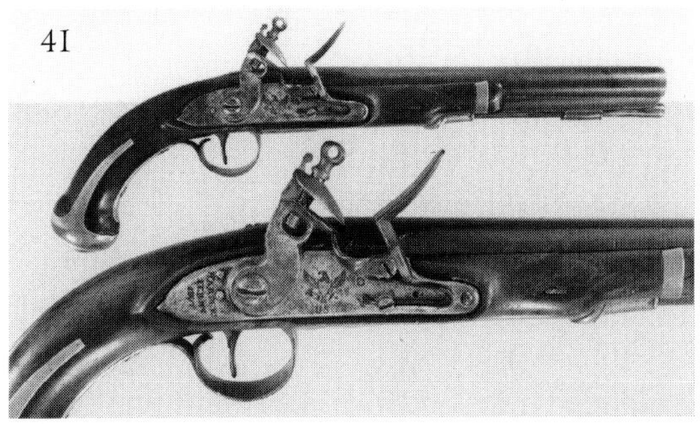

40. U.S. Model 1816 pistol by Simeon North. **41.** Harpers Ferry Model 1805 pistol, one of 4,096 produced. (*Courtesy: C. W. Slagle, photo by Ron Paxton*)

Riflemen had performed significant although specialized service during the Revolutionary War, but between the peace of 1783 and 1792, the U.S. Army had no rifle units. This emphasis on the musket continued in the immediate post war years and would do so well into the next century. Warfare between the army and American Indians in the Northwest Territory of Ohio and Indiana in the 1790s involved militia units as well as regulars and many of these militiamen who participated in the campaigns waged by generals Arthur St. Clair, Josiah Harmar, and Anthony Wayne carried their "long rifles" to war. Recognition of the rifle's usefulness was gradually growing and in 1792 President Washington ordered the army to be reorganized into the Legion with four sub-legions. Each of the latter was to include one company of artillery, two battalions of infantry, one troop of mounted dragoons, and a battalion of riflemen. Following this move, the government ordered approximately 436 rather plain but sturdy Pennsylvania-style "rifle guns" at about $10 each

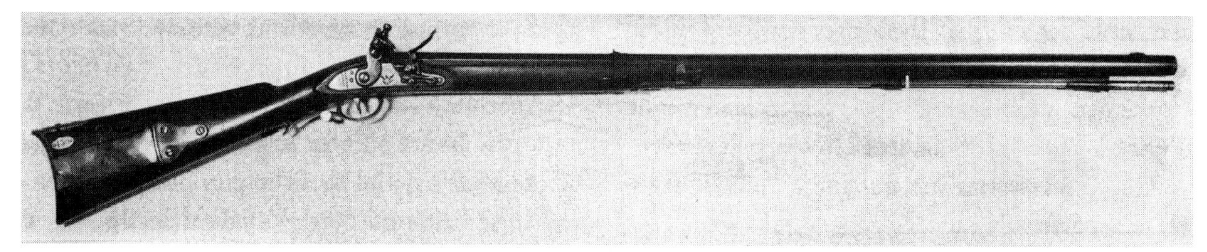

42. U.S. Model 1803 .54 Harpers Ferry rifle, considered by many to be one of the handsomest of U.S. martial long arms. (*Courtesy: National Park Service, Fuller Gun Collection, Chickamauga and Chattanooga National Military Park*)

from such Pennsylvania riflemakers as John Groff (80 rifles), Christopher Peter Gonter, Jr. (120), and Jacob Dickert (120), each of Lancaster. But riflemen disappeared from the regular army in 1796 and a rifle regiment didn't reappear until 1808.[28]

In 1803 in response to a shortage of rifles in public arsenals for issue to the militia, Secretary of War Henry Dearborn authorized the manufacture of the government's first regulation production rifle. It was .54 caliber with a barrel length of 33 inches, less than that of the typical 1792 contract rifles. It was a handsome halfstock weapon and the first rifled arm to be made at a national armory (Harpers Ferry). Its graceful lines and brass patch-box cover, trigger guard, and butt plate showed the influence of the Pennsylvania rifle on its design. There was no provision for a bayonet (Figure 42).[29]

The 1803 Harpers Ferry rifle for years was described as arming some of the members of Lewis and Clark's Corps of Discovery which journeyed up the Missouri River and on to the Pacific Ocean and back in 1804-6. It's more likely that the 15 rifles obtained at Harpers Ferry in addition to muskets for the remarkably successful expedition were from among those Pennsylvania rifles procured on contract beginning in 1792 or perhaps were prototypes of the Model 1803. Modifications made at Harpers Ferry to 1792 rifles intended for the corps could have involved such measures as reboring to the same caliber and perhaps altering the rifles to accept a common lock to facilitate repair and replacement of broken parts during the expedition.

One seemingly inconsistent entry in the annual Harpers Ferry production report for 1810 is for four harpoon guns. There seems to be no logical reason for a federal armory to be producing implements applicable only to the civilian whaling industry. A longtime friend of mine, Joseph M. Thatcher, Jr., seemingly solved the puzzle by discovering drawings of a torpedo or bomb patented by Robert Fulton in 1810. The device was attached by a line to a metal harpoon fired from a harpoon gun into the wooden hull of an enemy ship. Congress appropriated $5,000 to fund Fulton's experiments and these guns may well be associated with his project.

The infant U.S. Navy had ceased to exist after the Revolutionary War. But in the face of foreign depredations on American merchant shipping, Congress in 1794 began the rebuilding when it authorized funds for the construction of six frigates, including the famed *Constitution* and *Constellation.* The strength of the revitalized navy before the War of 1812 would ebb and flow with world situations which included the Quasi War with France and conflict with pirates along the Barbary Coast of North Africa (1801-5). Exact specifications for the muskets intended to be car-

28. Brown, *Firearms in Colonial America,* 362.

29. In a halfstock gun, the wooden forestock extends only about half way to the barrel's muzzle rather than close to the muzzle as in a fullstock. Halfstock rifles did not begin to achieve popularity with civilians until the 1830s or so.

ried on these new ships have eluded historians however the designation of "ship musket" appears frequently in navy documents.

Often these muskets for the frigates and the other new ships which followed were French style, many made by contractors in the Philadelphia area. In addition, perhaps as many as 500 British India pattern Brown Bess muskets with a .39-inch barrel were issued in 1805 to the Marines from stores at Harpers Ferry and were in use during the War of 1812. An arms requisition of 1797 for the first three U.S. frigates included 300 muskets, 250 pairs of pistols, and 44 blunderbusses in addition to 550 cutlasses, 300 boarding pikes and the same number of boarding axes.[30]

The War of 1812 and Its Aftermath

When the U.S. infantryman once again confronted his red-jacketed British antagonist in 1812-15, his musket was little different from that his predecessor carried during the War for Independence. Of course there now was a supply of American-made muskets available, but their production still required much hand filing and finishing. Substantial variations in tolerances in bore diameter did nothing to improve accuracy and the musket's only sight was a simple blade at the muzzle. Also hampering accuracy was the inconsistency of bullet weight and uniformity. The phrase was still appropriate when speaking of musket balls: "One went high, one went low, now where in Hell did the other one go?"

The powder charge of 160 grains of coarse black powder (minus the small amount placed in the pan as priming) produced a punishing recoil which quickly caused many soldiers to flinch; the resulting cloud of gray smoke could obscure one's vision on a windless day. However, the soldier who attempted to reduce that recoil by spilling some of the powder on the ground undoubtedly risked his sergeant's wrath if he was spotted. The .69 caliber ball weighed slightly less than an ounce and achieved a muzzle velocity of about 1,440 feet per second. The "buck and ball" cartridge retained its popularity, a result of the smoothbore musket's inaccuracy, as did a cartridge containing 12 buckshot and no ball. Loading and firing the flintlock musket (still referred to as a "firelock" in the 1814 infantry manual) continued to be done by command in a series of precise motions.[31]

Despite their inaccuracy, muskets of this period were relatively inexpensive and sturdy; they were proof tested with a powder charge of about two and one-half times the service load. In the hands of militia a musket was expected to last 10 years, a dozen years when issued to regular troops. As long as the enemy had no better weapons, the musket continued to be the dominant military firearm. Riflemen remained in the minority. Although the musket usually was loaded after first tearing open a paper cartridge, the rifleman normally loaded from powder horn and bullet pouch at this time.

30. William Gilkerson, *Boarders Away II* (Lincoln, R.I., 1993), 282, 284. An 1822 list of naval arms allowances included rifles and boarding pikes at a ratio of two per ten guns (cannon); cutlasses and muskets, three to every two guns; one blunderbuss to ten guns; and pistols to every two guns. Ammunition included fifty buckshot cartridges and one hundred ball cartridges per musket. Worth noting in this list is reference to one repeating pistol per five guns and one seven barrel "repeating swivel" to each ten guns. Rare today, these involve a family of multi-shot arms—long and short blunderbusses as well as pistols and muskets—developed by Joseph Chambers during the 1812-1815 period. They saw very limited use by the navy and their semi-secret development is well covered for the first time in *Boarders Away II.*

31. At this time the usual powder charge for a .69 pistol was about 80 grains, about 100 for a rifle. An 1834 memorandum listed the official charges as 144 grains for a .69 musket, 100 for a .54 rifle, 85 for a percussion Hall carbine, and 51 grains in a .54 flintlock pistol. The army's ordnance manual of 1849 continued to list three cartridges for the musket—single ball, buck and ball, and buckshot.

43. French eighteenth-century powder tester, resembling a pistol without a barrel and used to measure the strength and quality of gunpowder. A measured quantity of powder was placed in the chamber and when it was ignited, the relative strength of the sample was measured by the number of notches clicked off on the ratchet. Powder of that era was of inconsistent quality and was degraded by dampness. (*Exhibited at Fort Ticonderoga, New York*) **44.** Nineteenth-century one-pound can of DuPont "Summer Shooting Gun Powder." The many designs which appeared on powder containers by the various makers provide a colorful collecting opportunity. (*Courtesy: Neil and Julia Gutterman*)

Critical to the advance of military and civilian firearms technology was the availability of improved quality gunpowder. The census of 1810 showed that there were more than 200 powder mills in 16 states with Maryland leading in output. Powder manufacture was becoming an important industry in the new nation despite the hazards of operating a mill. Explosions caused by carelessness, lightning, sparks, or spontaneous combustion were common. One of the earliest mills to gain prominence in both quality and quantity produced was E. Irènée du Pont de Nemours located outside Wilmington, Delaware, identified with an eagle design as its logo. Alexander Wilson, a visiting Scottish ornithologist in 1804 described Pennsylvania frontiersmen in his poem "The Foresters" and paid tribute to du Pont powder.

From foaming Brandywine's rough shores it came,
To sportsmen dear its merits and its name;
Dupont's best Eagle, matchless for its power,
Strong, swift and fatal as the bird it bore.[32]

The national government was a major purchaser of powder, particularly during the War of 1812 period. Du Pont in 1812-13 provided at least 700,000 pounds of musket and cannon powder; other contractors included Whelen & Rogers, George Beidlemen & Co., and Randolph Ross of Richmond, Virginia. The most difficult to obtain of the three ingredients of gunpowder—saltpeter, sulfur, and charcoal—was the former, refined from deposits of nitrogenous wastes from animals and plants found in cellars, stables, caves, and elsewhere. More than 400,000 pounds during the war came from Kentucky caves alone. For two decades after the war, the army's powder supply built up during that conflict was sufficient for its needs even though by then it was overage. To test the strength of powder, two methods were used—either a spring tester or by firing a ball from a small eprouvette mortar at a set angle and with a fixed charge of powder. The distance the ball traveled determined the proof of the powder. The government also relied heavily on contractors to provide lead needed

32. Berkeley R. Lewis, *Small Arms and Ammunition in the United States Service* (Washington, D.C., 1956), 28.

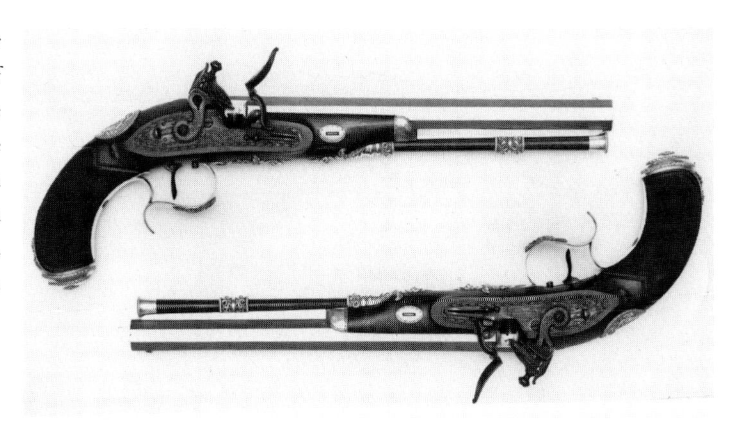

45. Pair of gold-mounted .52 caliber smoothbore pistols by Simeon North of Middletown, Connecticut, presented by the state's general assembly to Commodore Thomas Macdonough of the U.S. Navy in recognition of his victory over the British on Lake Champlain in September 1814. (*Courtesy: Smithsonian Institution, neg. no. 47085A*)

46. One of almost 40,000 Model 1817 "common rifles," all manufactured by contractors including this one by Henry Deringer, Jr. Some after conversion to percussion saw service in the Civil War. (*Courtesy: National Park Service, Fuller Gun Collection, Chickamauga and Chattanooga National Military Park*)

to make bullets, often cast at arsenals but occasionally by troops in the field (Figures 43, 44).[33]

During this second war with Britain, America's navy like the army relied on .69 caliber flintlock muskets for its seamen and marines, along with cutlasses and pistols. Private contractors accounted for many of these muskets, supplemented by production at Springfield Armory. Outfitters for privateers were welcomed by private contractors as buyers for muskets, for dealing with them meant no long wait for government payment of funds. Naval muskets were supplemented by a modest number of rifles, which apparently had not been authorized officially or in any significant numbers in the post-Revolutionary War sea service. But Captain John Rodgers of the famed frigate *Constitution* in 1809 requested a dozen for use in "Old Ironsides'" tops, and apparently some eventually were forthcoming. Wartime British documents confirm the presence of riflemen on American ships and these could have been armed with either examples of the Pennsylvania-style rifles for which the government contracted after 1792 or of the Model 1803 rifles being produced at Harpers Ferry (Figure 45).

Following the War of 1812 and into the late 1840s, the army musket remained predominant and basically unchanged except for minor improvements as well as better manufacturing techniques. Two models of flintlock rifles followed the 1803 Harpers Ferry, the short lived Model 1814 and that of 1817, both .54 caliber. These were full stocked and with an oval iron patchbox cover and

33. Ibid., 28. Civilians too sometimes relied on spring testers beginning in the early 1600s to gauge the quality of powder before purchasing. It helped reveal powder which had deteriorated or which dishonest sellers had deliberately adulterated.

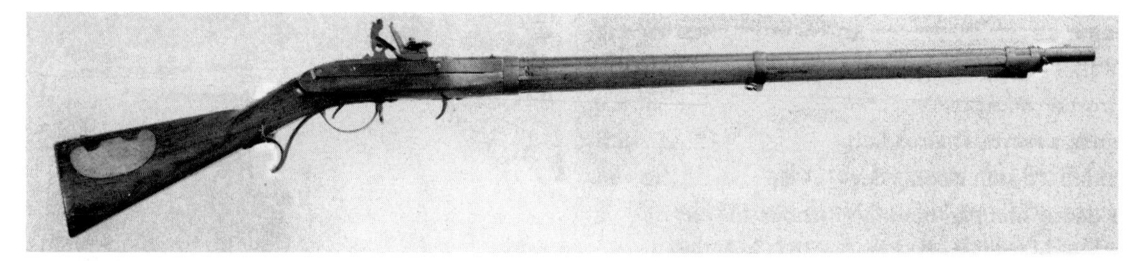

47. The silver plaque inscribed "By resolve of Congress, presented to Gustavus A. Bird, for his gallantry at the Siege of Plattsburg" identifies this Model 1819 Hall breech-loading rifle as one of fifteen presented to schoolboy heroes for their War of 1812 service. (*Courtesy: Smithsonian Institution, neg. no. 45072B*)

iron fittings rather than brass, guns not as aesthetically pleasing as their predecessor, but sturdier. All were made on contract, the Model 1817 accounting for almost 40,000 pieces, mostly produced in the 1820s, by Henry Deringer, Robert Johnson, pistol maker Simeon North, and Nathan Starr. Most of these rifles went to the states to equip their militia (Figure 46).[34]

Now a revolutionary new arm appeared, manufactured at Harpers Ferry (and later by Simeon North) under a patent issued to John H. Hall for a breech-loading rifle. He began production of his rifles in 1812 in Maine as rather expensive sporting arms (costing about $35) along with a handful of pistols. A spur lever protruding from the underside of the stock in front of the trigger guard pivoted the forward end of the breechblock upward to load. It had its faults, however. After extended firing powder fouling could hinder opening and closing the breech and gas leakage at the joint between the forward end of the breechblock and the barrel was objectionable. But it was faster to load than a muzzle loader and now a hunter or soldier could conveniently reload which lying or crouching behind cover.

The army approved an order for 100 rifles at $25 each in 1817 and, after testing, in 1819 ordered its first thousand, which didn't begin to reach the troops until 1824. Hall was hired at Harpers Ferry at $60 per month as assistant armorer and received $1 royalty for each rifle produced. Even though the Hall rifle was issued to the regular army and militia in both flintlock and later percussion form, the smoothbore musket remained the primary infantry arm until the mid-1850s. Yet Hall's rifle represented a milestone in the history of American military small arms, for it was the first breechloader to be issued in quantity and also was the first official arm which offered complete interchangeability of its parts. With the introduction of the Hall rifle into military service came the issue of metal powder flasks, which eventually replaced the army rifleman's powder horn. Adoption of the Hall inaugurated use of the term "common rifle" when referring to the army's various muzzle-loading rifles to distinguish them from this new weapon. Unlike the common rifles, which fired a patched ball, the Hall used a bare ball.

Prized among Hall rifles today is the handful of existing specimens of those 15 produced in 1824 and authorized by Congress to be presented to a group of men who as teenage schoolboys during the War of 1812 participated in the defense of a bridge against advancing British troops at Plattsburgh, New York, in 1814. Each gun was decorated with a large silver plaque set into the

34. Initial production of the Model 1803 Harpers Ferry rifle ceased about 1807 but resumed in 1814 in slightly modified form and continued until 1820. These later Harpers Ferry-made rifles sometimes today are referred to somewhat erroneously as the Model 1814 and should not be confused with the fullstock contractor-produced Model 1814 rifle.

buttstock commemorating the event and bearing the recipient's name. A smaller silver shield in the wrist bore the youth's initials and date of the battle (Figure 47).

The navy too obtained Hall flintlock rifles, and some were included among the arms issued in 1838 to Lieutenant Charles Wilkes's Pacific exploratory squadron along with muskets, flintlock pistols, and 150 of the unusual Elgin percussion cutlass pistols. In a way, this assemblage of flint and percussion arms marked the navy's transition to the percussion era beginning in the 1840s. In 1842 it adopted its first official percussion handgun, a "boxlock" in which the hammer was mounted on the inside of the lock plate rather than the outside. Less bulky than its predecessors—the barrel was only six inches long, These .54 caliber pistols lacked a belt hook and were produced by both Nathan P. Ames and Henry Deringer, Jr. Most were smoothbores except for a few by Deringer. The U.S. Revenue Cutter Service, forerunner of the Coast Guard, also obtained a limited number. Along with examples of earlier flintlock pistols converted to percussion ignition, these Model 1843s would serve onboard naval vessels until the Civil War era and the adoption of revolvers (Figure 48).[35]

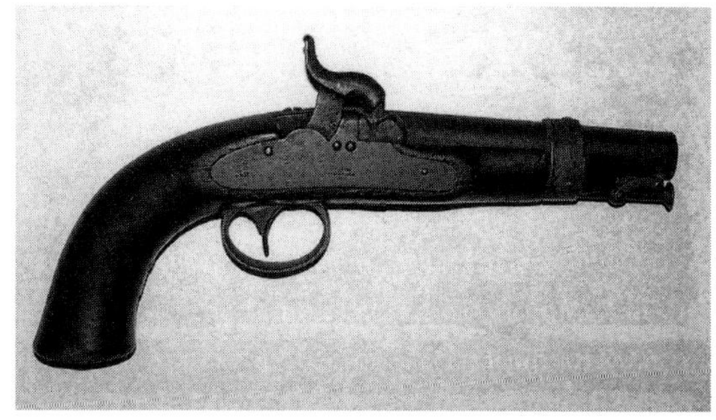

48. The first U.S. percussion martial pistol obtained on contract and issued, the Model 1843 made by two contractors, Henry Deringer and N. P. Ames. Most went to the U.S. Navy with a smaller number going to the U.S. Revenue Cutter Service.

35. These pistols by Ames and Deringer sometimes are called the Model 1843 and should not be confused with the slightly larger percussion Model 1842 pistols made by Henry Aston and Ira N. Johnson and at William Glaze's Palmetto Armory in South Carolina. In addition to its Model 1843 pistols, the Revenue Cutter Service also is known to have obtained some Jenks breech-loading carbines.

3 The Percussion Era and Advancing Technology

(1805–1866)

A Clergyman's Contribution

As the nineteenth century dawned, a Scottish Presbyterian clergyman, Alexander Forsyth, was conducting experiments which later would spark a giant leap forward in firearms technology. His work provided the basic theory for major advances in firearms ignition including the metallic cartridge of today. Prompted by his interest in hunting, he sought to speed the ignition time of the flintlock, that time between the trigger squeeze and the discharge of the ball or shot. His attention was attracted to fulminates, salts produced by dissolving metals in acid and which explode when struck. Attempts by others to use these salts rather than gunpowder as a propellant had been unsuccessful largely because of their instability and danger in working with them. Forsyth concentrated on using fulminates as a substitute for the fine priming powder in the flintlock's pan.

After much study and experimentation, in 1805 he demonstrated his first design, what commonly became known as his "scent bottle" gun lock. A pivoting magazine or reservoir deposited a small amount of fulminate in powder form in a channel leading to the main powder charge in the barrel, and a blow from the falling hammer exploded that fulminate. Once Forsyth's invention became known, other gun makers in Europe adopted the use of fulminate (often fulminate of mercury) as a primer, usually in the form of a pill, pellet, wafer, or tube. The most successful form of percussion primer was what's known as the percussion cap, a small copper cup with the fulminate inside its base covered with a tin foil disk and sealed with a bit of shellac to make it waterproof. The cap was placed on a hollow cone or nipple where it was crushed by the falling hammer, producing a jet of flame which was directed to the powder charge in the barrel. The cap was invented about 1816, probably in England. Various people claimed credit for the idea, and Joshua Shaw, an artist, obtained the first American patent but not until 1822.

Less prone to misfires in inclement weather, the percussion lock contained fewer parts and provided faster ignition than did the flintlock, eliminating the slight delay between the flash of the priming powder in the pan and the ignition of the main powder charge. One minor drawback was the difficulty in handling the small cap, as in winter when one's hands might be gloved or stiff with cold. By the late 1820s, the quality of percussion caps had improved and the new ignition system was gaining acceptance throughout much of the civilized world, including North America. But in the wilderness of the American west, some at first were reluctant to abandon the time-tested flintlock, for if a hunter or trapper was far from a source of percussion caps, his gun was useless when his supply was exhausted. However, it was often possible to locate and shape a piece of flint to fit between the jaws of a flintlock hammer.

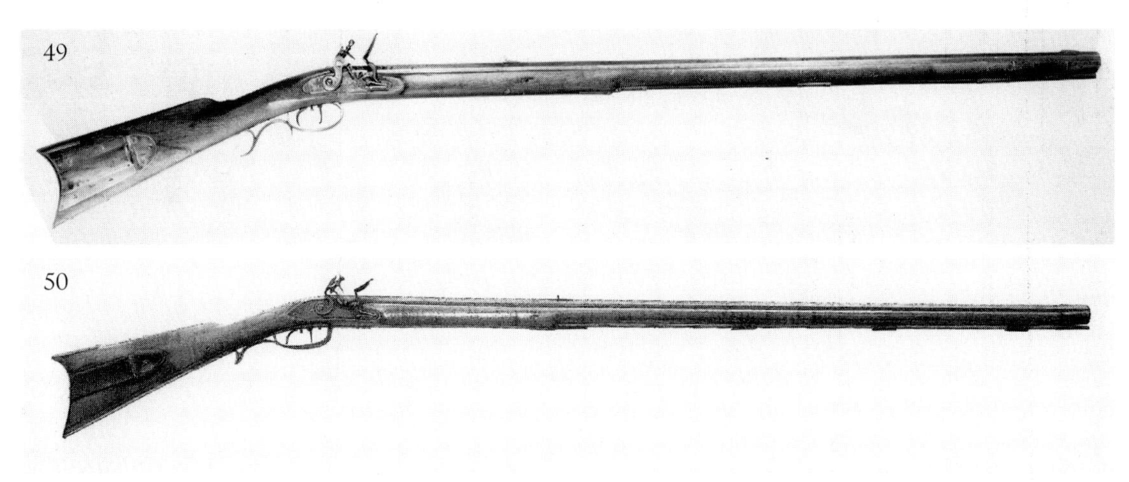

49. J. Henry & Son probably made this .42 caliber flintlock Pennsylvania rifle in the 1820s. (*Courtesy: Smithsonian Institution, neg. no. 72-4549*) **50.** A graceful .46 caliber southern rifle made by a Revolutionary War veteran from Pennsylvania and then Virginia, Joseph Bogle, Jr., for his friend and neighbor Capt. Robert McTeer. The rifle probably was made about 1795 after both men had moved to east Tennessee. The stock appears to be bird's-eye birch and mountings are of iron rather than brass. Instead of having a mere grease hole in the stock, it is fitted with a patch box as well as double-set triggers (pulling the forward trigger "sets" the rear one to fire the gun with just a slight touch). (*Courtesy: Wayne T. Elliott*)

The Changing Pennsylvania Rifle

The flintlock Pennsylvania rifle with its long octagonal barrel gained notoriety in the hands of frontiersmen who relied on its accuracy to put food on the table and to defend themselves and their families against marauders. By the 1790s some gunsmiths as far away as New England were turning out rifles similar to those by makers in the mid-Atlantic region. Poor roads and social isolation were factors leading to regional differences in style among rifles as is found with paintings in the world of art. Sometimes it's possible by careful examination of the various details of a Pennsylvania rifle to determine with substantial accuracy the locality, period, and perhaps the maker even if he didn't mark the barrel with his name or initials. However even though a name might be present on a barrel, that is no guarantee he made all of the rifle's components. A number of different wood and metal working skills went into making a rifle, and a rifle maker might have used a barrel blank obtained from another smith or a lock made by someone who specialized in such work or quite often an English or German lock perhaps acquired through a Philadelphia or other seaboard importer. Thus the name on a lock often has no bearing on who actually made the rifle (Figure 49).

The period from soon after the Revolutionary War to the 1820s, by which time the percussion ignition system was growing in popularity, is considered by many as the golden age of the Pennsylvania rifle. During this period many of the most elaborately decorated flintlock rifles appeared. Depending on a maker's skill and his customer's preferences and pocketbook, a rifle could vary from rather plain to an ornate specimen elaborately finished with raised carving in the stock; silver inlays of fish, Masonic designs, stars, teardrops, or other emblems; intricate patch box cover design; inlaid patterns in silver or brass wire; and other embellishments. In contrast, some of those rifles made farther to the southwest in the mountains of Tennessee and western North Carolina reflected a regional distinction. These often were built as strictly utilitarian arms with very plain stocks, few

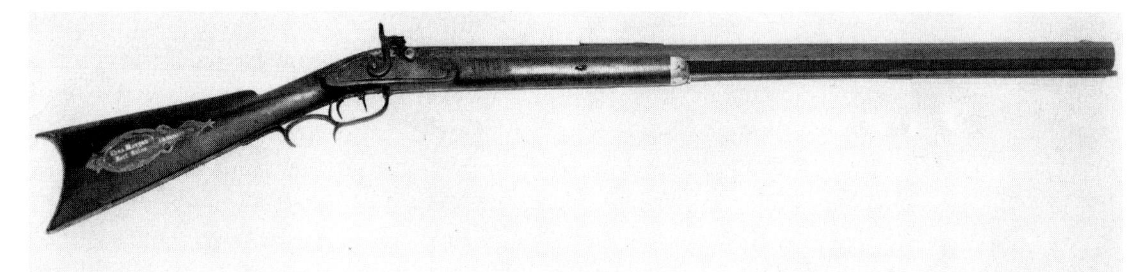

51. Heavy half-stock percussion .46 caliber rifle made by William Craig of Pittsburgh, Pennsylvania, in the late 1840s. The patch box lid is engraved "Chas. Morgan—Fort Union" and the patch box tang "A.D. 1851." Morgan was a hunter, scout, and supply caravan leader for Fort Union on the Missouri River in present-day Montana. (*Courtesy: Museum of the Fur Trade, Chadron, Nebraska*)

if any decorative touches, iron rather than brass fittings, and the substitution for a patch box of a simple uncovered cavity in the butt stock to hold grease to lubricate patches—a "grease hole." Sometimes even a butt plate was considered unnecessary (Figure 50).

A writer of historical fiction seeking the name of a rifle maker of the last quarter of the eighteenth century has scores from which to choose. Among the better known Pennsylvanians are Peter Gonter, Peter Angstadt, John Groff, and Jacob Dickert of Lancaster as well as Jacob Sell, Philip Heckert, and George Eyster of York County and Wolfgang Häga of Reading. William Henry served as armorer of General Braddock's ill-fated expedition into western Pennsylvania in 1755 and in 1777 became treasurer of Lancaster County, a position which he held until his death. He was the first of a distinguished line of Pennsylvania gun makers, which included his son John Joseph Henry, who accompanied Benedict Arnold on his epic march to attack the city of Quebec in the winter of 1775-76. In the spring of 1776 the Henry works employed fourteen hands.[1]

In business by the late 1790s, Marylanders Philip Creamer of Taneytown and John Armstrong of Emmitsburg would each establish a reputation as a maker of fine rifles. Creamer later moved to Illinois and Andrew Jackson owned a high-quality Creamer percussion pistol. Until the 1800s, New England riflesmiths were far less common than those in the mid-Atlantic region but Silas Allen, Welcome Matheson, and Hiram Slocum were three makers active in the northeast as early as the 1790s. As the end of the century approached, existing records reveal the average rifle cost between $12 and $18.

As settlers moved west over the mountains in increasing numbers, gunsmiths often set up shop in the newly established towns in Ohio, Kentucky, and elsewhere, repairing and cleaning guns and sometimes producing new ones. Kentucky reportedly was home to more than thirty gunsmiths by 1806. Gunsmiths' activities weren't always confined to arms making or repair but sometimes included such trades as silversmithing and general blacksmithing. Daniel Peck of Raleigh, North Carolina, in an 1809 advertisement stated he could make or repair "Gentlemen's fine Guns, Rifles" but also made door and furniture locks and keys, branding irons, and repaired or ground surgical instruments and swords. "Elastic Trusses Made to suit any size, age, or constitution, and warranted to afford relief," his announcement added. However, Pennsylvania continued to be home to many of the better known rifle makers in the early 1800s. Just a few names include Peter Angstadt, Frederick Sell, and

1. In 1824 Gonter had moved to Reading, Pennsylvania, and advertised his workshop at the sign of the Black Horse where he "makes every kind of new weapon, repairs old ones, and especially rifles."

Melchoir Fordney.[2] Regardless of where his rifle was made, the owner sometimes gave it a nickname such as "Deer Killer," or as Daniel Boone reportedly christened one of his rifles made by his gunsmith brother Squire Boone, "Ol' Ticklicker."

Following the gradual acceptance of the percussion ignition system, rifle makers in the 1820s often were busy converting or "percussioning" flintlock guns in addition to making new percussion ("cap and ball") arms during this transitional period. A common conversion method involved discarding the priming pan and frizzen, drilling out the vent hole and screwing in a drum containing the nipple, then replacing the flint hammer with one of percussion form. William Border had a lengthy career as one of the premier gunsmiths in Bedford County, Pennsylvania. His daybook showed he was charging $1.25 to convert or "pucution" [percussion] a rifle in the 1830s and 50 cents to "freshen" or recut rifling. To make a new rifle, he charged one customer $15 in 1846, but there was no indication as to how elaborate the gun might have been.[3]

Gradually barrels became shorter than on earlier guns—34 to 38 inches was a convenient length. At about the same time, mass production methods brought competition to the individual makers as such firms as the Henry family works at Boulton, Pennsylvania, and Henry Leman's factory in Lancaster hired large numbers of gunsmiths. This was a major change from the earlier days when a gunsmith worked in a small shop, often at the rear of his home, perhaps working alone except for an indentured apprentice or two.[4]

Percussion muzzle-loading rifles were made and were still in use in some backwoods areas in the 1890s, but they often were a far cry from the handsome Pennsylvania rifles of the early years of the century. By the 1840s the earlier graceful flowing lines were frequently replaced by the coarser lines of later factory-made specimens. The factory system often resulted in increased uniformity in design and somewhat less individualistic styling and ornamentation among percussion muzzle-loading rifles. There were exceptions, of course, and handsome ornate "cap and ball" target or hunting rifles exist, sometimes with several dozen inlays. Some such rifles appeared as late as the last quarter of the century.

Gunsmiths might establish their business in any population center where sufficient demand for their services as a gun repairer and maker existed. Walnut rather than maple wood for the gunstock was often used on percussion rifles, and while the patch box still was handy for holding precut patches, now it contained percussion caps or perhaps a spare nipple rather than flints. Gradually the rectangular brass patch box cover was replaced by a smaller round or elliptical cover or it was eliminated entirely. West of the Appalachian Mountains there were few flintlock rifles made, and most flintlocks present were brought from the east. But by the mid-1800s in what's thought of today as the midwest the typical rifle was a percussion arm, frequently a halfstock, its wooden forestock extending only about half way to the muzzle (Figure 51).[5]

2. Fordney died in his gun shop in 1846 when a religious fanatic struck him in the head with an axe to cleanse him of living in sin with his common-law wife.

3. Dillin, *The Kentucky Rifle*, 136-37.

4. As an example, in 1804 17-year-old John Blackburn indentured himself to John Armstrong of Emmitsburg, Maryland, for three years. He was to receive room, board, two weeks off during harvest season, and $20 at the end of the apprenticeship. Armstrong was to teach him "the art and mystery of the trade of a Gun Smith in the best way and manner that he can and also the art and mystery of Lock making." The apprenticeship system continued at least as late as the 1850s.

5. Not all gunsmiths were white. Among the scores of Ohio gun makers, Meshol Moxley was an African American known to have been active in the mid-1800s. Selwyn Peters, another Ohio gunsmith, is thought to have been part French and part black and reportedly was more than 100 years old when he died in the early 1900s.

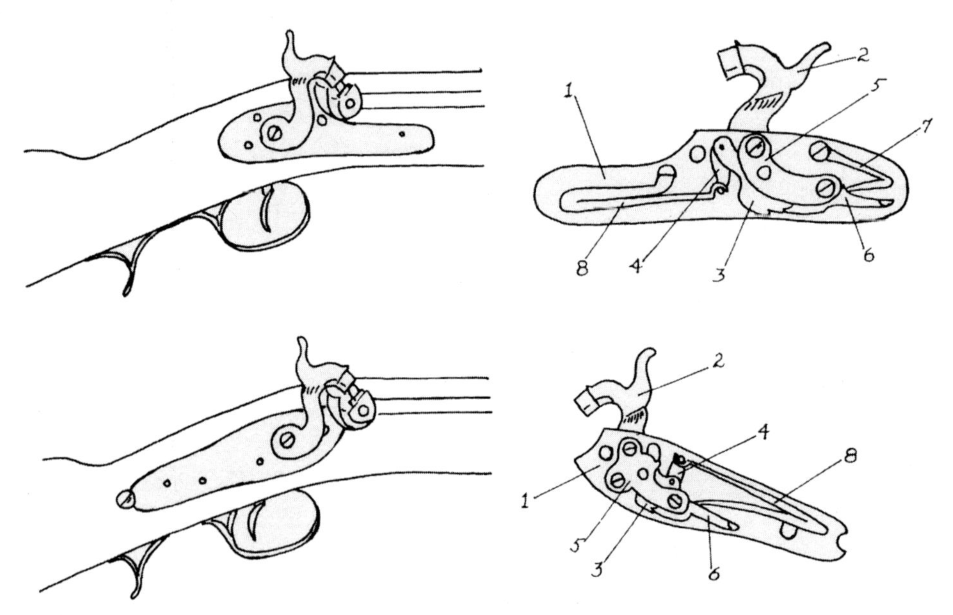

52. Design of bar and back-action percussion locks. 1. Lock plate. 2. Hammer. 3. Tumbler. 4. Swivel. 5. Bridle. 6. Scear. 7. Scear spring. 8. Mainspring. (*Courtesy: S. James Gooding, The Canadian Gunsmiths, 1608-1900*)

Two styles of percussion locks were common, the bar lock and the back action lock. In the former, the more common style, the mainspring was positioned in front of the hammer, as on most flintlocks. In the latter, the hammer was mounted at the forward end of the lock in front of the mainspring. The back action lock was inlet into the wrist of the stock, weakening it somewhat (Figure 52).

One somewhat unusual departure from the usual style of civilian gun lock appeared early in the percussion era, the side hammer or as it sometimes is called the "mule ear," a name perhaps derived from the shape of the spur on the hammer. On such rifles—and the less common side hammer pistols—the hammer was mounted so it operated in a sideways manner rather than falling forward and downward to strike the percussion nipple. Although never overly popular, the style retained appeal to some throughout the half century or so during which the percussion system reigned. It did offer several advantages, for its lock had fewer parts and could be made thinner, thus reducing the amount of wood removed from the stock to accommodate it. Just as important, the design allowed a straight route of travel for the flame from the percussion primer through the nipple to the main powder charge.

Throughout the percussion era the powder flask or horn was a necessary accessory to carry a supply of powder. Most of the former and some horns had a spring-loaded thumb piece and an adjustable spout (or charger) and dispensed a uniform amount of powder each time. Flasks were made in various shapes and of different materials, usually metal alloys of copper and zinc. Size varied from small flasks carried in the pocket to larger ones with carrying rings for suspension. Sporting flasks could be plain or decorated with patterns or images which might include dead game, hunting scenes, or the popular American eagle. Shotgunners often used similar flasks of leather, plain or decorated, to contain shot. The great variety of flasks produced over many decades offers a challenge to today's collector of these accessories (Figure 53).

Early Match Shooting, Hitting the Nail on the Head

By the early 1800s, target shooting (more commonly called match shooting then) had become an increasingly popular diversion, particularly in areas where the rifle was common, providing an escape from the drudgery of daily living. Sometimes the target was a live turkey, the objective being to shoot off the unfortunate fowl's head. The distance at which offhand matches were fired commonly was around 50 to 75 yards. In an 1835 offhand beef shoot, one man put up a cow worth $20, to be shot for at 25 cents per shot. Each shooter received a board on which he marked a cross in the center of the target. Two men were selected as impartial judges and determined which shots came closest to the center. Winner received the hide and tallow, considered the most desirable, next won his choice of hind quarters, third prize was the other hind quarter, fourth and fifth prizes were the fore quarters, while the sixth place finisher got the lead in the backstop against which the targets were placed.

An unusual match was advertised in a Maryland newspaper in 1810, and one has to assume it was proposed in jest. "A MAN to be shot at for the benefit of his wife and children—1 dollar a shot—100 yards distance, with rifles. . . . The above man is in a very low state of health, and wishes to leave his family snug."[6]

Naturalist John James Audubon lived in Louisville for several years just before the War of 1812 and wrote of several shooting contests he witnessed among Kentucky frontier hunters with their long rifles. One involved "hitting the nail on the head" at a distance of forty paces or so. "If he comes close to the nail his marksmanship is considered indifferent. If he bends the nail, this is . . . better; but nothing less than hitting it right on the head is satisfactory. One out of three shots generally hits the nail." Unfortunately Audubon didn't mention the size of the nail head. One night near a passenger pigeon roost along the Green River he investigated the sound of gunshots and found a dozen men practicing by torchlight to sharpen their skill in shooting at light reflected from the eyes of deer or wolves.

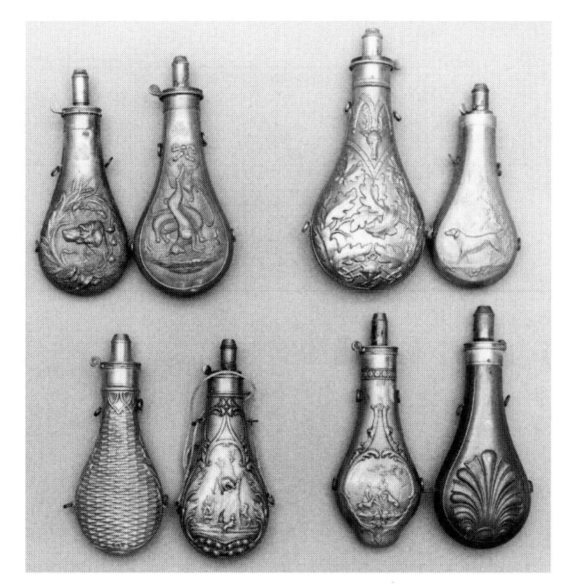

53. An array of sporting-type nineteenth-century powder flasks illustrating the variety of designs makers used.

> A burning candle. . . , almost indistinguishable at fifty yards, was the target. One man stood near it to watch the effects of the shots as well as to relight the candle if it went out, or to replace it if the shot cut it in two. . . . Some hit neither snuff nor candle and were congratulated with a loud laugh. One, particularly expert at this, was very fortunate and snuffed the candle three times out of seven. His other shots either put out the candle or cut it immediately beneath the flame.[7]

Just as impressive to Audubon was the skill of some hunters in "barking squirrels," as he first witnessed done by Daniel Boone.

6. Hartzler, *Arms Makers of Maryland,* 75.

7. Alice Ford, ed., *Audubon, By Himself* (Garden City, N.Y., 1969), 21-22.

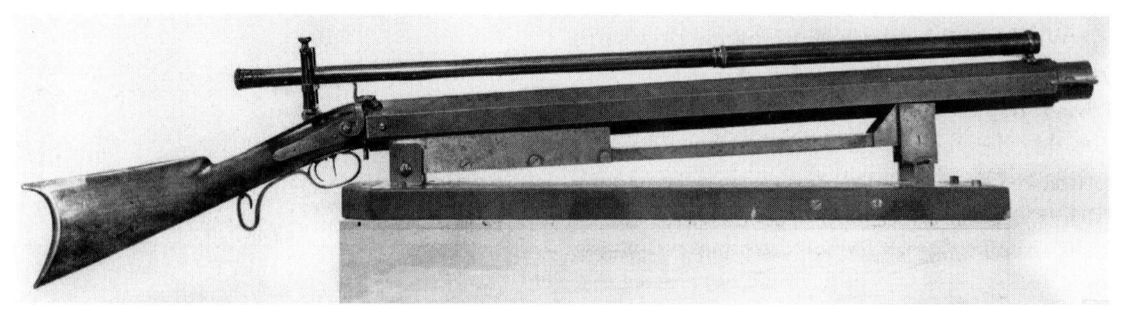

54. Percussion .47 caliber target bench rifle made by Carlos Gove in the 1850s when he was living in what today is Council Bluffs, Iowa. The gun weighs forty-two pounds and the bench rest from which it was fired weighs another twenty. (*Courtesy: Colorado Historical Society*)

> He wiped it [his rifle], measured powder, patched the ball with six-hundred-thread linen, and sent the charge home with a hickory rod. . . . Boone pointed to one crouched on a branch about fifty paces off, watching us. He bade me mark the spot well, and gradually raised his rifle until the bead or sight was in line with the spot. A whip-like report reverberated through the woods and along the hills. Judge of my surprise when I saw that the ball had hit the piece of bark just beneath the Squirrel and shivered it into splinters. The concussion killed the animal and sent it whirling through the air as if it had been blown up by a powder-magazine explosion. Since that first interview with our veteran Boone, I have seen many others perform the feat.[8]

As the public's acceptance of the percussion ignition system increased and time for sport gradually became more readily available, there was growing demand for rifles intended specifically for use in target matches, sometimes firing a conical bullet wrapped in a paper patch. This interest in competitive shooting with muzzle loaders continued throughout much of the remainder of the century. This was so even after their manufacture diminished in the face of the increasing availability by the late 1870s and 1880s of such high quality breech-loading target rifles as those by Wesson, Ballard, Stevens, Sharps, Remington, and others.

Muzzle-loading match rifles varied from those intended for unsupported "offhand" firing from the shoulder to those which became popular in percussion form and might weigh 20 pounds or substantially more and could be fired only from a bench rest. These monstrous and often rather ungainly-looking rifles were loaded with precision and minimized the human factor in achieving maximum accuracy. Adjustable and sometimes telescopic sights allowed precision sighting and distances for some matches increased. These target rifles might be sold in a fitted wooden case containing various loading and cleaning accessories. New Yorkers Morgan James of Utica and Nelson Lewis of Troy were well-known mid-century makers of such rifles, as was James Marker of Maryland. The Remington plant in Ilion, New York, became a major source of barrel blanks to those gun makers who chose to begin with a semi-finished barrel when fabricating a rifle for hunting or target shooting use.

A frequent method of determining the winner of a shooting match was by means of a string. Each competitor's bullet holes were measured from the center of the target by a piece of string. The shooter with the shortest string won the beef, cash, whiskey, or whatever the first prize might be.

An accessory often found with muzzle-loading target rifles was a false muzzle, intended to facilitate loading the conical or elongated paper patched bullet and to minimize wear to the rifling at the

8. Ibid., 21.

end of the barrel caused by loading and cleaning, thus extending the rifle's accurate life. Alvan Clark of Cambridge, Massachusetts, in 1840 patented the device and assigned rights to its use to the famed maker of high-quality target rifles, Edwin Wesson of Worcester, Massachusetts, who licensed their manufacture to others. The most effective means of producing the false muzzle was to cut off several inches from the unfinished barrel, insert pins which slipped into matching holes in the rifle's muzzle, then rifle the barrel with the false muzzle in place. The process required precision work, but when completed the rifling lands and grooves in the two units matched perfectly whenever the false muzzle was attached. After loading, the false muzzle was removed before firing. Some such target rifles saw Civil War use in the hands of sharpshooters, such as those organized by Col. Hiram Berdan, for long-range firing (Figures 54, 55).

55. The false muzzle partially removed from the Gove target rifle. (*Courtesy: Colorado Historical Society*)

Trappers, Explorers, and Other Western Travelers: Guns of the Mountain Men

Soon after Lewis and Clark's Corps of Discovery returned from the Pacific in 1806, a growing number of energetic Americans was crossing the Mississippi and venturing west in search of adventure and fur trade opportunities. Many remained for much of their life and became the "mountain men" of history. Most initially carried eastern-made flintlock Pennsylvania-style rifles, but by the 1830s the conditions found on the plains, in the mountains, and on the trail to Santa Fe had dictated a move to a sturdy rather plain rifle with few if any decorations, often iron rather than brass furniture, a stock of walnut or plain maple that extended only half way to the muzzle, an average weight of around 10 or 11 pounds, and by now usually a percussion lock. Barrel lengths for more convenient carriage on horseback were reduced to around 36 to 40 inches, and calibers of .50 or larger were prevalent, large enough if hunting bison or if one encountered *Ursus horribilis,* the mighty grizzly bear. These guns were built to withstand hard use, such as a fall from a horse, and their heavy octagonal barrel could accept an emergency double charge of powder to increase velocity if confronted by an angry grizzly. Simple non-adjustable open sights were the order, not fancy and fragile target sights.

These guns sometimes were advertised as Rocky Mountain rifles but the term "plains rifle" is in common use in the twenty-first century. Some people today are quite liberal in what they describe as a plains rifle and in the extreme are anxious to include almost any muzzle-loading sporting rifle of perhaps .40 caliber or larger. They forget that there was still large game such as bears and elk in New York, Tennessee, and other areas of the east for which a larger caliber rifle was appropriate. In short, not all big-bore rifles went west and warrant the term "plains rifle."

St. Louis became a major outfitting and jumping off point for many heading west, as it had been for Lewis and Clark, and by 1830 its population was close to 5,000. Gun making and repair was an important element of the city's economy. Today the best-known makers of true plains or mountain rifles are Jacob and Samuel Hawken of St. Louis, the most prominent among the various gunsmith sons of Christian Hawken, Sr., of Hagerstown, Maryland. Jake arrived in St. Louis at least as early

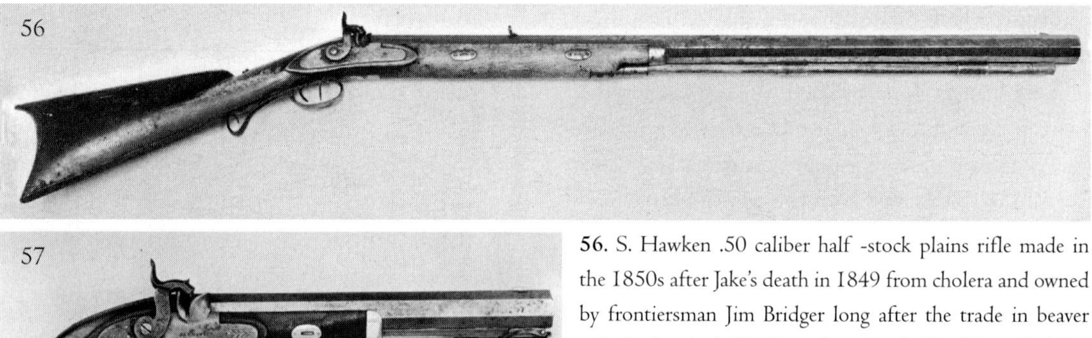

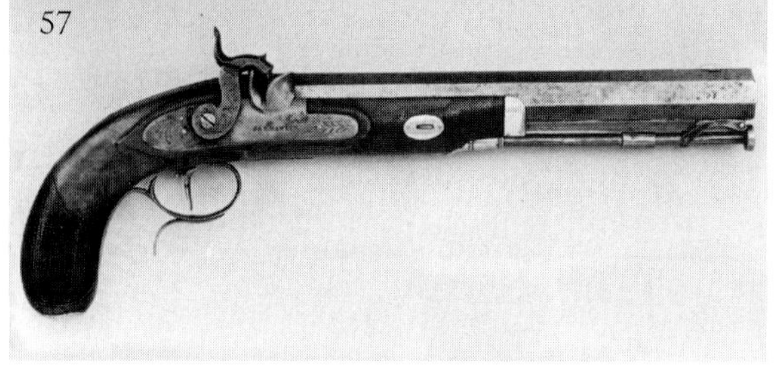

56. S. Hawken .50 caliber half -stock plains rifle made in the 1850s after Jake's death in 1849 from cholera and owned by frontiersman Jim Bridger long after the trade in beaver pelts had peaked. **57.** One of a pair of .65 caliber rifled J. & S. Hawken pistols with a swivel iron ramrod and a barrel almost 11 inches long. On the frontier, such pistols would have been suitable for "buffalo running"—hunting from horseback.

as 1818. Sam worked in Xenia, Ohio, for six years before joining his brother in St. Louis in 1822 and later forming a partnership with him which lasted until Jake's death from cholera in 1849.

While heavy percussion rifles by Jake and Sam Hawken were well publicized and their quality undeniable, evidence is strong that their rifles were better known in the 1840s and 1850s rather than during the several decades before which marked the zenith of the trade in beaver pelts. Their "mountain rifles" were rather expensive by frontier standards ($25 or so) and annual production probably didn't exceed many more than about 100 heavy rifles since the Hawken shop also was engaged in making pistols, shotguns, and lighter rifles for the local trade, and repairing or reboring or otherwise modifying guns by other makers. Nevertheless, among modern collectors the stamping "J. & S. HAWKEN ST. LOUIS" or the later mark "S. HAWKEN ST. LOUIS" on the barrel of a heavy Rocky Mountain rifle is a prize (Figures 56, 57).[9]

Eastern rifles carried west by trappers and traders included ones by makers in Pennsylvania and Maryland, Virginia and Tennessee, and later elsewhere. Along the route west were gun makers such as Benjamin Mills of Kentucky, Phillip Creamer of Illinois, and Stephen O'Dell of Natchez, Mississippi. Rifles suitable for the western trade by Pennsylvanians John Joseph and James of the prolific Henry family, George W. Tryon and later his son Edward K., Andrew Wurfflein, and John Krider, as well as the Golcher family and Henry Leman were commonly found in the west in the late 1830s and beyond. But in the 1840s and 1850s the Hawken brothers had competition from numerous other gunsmiths much closer, in St. Louis—Horace Dimick, Reno Beauvais, Baptiste Lebeau, T. J. Albright, Frederick Hellinghaus, and others plus competitors elsewhere as western towns grew in number and stature. As among makers of flintlock Pennsylvania rifles, all of a plains rifle's components might not have been made by the man whose name might be stamped prominently on the top of the barrel. It could be assembled incorporating a barrel by Remington, a lock by Golcher of Philadelphia, and fittings by Tryon. With the advent

9. A myth about the mountain men is that their reign ended with the increasing popularity of the silk hat and decline in demand by fur buyers for beaver pelts. In reality, trappers by 1840 had almost wiped out the beaver population in the central Rockies. At the trappers' 1838 rendezvous, 150 trappers brought in an average of fewer than 14 beaver pelts per man after a season's work and it would be a decade before the creatures' population recovered. "The Mountain Men, An Exhibit Catalog," *Museum of the Fur Trade Quarterly,* Fall–Winter 2006, 31.

of early breechloaders in the 1850s, particularly those by Sharps, the muzzle-loading plains rifle's popularity began its gradual decline, although it would serve some diehards well into the 1870s.

Throughout the early percussion era, a frontiersman (or anyone for that matter) had options if he wanted a rifle capable of firing more than one shot without reloading. Double-barrel rifles were available by the late 1700s, but they had the disadvantages of extra weight and the fact that each barrel didn't shoot to exactly the same spot. After the late 1830s there were revolving-cylinder rifles by Colt, Billinghurst, and others on the market, along with a few "harmonica" rifles. However in the west a frontiersman through the early 1850s generally relied on a single-shot muzzle-loading rifle.

With any muzzle loader, a ramrod was a necessity, and if one were to lose or break it, the gun was virtually useless unless one loaded only with a rather dramatically undersize ball which could be dropped down the barrel, with a resulting loss of accuracy and velocity. Thus it was not unusual for a trapper or other western traveler to carry a spare ramrod thrust into the bore. It also could be used as a form of "unipod" or shooting stick to help steady the gun.

While the rifle with its greater range and accuracy was the primary weapon among mountain men, pistols and shotguns were in evidence as well. In his journal covering the period between 1834 and 1843, Osborne Russell left one of the best accounts of the trapper's life in the Rockies. He wrote that during one encounter with hostile Indians, he kept a German military-style "horse pistol" handy in case the enemy attacked while his rifle was empty. A pistol thrust into one's belt was useful if needed for close-range fighting. Henry Deringer, Jr., is well known for his pocket pistols, but he was making larger belt-size pistols as early as 1825, he stated. But until the advent of revolvers and pepperboxes in the late 1830s, any multi-shot pistol would have had more than one barrel (Figures 58, 59).[10]

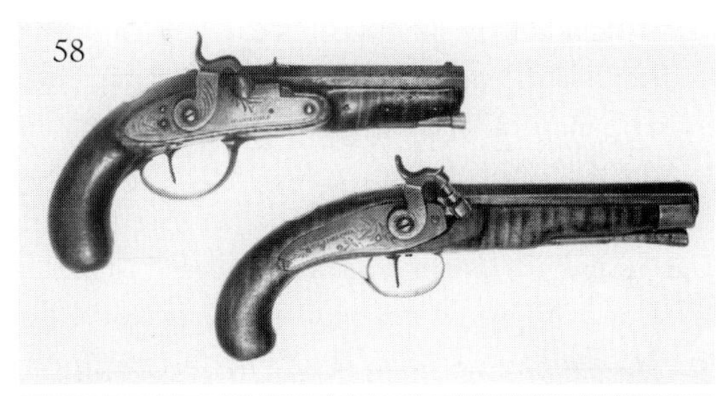

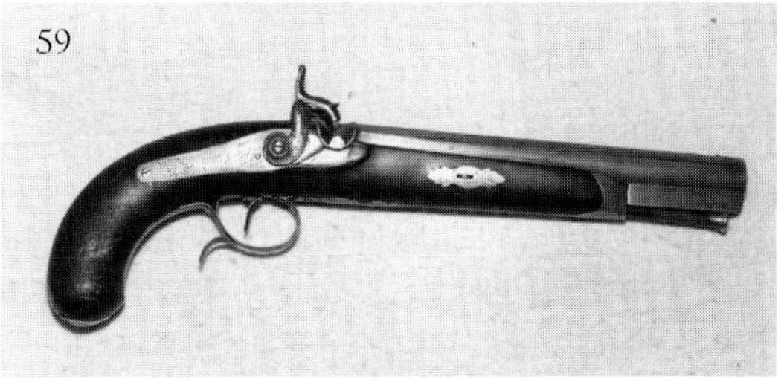

58. Percussion "Kentucky" or "Pennsylvania" belt-size pistols of 1830s vintage, each without a maker's name on the barrel. Although the locks may have been imported from England, each is marked with the name of a Philadelphia maker or merchant. **59.** What may be the largest Henry Deringer, Jr., pistol known, 16 inches in length. The bottom of the butt contains a small compartment for spare percussion caps, a feature found on some of his pocket pistols as well.

A shotgun too was just the thing for hunting smaller game for the cooking pot, but as some frontiersmen expressed, it was just as effective in countering an enemy night attack on one's camp. If hunting bison on horseback, a large caliber pistol was more convenient than a rifle, and a heavy barreled shotgun loaded with a single ball could serve well as a bison killer at close range. The versatility of a smoothbore was shown in a comment by an Italian count who visited America in 1837-38.

10. The mountain man is sometimes portrayed as an uncouth, uneducated individual and indeed some were. But trapper Osborne Russell when describing a winter encampment noted that they had books by Byron, Shakespeare, and Sir Walter Scott as well as the Bible and small volumes on chemistry, geology, and philosophy to read. Osborne Russell, *Journal of a Trapper*, Aubrey L. Haines, ed. (Lincoln, Neb., 1955), 109.

One night on the prairie on their way to visit a Sioux Indian village, they were cooking their evening meal of salt pork when they observed far-off smoke. Although they agreed there was nothing to fear, without anyone suggesting it, "we put tinder [priming] in our muskets, bullets instead of shot, and in spite of mosquitoes . . . , we let the fire go out, so that our blaze should not be reciprocally seen."[11]

By the 1840s, the lure of the west was attracting wealthy adventurers from other countries, particularly the British Isles. These travelers often brought with them hunting rifles and shotguns from home. Double-barrel rifles with a bore of a half inch (.50 caliber) or larger were common among such visitors, and the names of prominent English makers of this period include James Purdey and Joseph Manton. The most ostentatious such "sportsman" undoubtedly was Sir St. George Gore, an Irish baronet who between 1854 and 1857 hunted on the northern plains. He and his entourage and their equipment, which included his collapsible brass bedstead, traveled in almost two dozen modified two-wheel Red River carts—appropriately painted red—plus his private carriage and four freight wagons followed by a herd of spare horses, oxen, and milk cows. Gore's valet had the additional duty of tying his fishing flies, and others loaded and undoubtedly cleaned whatever guns Gore used from his arsenal of some 75 pistols, rifles, and shotguns. His name was appropriate indeed for the Gore party slaughtered some 2,000 bison, 1,800 deer and elk, and 105 bears in addition to uncounted numbers of birds and smaller game. His excesses angered many whites as well as American Indians.

Percussion Pocket and Belt Pistols by Deringer and Others

The man or woman in need of a handgun by the 1830s probably would have chosen a percussion rather than an obsolete flintlock weapon. Its size would have depended upon the intended use: whether it would be a compact pocket pistol or a larger belt or holster size. To achieve more than one shot, it would have had more than one barrel. American makers who gained prominence as makers of quality percussion single-shot pistols in this era included Ethan Allen, Henry Deringer, Jr., Edward Tryon, and others, the last two being Philadelphians. Imported pistols from England and the mainland European continent including France and Belgium also were common.[12]

Games of chance were a popular form of entertainment for some and a profession for others throughout the nineteenth century. In 1806, a visitor to the Ohio River town of Louisville noted that the inhabitants "are universally addicted to gambling and drinking." Compact pocket pistols were well suited to those who made their living as gamblers whether practicing their trade on riverboats or in gambling establishments built on solid ground, just as they were for any men or women who feared for their personal safety. In the 1840s, an English traveler stopping in Louisville, Kentucky commented: "Many of the males of the town carried pistols and bowie knives, more particularly the gamblers who flooded the place at this season, stopping here for a month or two on their way home from New Orleans. Later they would scatter themselves among the fashionable watering places to lure their game."[13]

Small pistols suitable for a man's pocket or a lady's purse were available in the United States as early as the flintlock period and often were of European manufacture. But it was Henry Deringer,

11. Count Francesco Arese, *A Trip to the Prairies and in the Interior of North America, 1837-1838* (New York, 1975), 76.

12. The use of a pineapple design on the forward end of the trigger guard is often found on pistols by Deringer and other Philadelphia makers, a period symbol representing friendship.

13. Isabel McLennan, *Louisville: The Gateway City* (New York, 1946), 55, 102-3.

Jr., of Philadelphia whose pistols put his name in the dictionary. Like the Pennsylvania rifle, the Deringer-style pistol was distinctly American in origin. He was a prolific maker of rifles and pistols for both the government and the civilian market, as well as a producer of rifles and smoothbore "Northwest guns" for the Indian trade. Deringer's initial civilian pistols were larger belt or holster size. In fact, the largest Deringer pistol known is .58 caliber with a 9 1/4-inch barrel and an overall length of slightly more than 16 inches, certainly nothing to carry concealed about one's person (Figures 60, 61).[14]

It isn't known definitely when Deringer began to produce his small, distinctive percussion pocket pistols in quantity but it appears to have been about 1852. Dimensions varied but they ranged from .33 to .50 caliber with a barrel as short as about an inch. However about .41 caliber with a 1 1/2- to 2-inch barrel length was typical, as was the use of German silver for the trigger guard and other mountings. Undoubtedly the best known Deringer was that used by John Wilkes Booth to assassinate President Lincoln in April 1865, a pistol which today is preserved by the National Park Service. Like some of Deringer's other pocket pistols, it contains a small compartment with a hinged German silver cover in the bottom of the butt to hold a few spare percussion caps. Deringer's pocket pistols usually were sold in pairs and they would prove particularly popular in the south and west. Some specimens found today bear a sales agent's stamp on top of the barrel such as Charles Curry of San Francisco, F. H. Clark & Co. of Memphis, or Wolf & Durringer of Louisville. Two years before Deringer's death in 1868 at age 82, he stated that in the previous 10 years he'd sold 5,280 pairs.

But thousands of other Deringer-style pistols were produced by copyists in the United States and abroad. Many of these were unmarked as to maker or sometimes bore the name "Deerringer" or "Deringe" to confuse potential buyers. The

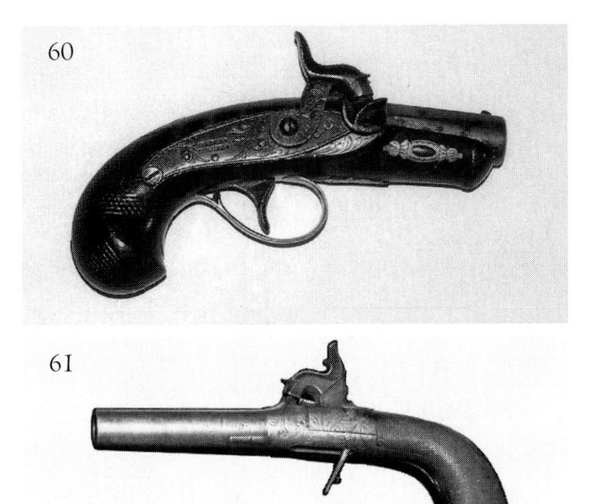

60. The genuine article, a Henry Deringer pocket pistol with the barrel marking of F. H. Clark & Co. of Memphis, Tennessee, agents for Deringer pistols. 61. About 1850 U.S. Senator Thomas Hart Benton gave this Belgian .36 caliber percussion pocket pistol to his daughter Mrs. Jesse Benton Frémont, the wife of western explorer Capt. Charles Fremont. The concealed trigger pops out when the hammer is cocked and the barrel screws off to load. Similar pocket pistols of British or often Belgian manufacture were common imports into the United States before the Civil War. (*Exhibited by the Arizona Historical Society, Tucson*)

Philadelphia firm of J. Slotter sometimes marked their pistols "J. Deringer" after hiring a tailor by that name to legitimize the procedure. Others such as Richard P. Bruff, Tryon, John Krider, and John and Andrew Wurfflein were more above board and marked their pistols with their correct name. A Prussian immigrant to San Francisco, A. J. Plate, sold both genuine Deringer pistols as well as

14. There is no uniformity in spelling when discussing these pocket pistols. *Webster's New World Dictionary* uses the spelling "derringer," and the use of the double "r" had became rather common by the 1860s. Today many gun collectors, including this author, use the spelling "Deringer" when referring to the genuine article by Deringer himself and "derringer" for others.

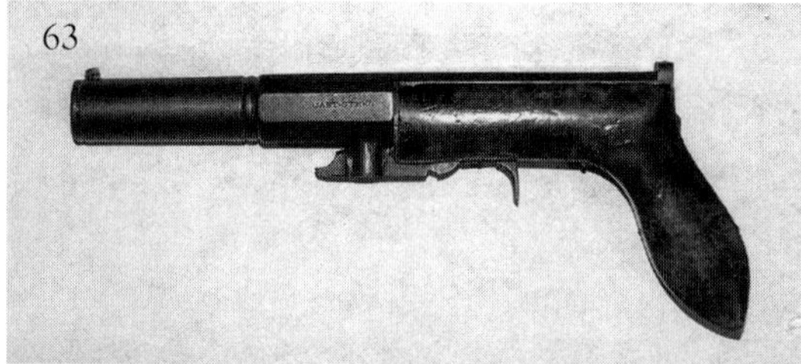

62. Counterfeit Deringer pistol made by Slotter but marked "J.DERINGER/PHILADELA" on the lock plate to confuse buyers. The barrel carries the stamp of San Francisco dealer A. J. Plate. Henry Deringer eventually filed a lawsuit in an attempt to counter such actions. **63.** New England underhammer "boot pistol" made by Gibbs Tiffany & Co. of Sturbridge, Massachusetts, and marketed through H. Hutchings & Co. of Baltimore.

Slotter counterfeits and in 1863 was sued by Deringer for trademark infringement in a suit that wasn't settled until 1870. In an 1883 gun repair manual the authors wrote: "The old [percussion] Derringer, though not now much manufactured in this country, is still among the people in considerable numbers. It is a muzzle loader . . . and, by the way, it is a very good pistol of its kind (Figure 62)."[15]

One style of single-shot handgun distinctly New England in origin was the percussion underhammer "bootleg" pistol made in substantial numbers from the mid-1830s to about 1860. Simple in design and rather inexpensive to purchase, these guns were shaped somewhat like a boot and they could be thrust in the top of a boot. Caliber usually was between .31 and .36, there rarely was a trigger guard, and the barrel length typically was between 3 and 6 inches, longer if target shooting or small game hunting was the purpose. Those with a long barrel fell into the category of a "pocket rifle." Locating the hammer on the underside of the barrel provided the shooter with an unobstructed view when sighting and precluded any fragment of percussion cap from striking the face when firing. Makers of these "boot pistols" included S. Osborn (Canton, Connecticut), Nicanor Kendall (Windsor, Vermont), Gibbs Tiffany & Co. (Sturbridge, Massachusetts), A. Ruggles (Stafford, Connecticut), and a score or more others. Some makers in addition to stamping their name at the rear of the barrel added a small American eagle. The underhammer design was applied to some rifles as well, often by New England makers. These ranged in size and style from light boys' rifles and "buggy rifles," small enough to fit beneath a buggy seat or in a wagon's tool box, to massive target rifles fired from a bench rest (Figure 63).

Early Repeating Pistols by Allen, Colt, and Others (1836–1857)

The idea of a revolver with a hand-rotated, multichambered cylinder can be traced back at least to the mid-1500s. Artemus Wheeler of Boston in 1818 patented a flintlock revolver, the same year in which Elisha H. Collier obtained an English patent for one. However the development of the percussion ignition system permitted Ethan Allen and young Samuel Colt independent of each other to introduce the first practical repeating handguns made in the United States, making each inventor an icon among nineteenth-century American arms manufacturers (Figure 64).

Allen was already established in Massachusetts manufacturing cutlery when about 1832 he turned to pistol making, specifically underhammer pistols and "pocket rifles." His use of cast steel barrels,

15. J. R. Steele and William R. Harrison, *The Gunsmith's Manual: A Complete Handbook for the American Gunsmith* (New York, 1883), 38.

drilled from a solid bar rather than of hammer-welded iron, encouraged others to use the process. The popularity of his pistols was such that in early 1837 Allen was advertising for additional help in his factory. But in November of that year he received a patent on a double-action "self-cocking" mechanism in which a long pull on the trigger raised the hammer and then allowed it to fall. He applied the system using a top-mounted hammer to his increasingly popular line of single-shot pistols, particularly to those with a barrel length of only two or three inches for handy pocket use (Figure 65).

In 1838 or 1839 Allen began production of the guns that ensured his fame, his "pepperbox" pistols. These consisted of a cluster of four to six barrels (five and six being most common) mounted on a center pin. The multi-barrel concept wasn't new and pepperboxes had been fabricated in Great Britain in both flintlock and percussion form. But Allen's guns were unique in America in that most incorporated his patented double-action mechanism with a bar hammer mounted on top of the frame. In size they ranged from small .28 caliber specimens with barrels about three inches long to the largest "dragoon size," a .36 caliber six-shot with barrels about six inches in length. Although never considered for adoption by the army, some Allen pepperboxes did see service as privately owned weapons during the Mexican-American War of 1846-48 and the Civil War. One made by Allen & Thurber, for example, armed a Civil War chaplain, Samuel McDaniel of the 4th Pennsylvania Infantry (Figures 66, 67).

Allen pistols were reliable and moderately priced for the period. An Allen company day book showed prices quoted in 1846 to dealer Samuel Sutherland of Richmond, Virginia as $5.25 per pair for bar hammer single-shot pocket pistols with a two-inch barrel and $8.25 for a standard six-shot pepperbox. If a dealer wanted more elaborate grips than walnut on a pepperbox, silver or ivory was available at $1.25 to $2.25 more, the latter being

64. Flintlock revolver by Elisha Collier. **65.** Two items in this image catch the viewer's eye—the tie and the double-action bar hammer pistol made by Allen or one of his competitors such as Blunt & Syms or W. W. Marston. (*Courtesy: Paul Henry*)

most common, although specimens with grips of either material are scarce today. The term "pepperbox" was in occasional use by the late 1840s, but where it originated is anyone's guess for they more often were called "repeating pistols" or "revolving pistols" in contemporary advertising. Presumably it evolved from a pepper shaker with numerous holes in its top.

The Allen pepperbox provided Colt revolvers with their stiffest sales competition among civilians in the 1840s and into the 1850s throughout the country. An advertisement by dealer J. G. Bolen of New York City illustrated a pepperbox and called

66. Cased Mexican War presentation Allen & Thurber .36 pepperbox with silver grips and engraved on the backstrap "Presented by the Buffalo Light Artillery to their late commandant Lieu. Col. John J. Fay 10th Reg. U.S. Infantry April 8, 1847." (*Courtesy: Charles Rollins*) **67.** Large "dragoon" size .36 caliber Allen & Thurber pepperbox.

them "Life And Property Preservers, For Housekeepers, Travelers, Captains, And Others. Patent Self-Cocking & Self-Revolving Pocket Pistols!!" The popularity of pepperboxes as well as Colt revolvers was particularly evident in the California gold fields following the rush of 1849 and is reflected in writings of that period. One gold seeker who arrived in San Francisco in 1850 wrote home that since his only weapon was a large California knife he had purchased for $8, "I now bought a small Allen's revolver; it was easy to handle, but was not very effective." An Allen was involved in a tragic fatal accident in California when a settler's younger brother playfully was snapping caps on it without realizing that one barrel was loaded and shot himself (Figure 68).

Not only were the double-action Allens a little faster firing than Colt handguns but in their plainest form they were less expensive. In addition, Allen's multibarrel design avoided the necessity that the chamber be in precise alignment with the barrel to avoid "shaving lead" as the bullet passed from cylinder to barrel. It was enough if a barrel on an Allen was sufficiently aligned to allow the hammer to strike the percussion cap. The Allen design, and subsequent similar pepperboxes by other makers, also eliminated the problem of a buildup of powder fouling between cylinder and barrel which could restrict rotation of a revolver cylinder. Despite these positives, a double-action pepperbox lacked accuracy. Not only were the barrels usually unrifled but the trigger pull was heavy and the barrel group as it rotated into firing position was difficult to hold steady on target. Accuracy with a self-cocking pepperbox against a man-size target for most shooters didn't extend very far beyond the width of a gambling table (Figure 69).

Samuel "Mark Twain" Clemens with typical exaggerated humor described his trip by stage coach across the central plains in 1861 during which one of his travel companions, Bemis, was armed with a pepperbox. Bemis came in for ridicule because of the Allen's inaccuracy. According to Clemens, when it was fired at a deuce of spades nailed to a tree the ball struck a mule about thirty yards to the left! Bemis had no use for the mule, but the animal's owner appeared with a shotgun and changed Bemis's mind, the humorist wrote in his classic volume *Roughing It*. Preparing for a journey to California via water and the Isthmus of Panama in 1849, Howard Gardiner purchased a pepperbox. It was "one of those newly patented pistols known as an 'Allen' which for 'comprehensive shooting' was truly a wonder for no matter in what direction it was pointed, there was no telling where the bullet might strike" (Figure 70).[16]

16. Mark Twain, *Roughing It* (New York, 1913), I: 5-6. Howard C. Gardiner, *In Pursuit of the Golden Dream*, Dale L. Morgan, ed. (Stoughton, Mass., 1970), 15.

68. A neatly dressed owner of a six-shot Allen & Wheelock pepperbox. (*Courtesy: Paul Henry*)

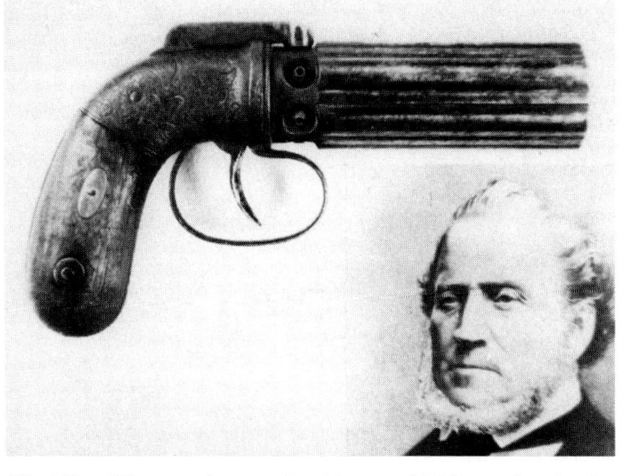

69. Allen .31 pepperbox made without a shield over the nipples and once the property of Mormon president Brigham Young, shown here. He later presented it to his colorful and controversial bodyguard, Porter Rockwell. (*Courtesy: Harold R. Mouillesseaux*)

Despite the preeminence of Allen pepperboxes, the firm of Blunt & Syms soon was marketing double-action ring trigger pepperboxes with a concealed underhammer. Stocking & Company and Thomas Bacon produced single-action models that could be fired with somewhat greater accuracy than Allen's "self-cockers." When the Allen double-action patent expired about 1857, the Manhattan Firearms Company and William Marston began offering double-action pepperboxes that in appearance duplicated those by Allen. Since pepperboxes weren't fitted with an attached loading lever, they had to be loaded using a separate ramrod to push the snug-fitting round ball down each barrel after the powder had been poured in. Some gun dealers loaded pistols for customers and advertised this service and sometimes maintained shooting galleries where patrons could practice with handguns (Figure 71).

Competing with American-made pepperboxes were those imported from Britain and continental Europe where they were equally as popular in the 1840s and 1850s. Among the numerous English makers were Parker, Field & Co.; J. W. Edge; and J. R. Cooper. Most widespread of those brought over from the continent to America appear to have been those marked "Mariette Brevete," most of which had a ring trigger and up to twenty-four barrels although any more than eight was rare indeed. Despite the popularity of pepperboxes in this country, there was still a market for single- and double-barrel pistols from the simplest inexpensive examples to better-quality arms in various sizes. But by the mid-1850s Colt revolvers were gaining preeminence over their less expensive pepperbox rivals and the Colt name was becoming largely synonymous with revolver (Figure 72).

Samuel Colt's Early Firearms

Young Sam Colt received his first American patents on a revolver in 1836, after five years or so of design tinkering which included his initial concept of a multi-barrel pepperbox-type revolver. Key features which were covered by his patents and used in his revolvers included cylinder rotation by cocking the hammer, actuation of the cylinder locking bolt by the hammer, and a partition between each nipple. With such protection, Colt cornered the mar-

70. Contemporary with Allen pepperboxes was the firm's popular line of pocket pistols such as this double-barrel specimen. (*Courtesy: Paul Henry*) **71.** Six-shot single-action pepperbox by Stocking & Co. of Worcester, Massachusetts. Being single action, it could be fired with somewhat more accuracy than a double-action Allen.

ket on the production of effective single-action revolvers, those in which the hammer had to be cocked manually between each shot. Between 1831 and 1835, he had invested the modern equivalent of almost $30,000 in having various machinists produce eighteen model hand and long guns. When quantity production began in late 1836 in Paterson, New Jersey, it wasn't of revolvers but revolving rifles with a concealed hammer, unusual in that the cylinder was rotated by a pull on a ring lever in front of the trigger guard. By 1841, Colt's Patent Arms Manufacturing Company produced perhaps 700 or so, most in .44 caliber with an eight-shot cylinder. The U.S. War Department in 1838 purchased 50 of the rifles for use by troops battling Seminole Indians in Florida, and the following year the young Republic of Texas bought 100 similar repeaters for its army, certainly not the larger quantities which the ambitious young inventor had anticipated.[17]

It's often written that young Sam Colt developed the concept for his revolver by watching the operation of the ship's wheel during an 1830-31 ocean voyage. He later claimed he had no early knowledge of such revolving arms as those flintlock and later percussion revolvers by Collier, but this probably is false.

Colt followed the first Paterson shoulder guns with a more conventional appearing .525 caliber smoothbore carbine with an external hammer in 1839. Almost 1,000 left the factory, along with several hundred as 16-gauge six-shot shotguns. About two-thirds of the carbines were purchased by the U.S. and the Texas republic for each nation's army and navy. Except for those later fitted with an attached loading lever, it was necessary to remove the cylinder to reload, after knocking out the transverse wedge which passed through the cylinder pin and which wed barrel and frame.

Handgun production got under way in January 1837, first a small pocket model followed by the larger and better selling .36 caliber No. 3 belt size. Each Paterson-made revolver, whether one of the small .28 or .31 caliber "Baby" Patersons or the largest and most popular No. 5 .36 caliber holster size (sometimes called the Texas model), had a five-shot cylinder, no trigger guard, and a concealed trigger which popped out from the underside of the frame when the hammer was cocked. In April 1839, the debt-ridden Republic of Texas purchased 180 of the No. 5s with a nine-inch barrel for its

17. Herbert G. Houze, *Samuel Colt: Arms, Art, and Invention* (New Haven, Conn., 2006), 41.

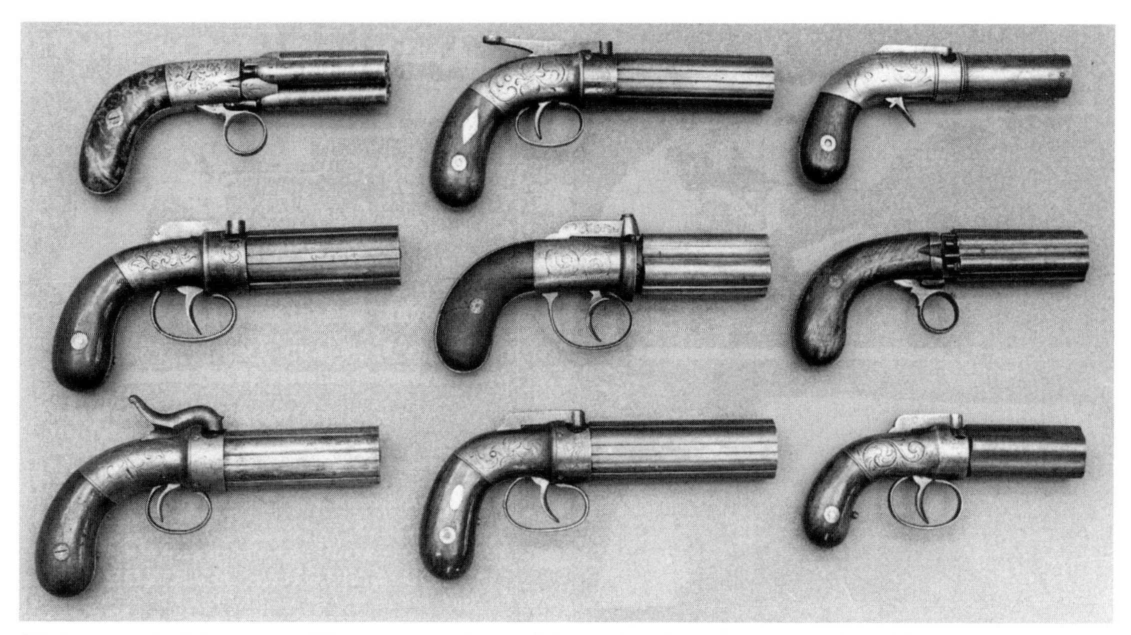

72. A potpourri of American and European pepperboxes of the 1840s and 1850s. From top down, left to right; Left column: Belgian Mariette underhammer; W. W. Marston of New York City; Allen single action. Center column: single action Stocking & Co.; an English double action unmarked as to maker; Allen & Thurber. Right column: rare four-shot Pecare & Smith of New York; English ring trigger Cooper; Allen & Thurber double action.

navy. In 1844, a company of Texas Rangers under Captain John Coffee Hayes obtained some used Paterson revolvers from the navy depot at Galveston. The guns had their baptism of fire in ranger hands on June 8, 1844, when Hayes and fifteen men including Samuel Walker routed a force of more than sixty Comanche Indians. The rangers' success with their revolvers is a frequently told tale yet it came too late for accounts of the incident to revive the now defunct Paterson firm (Figures 73, 74, 75).

By 1841, production of Colts had come to a virtual end. Civilian sales lagged and anticipated sizable government orders hadn't materialized. Only about 3,000 handguns had been manufactured and in 1842 the company went into insolvency. But young Samuel Colt was not an idler. In 1832 to help finance his revolver design efforts, he had advertised himself as a practical chemist "Doctor S. Coult," and for a fifty-cent ticket was demonstrating to crowds the effects of nitrous oxide or "laughing gas" or, as he called it, "exhilarating gas." He assured ladies that "not a shadow of impropriety attends the Exhibition to shock the most modest." Now as his initial venture into gun making faltered, he busied himself with various projects including experiments in underwater explosives for harbor defense, the telegraph, and tinfoil cannon cartridges.

But Colt persisted in his efforts to improve upon his Paterson arms, and in the fall of 1846 engaged the New York firm of Blunt & Syms to build a prototype of another revolver, a .47 caliber handgun weighing almost four pounds with a conventional trigger and trigger guard but no attached loading lever. He submitted it to the secretary of war for consideration and in December with the nation already six months into a war with Mexico showed it to former Texas Ranger Samuel Walker, now a captain in the U.S. Mounted Rifles. Walker's familiarity with Paterson Colts enabled him to suggest some improvements to the sample revolver—

73

74

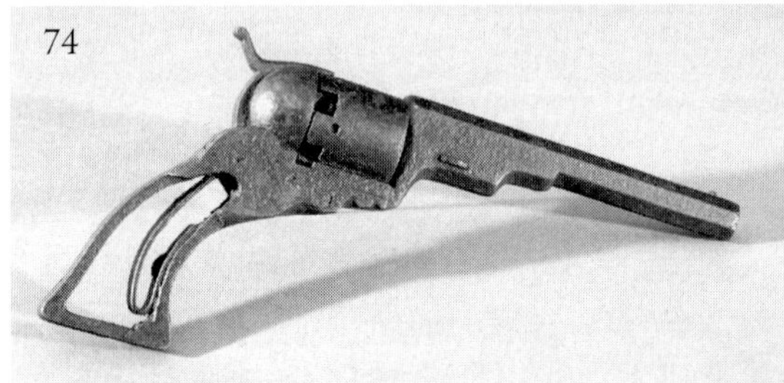

75

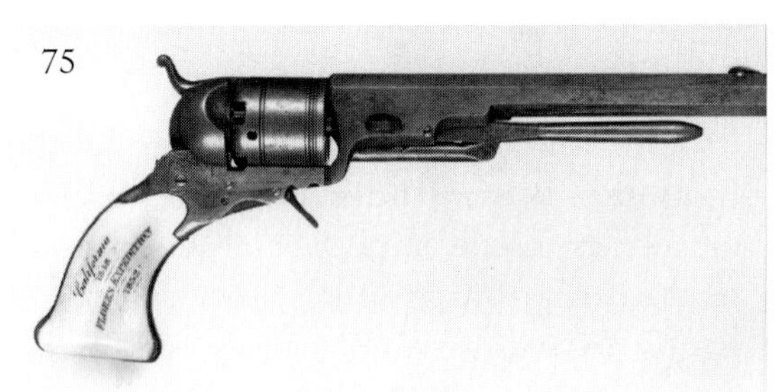

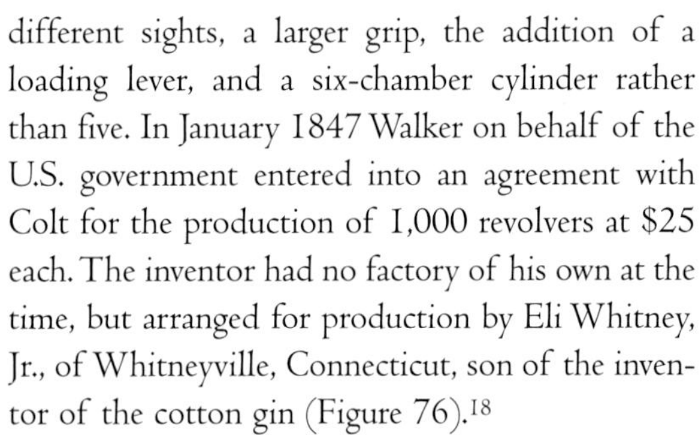

76

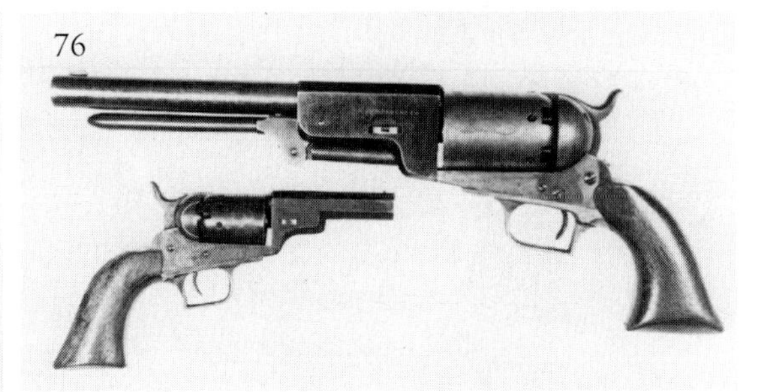

73. A pair of Colt Patersons, a .36 caliber belt model and a smaller .28 Baby Paterson with an attached loading lever. **74.** In 1929 a well-digging crew found this .36 Colt Paterson in a collapsed adobe building northeast of Amarillo, Texas. Such guns in "dug up" or relic condition have substantial collector interest and their condition should be stabilized rather than attempting to clean or restore them. (*Photo courtesy Leland E. DeFord III, gun courtesy Edward F. Cornett*) **75.** Judging by the inscriptions on the ivory grip of this Colt Texas Model Paterson, the gun probably could tell a fascinating story if it could talk—"Texas Rangers/Mexican War/1846" and "California/1848/Flores Expedition/1852." The latter inscription may refer to an effort by Ecuador's former president Juan José Flores and a mercenary force including some Americans to capture the city of Guayaquil in 1852. This revolver has a factory-installed loading lever, a feature found on only about one-third of the No. 5 size holster models. (*Courtesy: U.S. Military Academy, West Point Museum*) **76.** A study in contrasting Colts—a mammoth .44 Walker weighing almost five pounds and a Model 1848 .31 Baby Dragoon pocket model.

different sights, a larger grip, the addition of a loading lever, and a six-chamber cylinder rather than five. In January 1847 Walker on behalf of the U.S. government entered into an agreement with Colt for the production of 1,000 revolvers at $25 each. The inventor had no factory of his own at the time, but arranged for production by Eli Whitney, Jr., of Whitneyville, Connecticut, son of the inventor of the cotton gin (Figure 76).[18]

The new Colt was a monster, a six-shot .44 weighing four and a half pounds with a nine-inch barrel. The cylinder was engraved with a scene designed by New York bank note engraver Waterman Ormsby commemorating Hayes' 1844 battle. Deliveries for the Mounted Rifles began in June 1847 and, true to form as a promoter, Colt arranged for the shipment of additional revolvers to Mexico as gifts to army and navy officers, including a pair for Walker and other .44s for generals Winfield Scott, Zachary Taylor, and William Worth as well as several colonels including John Coffee Hayes. In Colt's final draft of the letter to Hayes presenting the pair of Walkers, he wrote:

> In return for the great service you have been in giving to my armes there present fair reputation I have

18. Ibid., 73.

77

78

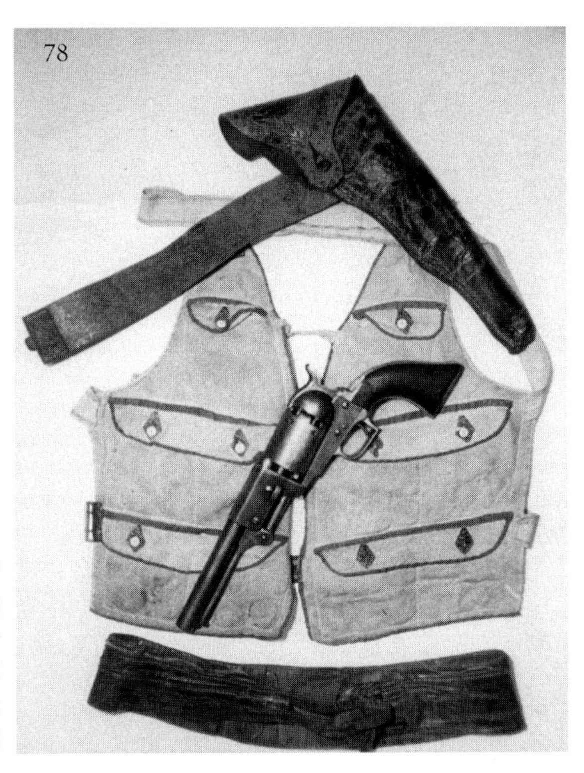

77. Colt .31 Model 1849 pocket model carried westward to the California gold fields in 1853 by an eager young Pennsylvania leather worker. Like most others who arrived by that date, he found gold scarce and prices high and he returned to the east disillusioned. (*Photo by R. K. Halter*) **78.** A California '49er's Colt .44 Dragoon, waist belt and holster, vest, and money belt.

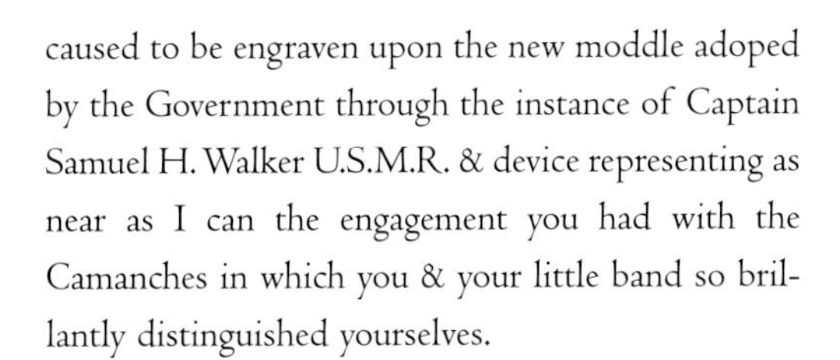

> caused to be engraven upon the new moddle adoped by the Government through the instance of Captain Samuel H. Walker U.S.M.R. & device representing as near as I can the engagement you had with the Camanches in which you & your little band so brillantly distinguished yourselves.

Unfortunately Walker didn't survive the war, but died in battle in October 1847.[19]

Two hundred eighty of the Walker Colts went to the Texas Rangers under Col. Hayes in August 1847. The Texans encountered difficulties almost immediately as some of the cylinders burst within the first few days of use. When 191 remaining Colts were turned in less than a year later, Captain John Williamson of the army's ordnance department on May 8, 1848, wrote:

> The remainder chiefly bursted in their hands—tho a few were lost in skirmishing with the enemy. Of the 181 [191?] turned in only 82 can be considered serviceable. All of the others have been more or less damaged by firing, and the wear and tear of some eight months use in the field. In some cases the cylinders are entirely destroyed—in others the barrels are irreparably injured where they join the cylinder—and again having bursted at the muzzle, the barrels have been cut off to one half their original length. As the pistols were turned in loaded in attempting to discharge them the Snaps were found fully equal to the explosion.[20]

Even though the arrival of these Colts in Mexico had no significant effect on the war's outcome and despite problems with failures in the field, profit and confidence from the order allowed Colt to establish his own factory in Hartford. He

19. Ibid., 75. Herbert G. Houze, "'Walker Colt' Breakthrough!" *Man at Arms,* December 2006, 26.

20. Maj. B. R. Lewis, "Sam Colt's Repeating Pistols," *American Rifleman,* June 1947, 24.

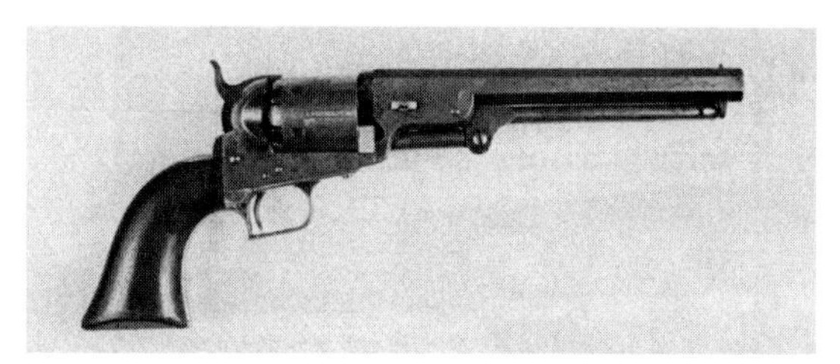

79. Colt .36 Model 1851 made during the first year of production. An early feature was its so-called "square back" trigger guard.

was back in the gun business and this time to stay. He wasn't oblivious to Walker's friendship and contributions, and in reply to a letter late in 1847 inquiring about availability included the sentence: "I can furnish a few of the 6 chambered Holster pistols of my construction...cal'd the Walker modle."[21]

The government's order for Colt Walkers was followed in November 1847 by another order for 1,000 additional holster revolvers, the first examples of a slightly lighter series of .44 so-called Dragoons, just over four pounds in weight. These solved the problem of cylinders being damaged by overloading, for the Dragoon cylinder was made of improved steel and the chambers' powder capacity was reduced. A much smaller .31 caliber Model 1848 "Baby Dragoon" pocket-size revolver intended for the civilian market followed in the fall of 1848 and in modified form as the Model 1849 Pocket became the most widely sold percussion revolver in this country, with some 330,000 made in both five- and six-shot versions and with an octagonal barrel in lengths of three, four, five, or six inches (Figures 77, 78).

Late in 1850, a .36 caliber six-shooter normally sold with a seven-and-a half-inch octagonal barrel appeared, commonly called the Model 1851 Navy, so named because its cylinder bore a roll engraved scene portraying an 1843 victory by the Texas navy over that of Mexico in which Patersons were used. Thus by 1851, Colt had established himself with the military and to civilians was marketing a .31 pocket size, a .36 "belt" revolver, and a heavy .44 "holster" pistol, a model to suit almost anyone's need, including the needs of those thousands eagerly seeking to strike it rich in California's newly discovered gold diggings (Figure 79).

The term "navy" as applied to the Model 1851 Colt was derived from the scene engraved on its cylinder, not from the model's use by the U.S. Navy, although the sea service did acquire almost 3,000 in the 1850s. A scene roll engraved on the cylinder was a decorative touch applied to most percussion Colt arms except those with a fluted cylinder, ones with longitudinal grooves cut in the exterior surface between the chambers to reduce weight. On specimens which were subjected to a substantial amount of use this engraving often is found to be worn off. On the first Paterson rifles the scene featured horsemen and a half-man, half-horse centaur hunting a deer, then on the 1839 carbines this was changed to a three-panel scene portraying a lion hunter, a ship, and a land battle. Paterson-made handguns bore an image of a centaur except for the largest size, the No. 5, some of which displayed a stagecoach holdup on the cylinder, a theme which was carried through on the pocket models of 1848 and 1849. The use of early Colts by the Texas Rangers was commemorated on Walker and Dragoon model cylinders, rangers chasing Indians. A reduced portion of the same view appeared on the early examples of the smaller Baby Dragoon. Many of these engraved cylinder designs were the work of W. L. Ormsby.

Occasionally percussion Colts, including some intended as presentation pieces, for an additional cost might be handsomely fitted with grips of carved or checkered ivory or other exotic material and metal surfaces engraved in varying degrees of coverage and detail in scroll, floral, or other design. Two masters of the gun engraving art at that time were Louis D. Nimschke and Gustave Young and genuine examples of their work are prized today.

21. Ibid., 79.

Not all engraving was done at the factory while the gun was still "in the white" before final finishing. Some guns were engraved later by skilled dealers or jewelers.[22]

Among the rarest of Colt presentation pieces are those few Model 1855 pocket revolvers and at least one Model 1851 with what are known as Charter Oak grips. Famed in Connecticut's history is a giant white oak tree which once stood in Hartford, still that state's official tree even though it doesn't exist today. In 1662, Connecticut received its official royal charter from King Charles II. A quarter century later the royal governor demanded the charter's return but the document was spirited away and hidden in a hollow in the aged oak tree thus preserving it. When the revered giant was blown down in 1856, it was draped with the American flag and Colt's armory band played dirges and patriotic music. Numerous mementos were made from its wood, including a cradle for Colt's son, veneer for several grand pianos, the governor's chair in the state's senate chamber, and grips for a few presentation Colt revolvers.[23]

Texas Independence and the War with Mexico (1835–1848)

A catalyst for the Mexican War was the annexation of Texas by the United States in 1845. Texans—or Texians—had won their independence from Mexico after a series of hard-fought engagements in 1835-36 against a Mexican army which, despite the bravery which it sometimes displayed, was not always well led or adequately armed and included many raw recruits within its numbers. When Texans, often short on munitions, sought to use a supply of captured Mexican gunpowder they called it little better than "pounded charcoal." Many Mexican infantrymen carried an obsolete flintlock Brown Bess musket, the same type British arm used at Waterloo twenty years before.

Some Texas army volunteers relied on shotguns and other smoothbores, although some did have rifles which far outranged their Mexican opponents' muskets. But there was a sprinkling of Mexican riflemen (*cazadores*) equipped with the famed Baker rifle adopted at the beginning of the century by the British army and which was accurate against an individual at several hundred yards. These *cazadores* could compete on equal terms with the independent and often ill-disciplined Texan riflemen. Some Mexican mounted troopers relied on the lance, but others were equipped again with British muzzle-loading arms, stubby Paget carbines with a 16-inch barrel. Reference to the modest number of Colt Paterson revolving hand and shoulder arms acquired by the Texas Navy and Texas Rangers has been made earlier (Figure 80).

In December 1835 fewer than 400 Texans besieged Bexar (San Antonio), and after four days forced the withdrawal of General Martin Cos' army of 1,100. Mexican losses were some 400 dead and disabled against about 30 Texans killed and wounded. But one of these dead was Colonel Benjamin Rush Milam, who had inspired the attack with his shouted challenge: "Who will go with old Ben Milam into San Antonio?" Felix de la Garza, one of the best shots in the Mexican army, reportedly killed Milam with a ball from his Baker rifle. In retaliation, several Texas riflemen spotted Garza's position and shot him out of his tree.

War broke out between the U.S. and Mexico in the spring of 1846, a conflict which did not have universal support among Americans. The result of the American victory was an expanded United States border, and the war did serve as a training ground for some officers who had no or only modest combat experience but who soon would be test-

22. Another among the most talented nineteenth-century gun engravers was Conrad F. Ulrich, Jr., who was responsible for much of the engraving work done for Marlin after about 1881 and later for Winchester.

23. John E. Parsons, *The Metropolitan Museum of Art Catalogue of a Loan Exhibition of Percussion Colt Revolvers and Conversions, 1836-1873* (New York, 1942), no page numbers.

80. English flintlock Paget carbine with Mexican military markings imposed over the original English marks. A swivel secures the iron ramrod to the barrel to prevent loss. (*Courtesy: Arizona Historical Society*) **81.** U.S. Model 1842 .69 musket, the last of the infantry's smoothbores and its first percussion musket. About 275,000 were made at Harpers Ferry and Springfield before a new .58 caliber rifle musket was adopted in 1855. Some later were updated with the addition of rifling and a rear sight. It also was the first U.S. arm made at both national armories with interchangeable partsl regardless of location of manufacture. **82.** Model 1841 .54 "Mississippi" rifle, last of the army's rifles to fire a round ball. The M1841 originally was not designed to accept a bayonet although some later were modified to do so. (*Courtesy: National Park Service, Fuller Gun Collection, Chickamauga and Chattanooga National Military Park*)

ed again, this time in a civil war. Robert E. Lee, U.S. Grant, Thomas J. Jackson, George E. Pickett, Braxton Bragg, and George B. McClellan were but a few who led troops in Mexico before assuming higher command in the Civil War. The Mexican army still relied extensively on imported flintlock muskets, particularly the British India Pattern Brown Bess with a 39-inch barrel. Although the U.S. in 1842 had discarded the flintlock in exchange for a percussion smoothbore musket for its infantry, production had not been sufficient yet to allow extensive distribution of the new arm. The infantry fought largely with flintlocks and not until 1849 did the War Department issue a general order directing the conversion of all its serviceable flint muskets to percussion (Figure 81).[24]

Young Lieutenant Ulysses S. Grant served with the 4th Infantry during the Mexican War and in his

24. Of the approximately 105,000 regulars and volunteers who served in the U.S. army during the war, about 1,700 were killed in battle or died of wounds. In contrast, 11,000 died of disease. Francis B. Heitman, *Historical Register and Dictionary of the U.S. Army,* Vol. 2 (Washington, D.C., 1903), 282.

memoirs noted that the infantry under General Zachary Taylor was equipped with flintlock muskets and paper cartridges charged with powder, buckshot, and ball. "At the distance of a few hundred yards a man might fire at you all day without your finding it out," he recalled.[25]

Regardless of ignition system, the .69 musket had a bruising recoil. One tale that circulated among infantrymen during both the Mexican and Civil Wars was that the kick was caused by excessive space in the barrel behind the touch hole or nipple and could be reduced by eliminating this space. A dime was about the same diameter as the bore and to test this theory one soldier rammed six down the barrel of his musket. When he found this did nothing to soften the kick, he wanted his 60 cents back but found that the dimes had become firmly affixed in place and couldn't be recovered. Since it was common for soldiers to ascribe nicknames to their guns, such as "Mary Jane" or "Hannah," he christened his "Silver Sue."[26]

Colonel Jefferson Davis commanded the red-shirted 1st Mississippi Infantry during the Mexican War, and following his persistent efforts the regiment received a new percussion rifle adopted in 1841 (production actually began two years later). This was a handsome, accurate, sturdy rifle with patch box cover and other mountings of brass, firing a round .54 caliber patched ball, the last regulation U.S. rifle to do so. In tribute to the effectiveness of this rifle in the hands of Davis' regiment one of its most common nicknames was the Mississippi rifle, along with "Yager" or "Windsor" rifle. The new rifle also was issued to other federal and militia units including the Regiment of Mounted Rifles, organized in 1846 and mounted for mobility but intended to fight primarily on foot. Armed with this new rifle, single-shot pistols, and saber the rifle regiment's anticipated assignment was to guard emigrants along the Oregon Trail but service in the Mexican War delayed that. After 1855 on the order of now Secretary of War Jefferson Davis, many M1841 rifles would be fitted with an adjustable rear sight and rerifled to fire the new .58 caliber conical hollow-based Minie ball (Figure 82).[27]

The army's two regiments of mounted dragoons, organized in 1833 and 1836, during the Mexican War relied on a carbine, a heavy saber selected in 1840 ("old wristbreaker"), and .54 caliber single-shot flintlock and percussion pistols, the latter the new model of 1842 (not to be confused with the Navy's Model 1843 boxlock pistols). Being smoothbores like their predecessors, these new pistols still offered little use beyond a dozen or so yards yet would remain in service until the mid-1850s. The dragoons' carbine was an improved version of the breech-loading percussion Hall, inferior in range to the Mounted Rifles' Mississippis. Early model Hall carbines had been first issued to dragoons during the Seminole War in Florida in the mid-1830s and were the army's first official percussion weapons. An unusual feature of these early Hall carbines was a slim triangular rod bayonet mounted beneath the barrel and which could be slid forward and locked into place or

25. U. S. Grant, *Personal Memoirs of U. S. Grant* (New York, 1894), 61.

26. *Collections of the Kansas State Historical Society, 1913-1914* (Topeka, 1915), vol. 13, 26-27.

27. As originally intended, dragoons were trained to fight mounted or on foot while the Mounted Rifles fought as infantry. The two cavalry regiments organized in 1855 were lightly armed with revolvers and some carbines and were intended to fight on horseback. In time the distinction between these units became blurred and in August 1861, they were reorganized into one arm known as cavalry. The dragoons and rifles were bitter at losing their distinctive designations and trim colors on their uniforms (orange for dragoons, green for rifles) and now would have to adopt cavalry yellow. To pacify them, they were allowed to wear their former color stripes until their existing uniforms wore out. However some uniforms acquired a miraculous durability and orange and green stripes could still sometimes be seen as late as 1865. Stephen Z. Starr, *The Union Cavalry in the Civil War*, Vol. I (Baton Rouge, La., 1979), 59, n. 31.

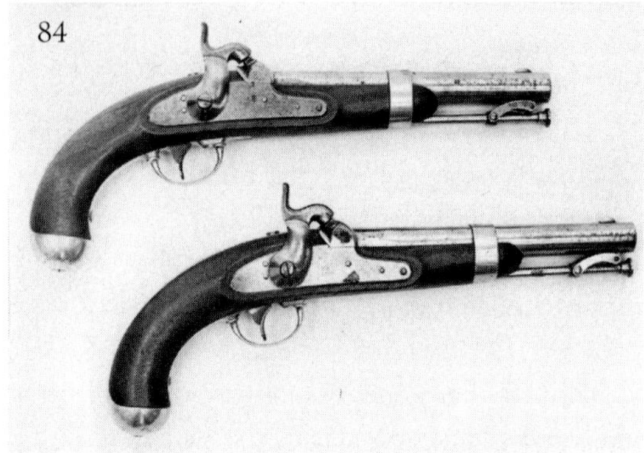

83. U.S. Model 1836 flintlock made by Asa Waters and carried in the Mexican War by Colonel Ashbel Fairchild. The government obtained about 41,000 Model 1836 pistols from contractors Waters and Robert Johnson. (*Courtesy: Smithsonian Institution*) **84.** Close on the heels of the Model 1836 flintlock martial pistol came the army's model of 1842, little changed except using the percussion ignition system. Henry Aston and Ira N. Johnson produced about 40,000 on government contracts. These are Astons, each dated 1846 and each inscribed "Lt Col. T. H. Seymour/U.S.A./from/Capt W. A. U.S.A./City of Mexico/Feby 1848" on the butt cap. Seymour later became governor of Connecticut and Colt's agent in Russia.

removed and used as a cleaning rod or even as a ramrod if by chance it became necessary to load from the muzzle. By the Mexican War the rod bayonet had been eliminated from those in production although the concept would be resurrected in 1880 and applied to some of the .45-70 Springfield "trapdoor" rifles when the supply of Civil War-style triangular bayonets became exhausted. (One problem encountered with the later rod bayonet was the difficulty in retracting it after the rifle had been fired for a while with it extended and black powder residue had built up around the rod) (Figures 83, 84).[28]

Another unusual feature, common to all Halls, was the ability by withdrawing a single screw to remove the breechblock, mainspring, hammer, trigger, and trigger spring as a single unit. Samuel Chamberlain, a dragoon who later wrote a colorful account of his Mexican War exploits, described carrying the block of his Hall carbine as a crude but functional pocket pistol when he went into town for rest and relaxation. I have examined three simple pistols made by securing a Hall action in a crude wooden stock, each obviously of nineteenth-century vintage, perhaps assembled by American Indians (Figures 85, 86).

The domestic repeating handguns to see service during the Mexican War initially would have been privately owned multi-barrel pepperboxes or Colt Patersons which had served in the hands of the Texas Rangers or as personal sidearms. However in January 1847 the government signed a contract with Colt for the 1,000 revolvers which Captain Samuel Walker of the Mounted Rifles had helped design and which bear his name today. Walker received his pair from Colt and additional guns reached the Mounted Rifles and Texas Rangers soon after. One of history's most powerful military handguns, it had a muzzle velocity of 1,300 feet per second. Walker described it as "effective as a common rifle at one hundred yards and superior to a musket, even at two hundred" (Figure 87).

Adopted in 1847 but produced too late for service in that war was a series of .69 smoothbore iron-mounted musketoons or short muskets, for use by artillerymen and sappers (engineers). Others with brass rather than iron fittings were used by dragoons and cavalry throughout most of the 1850s decade. Though these latter musketoons are

28. Even though a weapon was adopted in a particular year, it might be several years before production and issue to the field began as with the army's Model 1842 pistol. Actual production began in 1845.

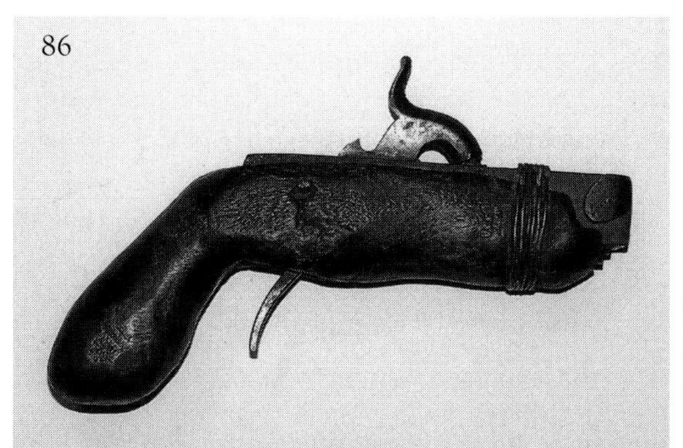

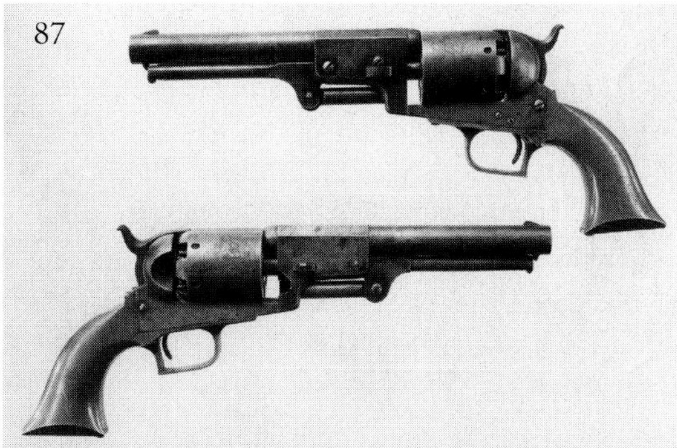

85. Model 1843 Hall carbine with the breechblock tipped up for loading. (*Courtesy: National Park Service, Fuller Gun Collection, Chickamauga and Chattanooga National Military Park*) **86.** Hall breech mechanism converted into a functional pistol by securing it to a crude wooden stock, perhaps by an American Indian. **87.** As a postwar expression of gratitude, the citizens of Knox County, Ohio, presented this pair of Colt .44 Dragoons with silver grips to Colonel George Washington Morgan for his gallantry in action during the Mexican War. (*Courtesy: Smithsonian Institution, neg. no. 33327-A*)

attractive pieces with their brass furniture, they were scorned as inaccurate and the recoil or "kick" was punishing. In an attempt to correct this latter fault, some had a slug of lead inserted in the butt stock. "Worthless" was a term often applied to these musketoons, which served the dragoons alongside the Hall breechloader and in the late 1850s the highly regarded Sharps carbine. To counter their musketoons' limited accuracy and range, it was common for a dragoon company to be issued one percussion rifle for hunting. Similarly, dragoon companies armed with the Hall carbine had relied on a few Hall rifles for longer range shooting (Figure 88).[29]

The decade of the 1850s was a period of experimentation and the testing and eventual adoption of a miscellany of new weapons. Unwittingly, the military was preparing itself for a bloody four years of civil war soon to come. The Model 1842 musket was being issued in increasing numbers and it did offer two significant advantages over proceeding models of flintlock infantry muskets. It used a percussion cap and had the distinction of being the first regulation arm produced in both national armories with complete interchangeability of parts. It also was the last of the army's official smoothbore muskets. The practicality of parts interchangeability was demonstrated dramatically after Harpers Ferry suffered a particularly severe

29. In response to complaints about the cavalry musketoon, army ordnance colonel Craig in 1851 defended the muzzle-loading piece with its "whole barrel." "If it be not a suitable arm for cavalry, I know not where to look for one that will answer." Claude E. Fuller, *Springfield Shoulder Arms, 1795-1865* (New York, 1931), 107.

88

88. Three versions of the Model 1847 .69 musketoon exist with minor variations—one for artillerymen, another for miners and sappers, and a brass-mounted version for dragoons and later cavalry. A few like this artillery model were later rifled and a rear sight added. "Worthless" was an adjective sometimes used to describe the cavalry musketoon. (*Courtesy: National Park Service, Fuller Gun Collection, Chickamauga and Chattanooga National Military Park*) **89.** Copper powder flask of the type issued to regular army and militia riflemen in the 1840s and 1850s. This style is commonly known as a "peace" flask because of its design with clasped hands within a circle of stars to represent union among the states. (*Courtesy: Neil and Julia Gutterman*)

periodic flood. In the process of removing mud from guns in storage, 9,000 of the new muskets were stripped with no effort made to keep parts from the same gun together. The parts were cleaned and then workers reassembled them without difficulty. Nevertheless the M1842 was still a .69 smoothbore, faster to load than a muzzle-loading rifle but with limited accuracy, not much improved over the flintlock muskets acquired from France in the 1770s (Figure 89).[30]

1855 and a New Series of Military Arms

Less than a decade after peace was established between the U.S. and Mexico, in 1855 under Secretary of War Jefferson Davis the army selected a new series of rifled arms to replace the Model 1842 and earlier smoothbores. These new weapons included a musket with a rifled 40-inch barrel (rifle musket as it came to be called), in substantially smaller quantities a rifle with a barrel seven inches shorter, a single-shot pistol-carbine with a twelve-inch barrel and a detachable wooden shoulder stock for use by artillery and mounted troops, and a .54 caliber rifled carbine (Figure 90).

Each of these new weapons was still a muzzle-loader and except for the latter was now .58 caliber and employed a mechanical priming device similar to the roll of paper caps adopted several generations later for use in children's repeating toy cap pistols. This mechanical priming system—ordinary individual musket caps still could be used instead—was the invention of a Washington, D.C., dentist, Dr. Edward Maynard. The roll containing fifty caps (twenty-five for the pistol-carbine) was contained in a compartment in the lock plate, and whenever the hammer was cocked, the tape of primers was advanced one position to place a fresh primer over the nipple. The system never achieved

30. An infantry private in 1850 was paid $7 a month, a dragoon private $1 more, as compared with a sergeant's earnings of $13 a month. Although soldiers were supposed to be paid every two months, in remote areas as on the frontier there might be a six-month delay between paymaster visits.

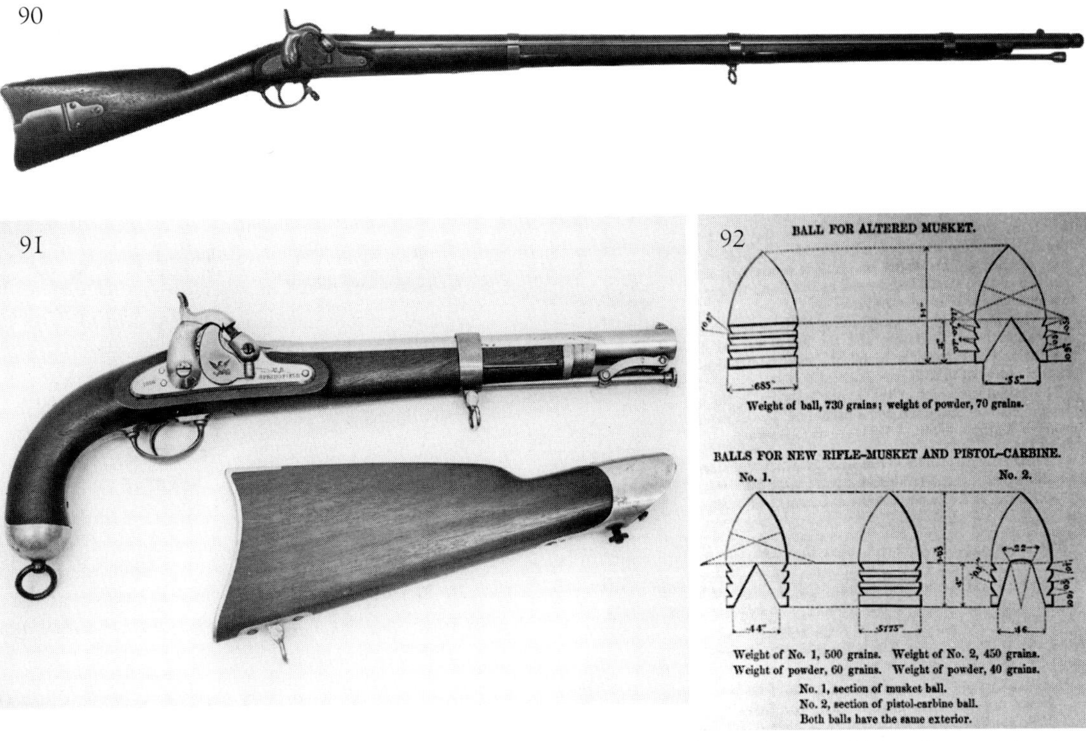

90. U.S. Model 1855 .58 rifle musket equipped with the Maynard tape priming device, a feature which was eliminated from later models. This was the first U.S. infantry arm to fire the .58 Minie ball. (Courtesy: National Park Service, Fuller Gun Collection, Chickamauga and Chattanooga National Military Park) **91.** Springfield Model 1855 pistol-carbine. **92.** Drawings of the new Minie bullets—.69 caliber (above) and .58.

much support from those who used it in the field, although it allowed the soldier to load without fumbling with a percussion cap. However the Maynard tape primer was not always reliable in damp weather and the system was dropped from subsequent models of government arms (Figure 91).[31]

This new rifle musket was expected to shoot sufficiently well to strike a target the size of a horseman at 600 yards, penetrate four inches of soft pine at 1,000 yards, and put ten consecutive shots in a nine-inch bullseye at 200 yards and in a twenty-seven-inch bullseye at 500 yards, a major improvement over what was expected from its smoothbore predecessors. This also was an improvement over the maximum effective range of about 300 yards for a round ball rifle such as the Model 1841 Mississippi. The secret to the new arm's success was a cylindro-conical bullet developed in 1847 by Captain Claude Minie of the French army. Its hollow base was filled with an iron expander plug which was driven forward by the force of the exploding gunpowder, causing the bullet base to expand and fill and grip the rifling grooves.

In 1854 at Harpers Ferry Armory, assistant master armorer James H. Burton found that by widening and deepening the cavity in the base the plug was unnecessary since the gases were sufficient

31. Maynard also was the inventor of a successful breechloader, some of which saw Civil War service as a cavalry carbine and others as various models of hunting and target rifles manufactured by the Massachusetts Arms Company as late as the 1880s in calibers from .22 to .50.

93. Despite the navy's satisfaction with Jenks breech-loading rifles and carbines such as this one, the army rejected the carbine because of the difficulty in loading it with loose powder and ball on horseback. Nevertheless, in an army test in the summer of 1842 a Jenks was fired 14,813 times before any significant damage occurred--the nipple split and the test was halted. (John D. McAulay, Civil War Carbines, Vol. II [Lincoln RI, 1991], 51.) (*Courtesy: National Park Service, Fuller Gun Collection, Chickamauga and Chattanooga National Military Park*)

to expand the undersize bullet's hollow base. History has often ignored Burton's contribution and consistently the bullet has borne the name "Minie ball." With the advent of civil war, Burton resigned his post at Harpers Ferry and soon was named superintendent of Confederate armories (Figure 92).[32]

The caliber chosen for the new rifle musket (.58) was a compromise between the less severe recoil of the .54 caliber Mississippi rifle and the better long-range accuracy of a .69 Minie ball. Now the infantry had a more accurate weapon which unless heavily fouled could be loaded as quickly as a smoothbore. Soon workers at government armories were busy altering Model 1842 smoothbores into percussion rifled muskets, rifling them to fire the new Minie ball in .69 caliber and fitting some with an adjustable rear sight. Similar modifications were made to substantial numbers of Mississippi rifles, reboring them to .58 caliber. The new .58 rifle musket would become the primary infantry weapon in the civil war which was to come, although its reign would only be for a decade.[33]

Like its predecessor the Model 1842 musket, the new rifle musket's bayonet had a fluted triangular cross section. It slipped over the muzzle and a rotating ring on the shank of the bayonet locked it in place. Wicked though it might appear, in the Civil War the bayonet inflicted far fewer wounds than might be anticipated. Not an enemy's blood but candle wax often dripped down its shank for it served as a convenient candle holder or like the rifle musket's iron ramrod a handy spit on which to cook meat or other rations. Some wartime doctors never even once were called upon to treat a bayonet wound.

Also in 1855, four new regiments were created within the army, two as infantry and two designated for the first time as cavalry rather than as dragoons or mounted rifles. These two new mounted regiments were sent west to join the existing three horse-borne regiments attempting to maintain order in the face of a growing problem of white versus American Indian. Over the next few years, cavalry and dragoon companies received a potpourri of weapons in addition to the Halls and muske-

32. In something of an ironic twist, Burton soon after war's end accepted the position of foreign sales agent attempting to market Spencer repeaters in Europe, a weapon which had helped speed the downfall of the Confederacy.

33. The term "rifled musket" describes those muskets which were originally smoothbore but were updated with the addition of rifling as opposed to the rifle musket which was rifled originally. The Springfield Armory alone in addition to manufacturing new guns reported that in 1850-52 it altered more than 86,000 flintlock muskets to percussion and in 1857-59 rifled and added rear sights to more than 7,000 Model 1842 percussion muskets (Fuller, *Springfield Shoulder Arms, 1795-1865*, 142).

toons already in service—the new Springfield muzzle-loading rifled carbine and the pistol-carbine (which proved unpopular) as well as breech-loading carbines for evaluation including those by Sharps, Maynard, Burnside, and several others of lesser renown.[34]

The handguns issued to the mounted regiments constituted a mixed bag as well. In addition to pistol-carbines, one found single-shot percussion pistols of the model chosen in 1842 and four-pound .44 Colt Dragoon six-shooters, each type carried in saddle holsters, .36 English Adams five-shooters (some made by Massachusetts Arms Company), plus the .36 caliber 1851 Colt. The two new cavalry regiments were the first army units to receive the Colt '51, one per trooper. There had been occasional "in the field" modifications to saddle holsters to allow the heavy Colt Dragoon to be worn on a waist belt and apparently some issue examples, but the Colt 1851 Navy was light enough that it was regularly carried in a belt holster, known then as a "pistol case."[35]

New Navy Small Arms (1828–1860)

Navy and Marine Corps personnel too were witnessing improvements in their arms. In 1827-28 they had received some Hall flintlock breech-loading rifles and beginning in 1841 a number of Hall percussion carbines. Also between 1843 and 1848 the navy procured Jenks percussion breechloaders, 1,000 rifles and 5,200 carbines, unusual in their side hammer or "mule ear" design. Seventy of the Jenks carbines were on board the famed frigate *Constitution* when at the end of 1852 she was outfitted for a cruise that would last for almost two and a half years. Other navy breechloaders including those by Sharps and Burnside. A few Colt Model 1855 revolving rifles and Maynard carbines were procured between 1856 and 1860 but as within the army, breechloaders did not displace the muzzleloader as the sea service's primary long gun (Figures 93, 94).

94. Copper naval powder flask with fouled anchor design dated 1845 for use with the Jenks breechloader. (*Courtesy: Neil and Julia Gutterman*)

The navy and marines—most of the latter personnel served as guard detachments on ships—by the late 1840s had begun the transition to the percussion age, converting their flintlock muskets to the percussion system, but it wasn't until 1853 that the marines received their first allotment of the new Model 1842 percussion muskets, a design chosen by the army a decade earlier. Three years later, in 1856, the navy and marines began receiving army .69 percussion rifled muskets, altered from smoothbore flintlocks and fitted with an adjustable rear sight, followed not long after by deliveries of Model 1842 muskets, updated with rifling. Not until 1860 did the marines begin rearming with the army's newly adopted Model 1855 .58 rifle muskets. One tragic accident hap-

34. The 1888-89 catalog of J. H. Johnston's Great Western Gun Works of Pittsburgh offered surplus Springfield pistol-carbines described as the Hunter's Companion for $2.50 complete or $2.25 without the detachable shoulder stock.

35. Perhaps disenchanted with word of the Walkers' performance, unlike the 2nd Dragoons and Mounted Rifles, the 1st Dragoons as of early 1849 had shown no official desire to receive Colt revolvers and instead appeared content to retain their single shot pistols. (Maj. B. R. Lewis, "Sam Colt's Repeating Pistol," *American Rifleman*, June 1947, 25.)

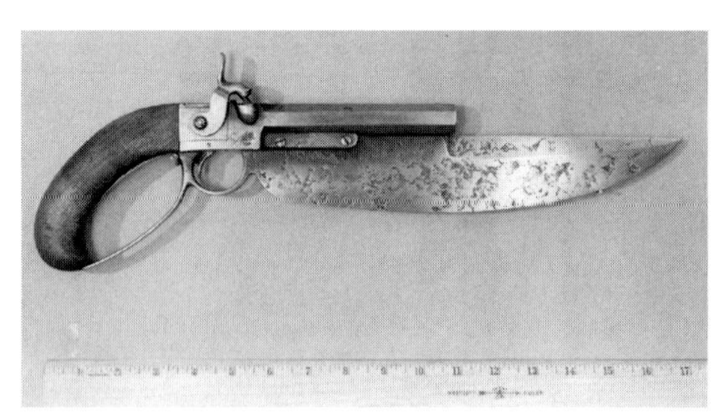

95. Elgin cutlass pistol, the U.S. Navy's first percussion weapon. (*Courtesy: John A. Williams*)

pened with a .69 musket when a detachment of marines went ashore in Shanghai, China, in 1855. As one of the men attempted to break down a door with the butt of his gun it discharged, killing the boatswain behind him.[36]

The navy had abandoned most of its remaining flintlock pistols by mid-century and was relying on .54 caliber single-shot percussion smoothbore handguns, either the short barrel box-lock model of 1842 made by N. P. Ames of Springfield, Massachusetts, or Deringer or the longer army model of the same year produced by Aston or Johnson. On the *Constitution*'s cruise mentioned above, records indicate her crew was partially armed with Ames pistols dated 1844. Quite rare today are those Ames pistols dated 1843 and marked "U.S.R." indicating purchase for the U.S. Revenue Cutter Service, forerunner of the U.S. Coast Guard.

In 1852, the navy did procure Colt revolvers—fifty Model 1851s, twenty-five .44 Dragoons, and twenty-five .31 caliber Model 1849s, the latter including four-, five-, and six-inch barrel lengths. Some of these were presented as gifts during Commodore Matthew Perry's visit to Japan in 1853. But despite the increasing popularity of Colt's arms in other circles, Chief of Naval Ordnance Commodore Morris considered them too costly (four times that of a single-shot pistol), more liable to get out of order, and more dangerous for a common seaman to use. Pistols, swords, and hatchets were more useful on the rare occasion a party boarded an enemy vessel to capture it, he felt. Not until 1857 did the navy place its first substantial order for Colts, 2,000 Model 1851 revolvers. The only other revolvers received came in late 1860 against a modest navy order for 300 of the clumsy appearing Savage-North .36s.[37]

One unique example of an exclusively navy handgun was a single-shot percussion pistol designed by George Elgin with a sturdy blade mounted beneath the barrel. The navy did obtain 150 of these seventeen-inch cutlass pistols in .54 caliber made by C. B. Allen for use during its 1838-42 U.S. South Seas Exploring Expedition led by the tyrannical Lieutenant Charles Wilkes. These hybrid pistols apparently were put to use in at least one struggle with natives on one of the islands. Not only was this the only eighteenth- or nineteenth-century combination knife and pistol purchased by one of the U.S. armed services, but it was the first percussion handgun the Navy procured even in limited numbers. A few smaller size Elgin knife pistols were marketed to civilians as well; all are rare (Figure 95).

Across the Plains by Wagon, Handcart, and Foot (1821–1870)

In 1821, William Becknell pioneered the Santa Fe Trail from Missouri to the Spanish town of Santa Fe, New Mexico. It would remain largely a trade route rather than a path for settlers, but to the north, in 1841 the first party of emigrants from the east to make their way by wagon across the plains to Oregon reached their destination. The trickle of wagon trains bringing settlers increased in volume with the resolution of the Oregon

36. John D. McAulay, *Civil War Small Arms of the U.S. Navy and Marine Corps* (Lincoln, R.I., 1999), 24.

37. Ibid., 30, 50, 53.

boundary dispute with Great Britain in 1846, but became a torrent after discovery of "something shining in the bottom of the ditch" by John Marshall in January 1848. For the next few years, the gold country of California became the destination for many of these travelers, often not families but parties of single men or those who had left wife and children at home (Figure 96).[38]

The non-Indian population of California before Marshall's discovery of gold was about 14,000 but word spread throughout the world. By 1851 as gold fever peaked it had jumped to almost 200,000. While the first arrivals on the scene in 1848 might find the wealth they sought, in the early 1850s gold no longer was easily found and it soon was available only to those organized companies with more elaborate equipment than a pan or a rocker and with substantial capital investment. As colorful a chapter in American history as the California gold rush was, it was accompanied by rape of the land as hydraulic mining with its high-pressure water hoses soon began to rip the earth apart, severe discrimination against Chinese and Mexicans, and by a policy of virtual genocide which eliminated three-fourths of California's American Indian population.

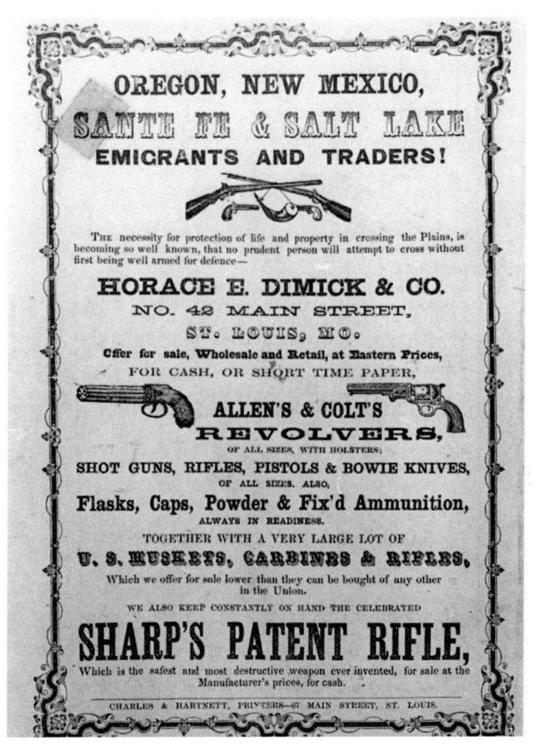

96. "The necessity for protection of life and property in crossing the Plains, is becoming so well known, that no prudent person will attempt to cross without first being well armed for defense." So advised this 1850s ad by Horace E. Dimick & Co. of St. Louis.

Regardless of whether they were seeking gold or a new family home, a journey across the plains in the 1840s and 1850s took about four months by wagon, and many dangers and hardships lay ahead of those bold enough to challenge terrain and weather. They couldn't start too early in the season or there wouldn't be sufficient grass for their sturdy but slow-moving oxen, the preferred draft animal for the trip. Too late a start put emigrants at risk of being trapped in the mountains by winter snows, as happened to the ill-fated Donner party in 1846, some of whom were forced to resort to cannibalism to survive.

The passing of time and hundreds of wagons gradually eased the difficulties of the trail somewhat but dangers persisted. Indians always might be a threat to run off carelessly herded livestock or engage in petty thievery, but before the mid-1850s there were few instances of attack if travelers were respectful and took reasonable precautions. In fact, American Indians sometimes provided aid in the form of food, directions, and guide services. One group of families was grateful to Crow Indians who ferried women and children across the Green River on horseback. Even if confronted by marauding Indians, a show of defiance and a bluff some-

38. Not all those heading to the west coast traveled by land. An alternative used by some was by ship in a hazardous passage around the storm-battered southern tip of South America. Others opted to travel by ship to Panama, cross the mosquito-infested isthmus on foot, then take passage on another ship to California.

97. Ready to depart for the California gold fields, a prospective '49er poses with a Colt .31 Baby Dragoon in his belt and what appears to be a Stocking pepperbox in his shirt front. (*Courtesy: Minnesota Historical Society*)

times was a good defense as shown by several recorded later incidents. Captain Luther North, who commanded a company of Pawnee scouts for the army, in 1870 found himself on foot unarmed and separated from cowboys with whom he was traveling. Six Sioux threatened him and in the growing dusk he grabbed a long stick, dropped to one knee, and pointed it at his enemies who scattered, thinking it was a rifle. He continued this deception whenever they approached until he came in sight of his comrades' campfire. Similarly in 1868 a Colorado farmer used the handle of a hoe to intimidate marauding Indians long enough for him and his sister and daughter to reach the safety of a neighbor's house. Generally on the overland trails, other dangers were more of a concern than Indians.[39]

Illness such as cholera, drowning during a river crossing, and children being run over by a wagon were not uncommon causes of death. But diaries reveal that injury or death due to the careless handling of firearms was one of the most often mentioned hazards of the trail. Frequently such accidents resulted when withdrawing a loaded and capped percussion rifle, shotgun, or musket from a wagon. If the hammer should catch on an obstacle, it might be drawn back far enough so that when it slipped free it fell on the percussion cap with sufficient force to detonate it. Some guidebooks for travelers warned never to place a cap on a gun until you were ready to shoot.[40]

This and most other forms of firearm-related accidents resulted from carelessness in some manner. A woman lost a lock of her hair when she jumped into a wagon and a rifle leaning against the vehicle discharged. When a man tossed a bundle of clothing from a wagon, it struck an upright rifle and he was shot in the knee. Expulsion from a Mormon wagon train bound for the Great Salt Lake was the sentence passed out to a young man who carelessly discharged a flintlock pistol loaded with three balls (an unusual load for a pistol), seriously wounding a companion. The injured man was expected to die but Brigham Young and others "prayed me out of danger" and he survived. A grazing horse's halter snagged on a loaded musket carelessly left lying on the ground, and when it discharged, the ball struck a boy in the knee forcing the amputation of his leg. Another party suffered its fifth gunshot accident when a youth playfully stalked a guard Indian-fashion a little too convincingly. Its sixth occurred when another member of the same group held up a trunk lid as a target for an inept marksman. In 1850 David Jones shot him-

39. Charles G. Worman, *Gunsmoke and Saddle Leather* (Albuquerque, N.M., 2006), 482.

40. Some clergymen proclaimed that among those seeking gold, cholera was God's retribution for their greed. Pawnee Indians are thought to have suffered 1,200 deaths in 1849 alone from this dreaded disease brought to them by white travelers.

self in the groin and thigh as he thrust a pistol into its holster as he was saddling his mule and the hammer caught on the holster's edge. He thought of taking his life with his other pistol when he saw the seriousness of his wound, but he reconsidered and did survive.

Emigrants whether carrying their possessions by wagon, hand cart, or on their back might arm themselves with any firearm available in the east or in St. Louis or other jumping-off place. Guidebooks and others recommended a good sturdy rifle or shotgun capable of withstanding the rigors of the trip—"no gingerbread work and sundry silver fixings, with a stock so slight that it must be laid down with great care and carrying a ball only suitable for squirreling." One party of gold seekers purchased eighty double-barrel shotguns in Baltimore before setting out. Another group of twenty-five eager men hoping for wealth was armed as though going to war. Each carried a revolver, sword, and a bowie knife, mounted men carried a pair of holster pistols and a rifle, and there were shotguns and more rifles suspended in their wagons (Figure 97).

Some emigrant parties relied in part on military arms, for the government since the 1820s had been selling its surplus worn or obsolete weapons. In addition, Congress in March 1849 authorized the sale of martial arms and ammunition, including some of the new Colt Dragoon revolvers, at cost to emigrants bound for Oregon, California, or New Mexico. There was no shortage of willing buyers and as one '49er wrote in his journal, his party secured new Mississippi rifles and percussion holster pistols from the government arsenal at Baton Rouge, Louisiana. But the perceived threat of Indian attacks lessened as an emigrant train progressed westward and often people gradually laid their guns aside. As one traveler wrote, by the time they reached Fort Laramie (Wyoming), guns were left stowed in wagons and the only weapons carried were a knife and perhaps a pistol peeking from a pocket (Figure 98).[41]

98. On the back of this daguerreotype's case is written "Charles G. Alexander, King George County, Virginia. Taken on his way to California. April 1849." He holds a Model 1841 Mississippi rifle and has what appears to be a pepperbox thrust in his belt. (Although the patch box appears to be on the left side of the rifle stock, the image is reversed due to the photographic process.) (*Courtesy: Herb Peck, Jr.*)

Passage all the way by steamboat obviously wasn't an option for those heading to the Pacific coast, but this means of travel for passengers and for the shipment of mining machinery and other goods up the Missouri River as far as Montana was by the late 1860s, the height of the gold rush to that territory. The threat of Indian attack against both the vessels and the wood yards that provided them with fuel was very real, and steamboats often were

41. Just outside Guernsey, Wyoming, one still can see visible but silent evidence of the Oregon Trail in the form of deep ruts worn in stone by the passage of thousands of emigrant wagon wheels.

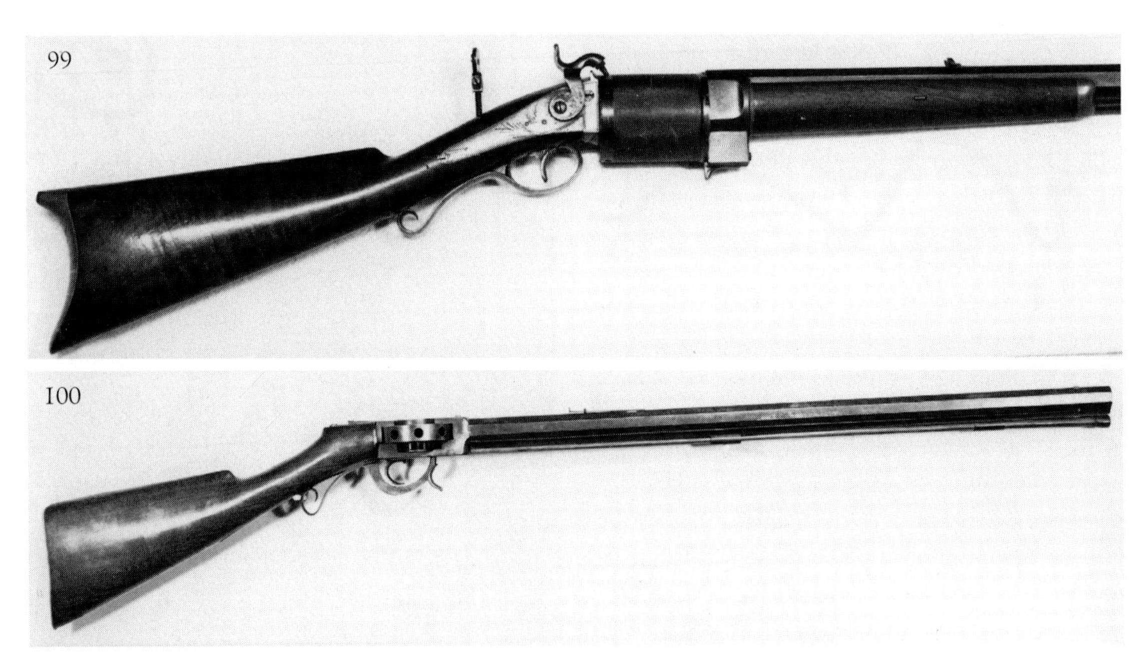

99. Pill lock revolving rifle by William Billinghurst of Rochester, New York. A finger-operated lever at the front end of the cylinder locks it in alignment with the barrel. A more advanced feature present in Sam Colt's revolver patents was an automatic cylinder locking arrangement. **100.** Nine-shot percussion Cochran "turret rifle." The fact that one chamber always was pointing toward the shooter may have been disconcerting to some potential buyers. (*Courtesy: National Park Service, Fuller Gun Collection, Chickamauga and Chattanooga National Military Park*)

equipped with artillery as well as small arms for protection. Army posts and government arsenals were the source for some of these firearms on board, just as they had been for some of the emigrants crossing the plains in earlier years. The steamer *Ida Rees* journeying from Pittsburgh to Fort Benton, Montana Territory, in the spring of 1868, for example, from the St. Louis arsenal obtained obsolete muzzle-loading muskets plus buck and ball cartridges and purchased one Spencer carbine and 150 cartridges for $22 there before steaming up river.[42]

Percussion Repeating Rifles

The invention of the percussion cap opened the door to firearms inventors for it made a number of technological advances practical. One of these was a gun with a revolving cylinder, best known in handgun form—the revolver. The concept of a multi-chamber rotating breech (cylinder) had been present as early as the matchlock era and it had been available in a very limited manner in flintlock form. Bostonian Elisha Collier had not found acceptance in America and from about 1818 until the early 1820s had produced revolving rifles, shotguns, and handguns in London in both flintlock and percussion form. The cylinder in these early attempts had to be turned by hand but did offer multi-shot (often five charges) capability before reloading.

Two- and three-barrel percussion hunting arms, with all rifled barrels or a combination of rifled and smooth barrels, were available throughout the percussion or "cap and ball" era although they were

42. Information provided by Dr. John J. Kudlik.

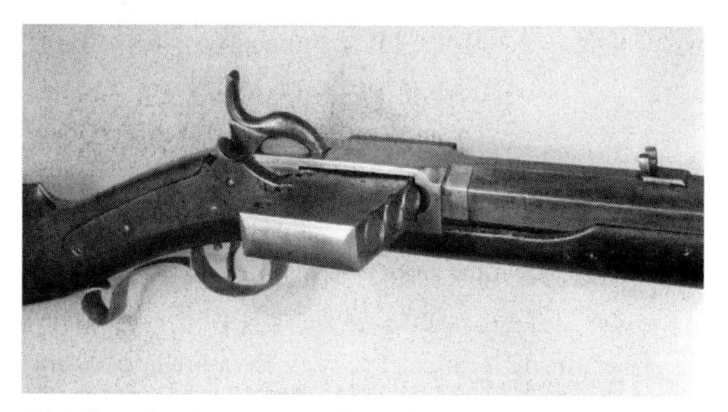

101. Five-shot harmonica rifle made by Jonathan Browning in 1853, a design he called his "slide gun." The bar is moved by hand to successively align each chamber with the barrel and the thumb lever locks the bar in place. To reduce gas leakage, the front of each chamber is countersunk to mate with a lip projection at the rear of the barrel. (*Courtesy: Norm Flayderman*)

heavy. One unusual example of such multi-shot arms was a four-barrel .38 rifle with four side hammers made by James Goodell of Olean, New York, in the 1850s. Earlier John Miller of Rochester, New York, in 1829 patented a repeating rifle with a hand-turned cylinder, and a few other makers followed suit between the 1830s and 1850s. Best known are those made by William Billinghurst, also of Rochester; Benjamin Bigelow of Marysville, California; and of course Samuel Colt with his mechanically rotated cylinder (Figure 99).

Many of these early revolving rifles, other than Colts, were pill locks in which the small pills of fulminate were held in place with beeswax in depressions in the cylinder. The wax could be contained in the patch box if one were present. Often they were converted to percussion as the more convenient caps became increasingly available and more reliable. Various means of locking a chamber in alignment with the barrel were used and gas leakage at that joint was a distraction as was the occasional discharge of more than one chamber at a time. For this reason, it was advisable to hold the weapon with the foremost hand behind the cylinder. With the exception of those by Colt, early revolving rifles were made in limited numbers, often only a few hundred at most by any maker.

C. B. Allen of Springfield, Massachusetts, in the late 1830s produced several hundred revolving turret rifles and pistols designed by John W. Cochran. In these underhammer guns the horizontally mounted radial or turret cylinder contained seven or nine chambers, arranged like the spokes of a wheel. Undoubtedly disconcerting to the shooter with a fully loaded cylinder was that one chamber pointed directly to the rear in line with his face while other chambers pointed at anyone standing at the side. Yet Josiah Gregg on an 1840 journey to New Mexico carried just such a gun, a nine-shot Cochran rifle. On one occasion while alone he encountered a band of bison and fired six shots in rather rapid succession. In a few moments some of his companions from the wagon train came galloping to his aid, forgetting he had a repeating rifle and assuming he had been attacked by Indians (Figure 100).

One uncommon yet practical example of a repeating percussion gun was the "harmonica rifle." Rather than a revolving cylinder, multiple charges were contained in a rectangular block which was slid horizontally by hand through an opening at the breech for each successive shot. It didn't solve the problem of escaping gas between breech and barrel, but if a shooter had several spare loaded and capped blocks in his pocket, he had what in the 1840s and 1850s was a substantial amount of firepower at his disposal. Known makers of harmonica rifles included C. B. Allen of Springfield, Massachusetts; Nicanor Kendall of Windsor, Vermont; Henry Gross of Tiffin, Ohio; and Mormon gunsmith Jonathan Browning, father of famed inventor John Moses Browning (Figure 101).

Possible references to what Browning advertised as his "slide repeaters" appear in accounts by several Mormons. When in June 1846 a mob from Carthage threatened the Mormon settlement at Nauvoo, Illinois, on the Mississippi River, they were ambushed by the "Spartan Band" of thirty men, mostly armed with fifteen-shooters as Hosea Stout later wrote in his diary. In describing an encounter with Ute Indians in 1850, Mormon Bill Hickman later wrote: "I had a slide rifle; six shots in a slide, and three slides, making eighteen shots on hand." John Doyle Lee in September 1846 was traveling to Santa Fe on a mission for Mormon leader Brigham Young when his party awoke surrounded by grazing bison. "I took [my] fifteen shooter and walked within 60 or 70 yards of them . . . and it was as much as I could do to refrain from shooting."[43]

Despite the advanced designs offered by some makers of early repeating rifles, the single-shot muzzleloader retained its popularity. It would remain the primary form of civilian rifled shoulder gun until the advent of successful breechloaders such as the Sharps in the early 1850s and improved Colt revolving shoulder guns after 1855 and even then retained some popularity as the era of the metallic cartridge dawned in the 1860s.

New Colts and Their Competitors

There had not always been agreement, particularly on the western frontier, whether the lighter and handier .36 Navy or heavier .44 Dragoon with its greater bullet weight and powder charge was the wiser choice. Nevertheless through 1859 production of the Navy far surpassed that of the Dragoon (about 90,000 to 20,000). The well-balanced Colt Navy achieved such prominence that it sometimes was referred to in nineteenth-century writings merely by the term "navy six" or just "navy" with no need for any other identification. One Kansas pioneer of the 1850s later described what he perceived as the typical pro-slavery "border ruffian" and noted: "He never sat his horse, he just hung there; his feet ornamented with spurs, a pair of navy revolvers and a large knife attached to his belt; a skillful rider, an excellent shot and I might say when in the minority, . . . the most sneaking, cringing coward of the plains. But when in the majority, the most braggardly, overbearing, insolent, tyrannical and brutal of creatures." Ultimately when production of the '51 in Hartford ended about 1872 it totaled about 215,000 (Figure 102).[44]

The decision for a buyer who wanted a .44 but didn't want to be weighted down by more than four pounds was eased in 1860 when in February Colt introduced what it called its New Model Holster revolver, replacing the Dragoon with a better balanced six-shot .44 which weighed just over two and a half pounds. It evolved from altering the frame of the '51 to accept what's known as a rebated cylinder in which the forward section is a little larger in diameter than the rear allowing an increase in caliber to .44. Some early cylinders were fluted or dished out longitudinally between the chambers to reduce weight, but this practice was halted in 1861 when it was found it weakened the cylinder walls and sometimes caused them to split. Unfluted 1860 .44 cylinders bore the same naval battle scene engraving as found on the Model 1851. Popular as this model might have been with civilians from the start, beginning in 1861 most of these went to fill government orders for more than 125,000 as the Model 1860 became the most widely used of the Federal cavalry's handguns. Production ended in 1872 after about 200,000 had been assembled. Colt charged the government an average of $25 each for these Model 1860s during the early Civil War years until competition forced a reduction in price to $14 (Figure 103).[45]

Other Colts of somewhat lesser production quantities than the '51 and '60 included in 1855 a solid-frame pocket revolver in .28 and .31 caliber, the only percussion Colt handgun to vary from the usual style of separate frame and barrel secured by a wedge. Designed in part by Elisha K. Root, it also differed in having a hammer mounted on the side

43. Worman, *Gunsmoke and Saddle Leather*, 74.

44. William W. Denison, "Early Days in Osage County," *Kansas State Historical Society Collections*, vol. 17 (1926-28), 379.

45. Houze, *Samuel Colt: Arms, Art and Invention*, 106.

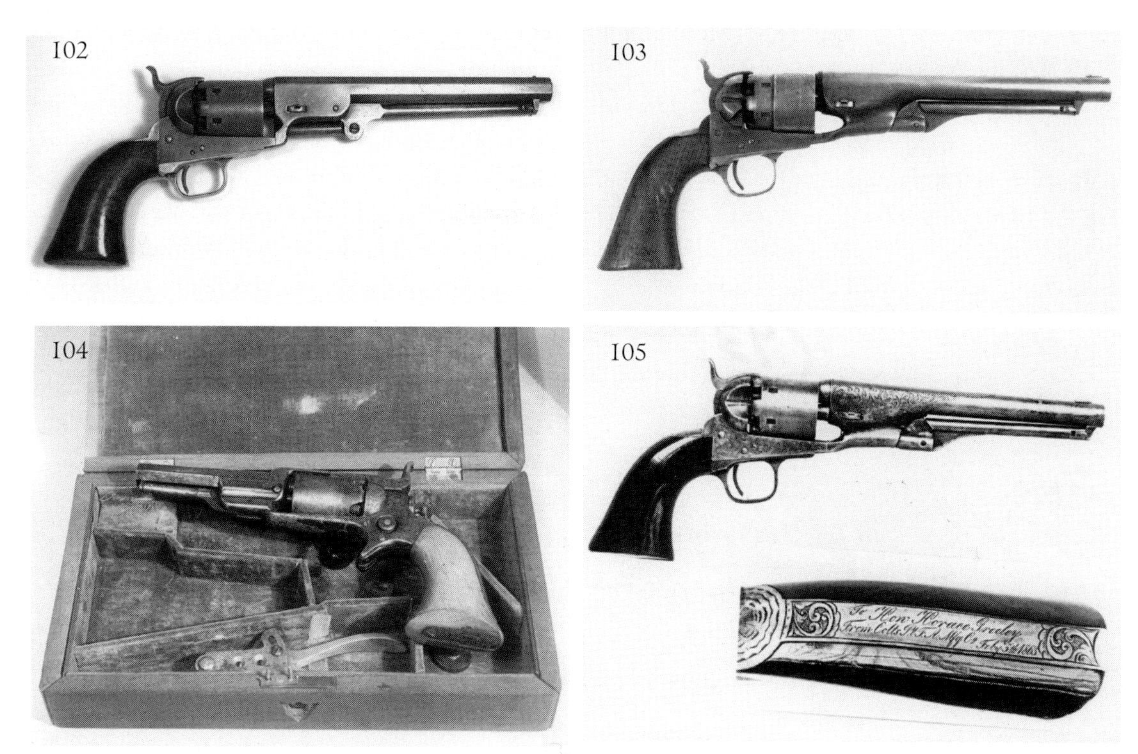

102. Christopher "Kit" Carson was one frontiersman who chose the Colt Model 1851 Navy over the heavier .44 Dragoon. This Navy was passed down through his descendants. Carson was apprenticed to a saddle maker when at age fifteen he went west in 1824 and remained there for the rest of his life as trapper, trader, guide, Indian agent, and army officer. (*Courtesy: Ron Peterson*) **103.** Colt .44 Model 1860 with a bloody history. In 1863, when the Espinosa brothers were terrorizing residents in southern Colorado with more than a dozen murders attributed to them, Colonel Sam Tappan at Fort Garland hired frontiersman Tom Tobin to track the outlaws. When Tobin and a party of soldiers caught up with them, Tobin shot the remaining brother Vivian Espinosa and nephew Felipe, then cut off their heads and delivered them in a gunny sack to Tappan. This Colt reportedly was taken from Felipe's body. (*Courtesy: Colorado Historical Society*) **104.** "R. E. Lee Col. U.S.A." is inscribed on the backstrap of this cased and engraved .28 caliber Colt Root 1855 pocket model. It isn't known when or how Lee acquired it, but he served as superintendent at West Point (1852-55) and then with the 1st and 2nd regiments of U.S. cavalry before the Civil War. It was found among Mrs. Lee's effects at the time of her death. (*Courtesy: National Park Service*) **105.** Newspaper editor Horace Greeley coined the popular expression "Go west, young man." This engraved .36 Model 1861 Navy was presented to him by the Colt firm in 1863. Colt produced about 39,000 of these streamlined "round barrel Navies" between 1861 and 1873 but rather surprisingly they never sold as well as the earlier octagonal barrel Model 1851 Navy. (*Courtesy: Connecticut State Library*)

of the frame rather than the center and a "spur trigger" with no trigger guard. Two other .36 caliber models appeared, a round barrel six-shot New Model Navy or Belt pistol, known by collectors today as the Model 1861 Navy, and in December 1860 a smaller frame five-shooter, the streamlined New Model Police with a round barrel and a semi-fluted cylinder, today incorrectly called the Model 1862 Police. The five-shot .36 Pocket Navy (sometimes erroneously referred to as the Model 1853) with octagonal barrel apparently wasn't introduced until sometime after a disastrous fire seriously damaged the Colt factory in Hartford in February 1864 (Figures 104, 105, 106).

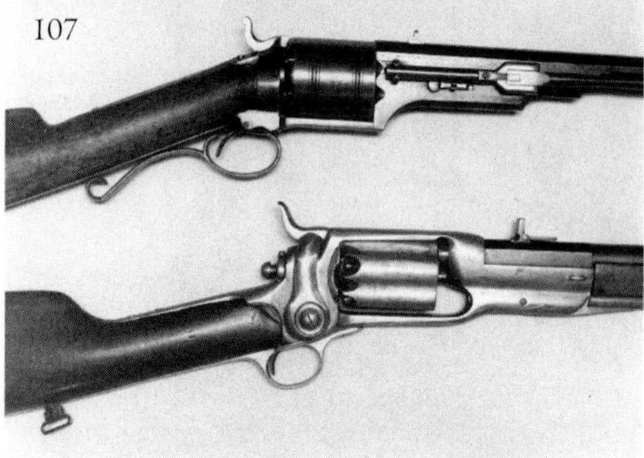

106. This Colt .36 five-shot is called the Model 1862 Police but it actually was introduced at the end of 1860. **107.** A pair of Colt revolving shoulder arms. Above, a Model 1839 .525 smoothbore Paterson-made carbine with a loading lever attached to the side of the barrel. Below, a Hartford-made Model 1855 rifle.

In 1856 Colt also had introduced a solid-frame revolving rifle of the general design as the Root pocket revolver, initially a six-shot in .36 caliber for hunting moderate-size game. Aware of the potential risk to a shooter's hand in the event of a multiple discharge, these early sporters had a finger spur behind and another in front of the trigger guard indicating where one's supporting hand should be placed, beneath rather than in front of the cylinder. More popular were the larger frame carbines, rifles, and muskets available in 1858 in .44 or .56 caliber. In late 1859 Colt introduced five-shot revolving shotguns in either 10- or 20-gauge (.75 and .60 caliber, respectively), but these had little appeal to sportsmen as limited sales figures showed (Figure 107).

Sam Colt was an enthusiastic promoter of his guns, often making use of letters of endorsement for advertising purposes. Beginning in 1854 he had commissioned ten original paintings in oil by George Catlin, a financially challenged artist famed for his portrayals of American Indians, of hunting scenes with Colt arms in use in the American west and South America. Colt also felt it well worthwhile to make presentations of highly finished examples of his arms to both military officers and civilians, including members of Congress and others in positions which warranted such gifts. How much such pieces influenced the award of government contracts and other orders for Colt arms can't be determined.

Often these gifts before Colt's death in January 1862 were inscribed on the backstrap (the strip of metal encircling the rear of the grip) with such phrases as "From the Inventor, Col. Colt" or "Compliments of Col. Colt." Just a few of the recipients of such gifts included England's Earl of Cardigan; the King of Siam; Democratic presidential nominee General Franklin Pierce; Simon Cameron, Lincoln's first Secretary of War; General George B. McClellan; General Ambrose Burnside; and Colonel James W. Ripley, then superintendent of the Springfield Armory and later during the Civil War the army's chief of ordnance. One of the foremost such presentations was a double set of double-cased revolvers with accessories, a pair of consecutively numbered Model 1860s and a second cased set of a Model 1861 Navy and a New Model Police, each of the four Colts inscribed on the backstrap "Gen'l. Irvin Mc.Dowell/With Compliments of Col. Colt." McDowell commanded the Army of the Potomac at the beginning of the war (Figure 108, 109).[46]

46. In the spring of 1851 the governor of Connecticut appointed Colt as a lieutenant colonel in the state militia. Thereafter Colt used the title "colonel," helpful when dealing with military brass.

108. One of the George Catlin scenes promoting Colt firearms, entitled "Catlin the Artist Shooting Buffalos with Colt's Revolving Pistol, 1855." **109.** Cased pair of Colt Model 1860 .44 revolvers presented by Colonel Colt to General James W. Ripley, chief of the army's ordnance department. (*Exhibited at Springfield Armory National Historic Site*)

Earlier in the Paterson era, Sam Colt had incurred the displeasure of Dudley Selden, his cousin and treasurer of the Patent Fire Arms Manufacturing Company, when he became aware of the inventor's efforts to influence the army's ordnance chief to approve a purchase order for guns. In January 1839 Selden wrote: "I will not become a party to a negotiation with a public officer to allow him compensation for aid in securing a contract with the government. The suggestion with respect to Col. Bomford is dishonorable in every way and if you write me I trust it will relate to other topics."[47]

Directions which the Colt factory provided for loading its percussion arms applied to most percussion revolvers regardless of make. Before loading, the shooter was advised to "snap off a round of percussion caps to blow the oil and dirt out of the nipples. Great care should be taken when Colt's Cartridges are not used, that all the Balls are perfect and fit the chambers snugly, otherwise the charges may jar out, and more than one chamber be discharged at once." The factory recognized a problem facing most percussion revolvers—that of multiple discharge. Such happened when flame passed from one chamber to another, often through the use of loose fitting caps or imperfectly shaped or undersize balls. One loaded by drawing the hammer back to half cock to allow the cylinder to rotate freely, pointed the gun upward and inserted a self-consuming cartridge containing powder and bullet into each chamber or loaded the chambers with loose powder then a snug fitting round ball or conical bullet in the mouth of the chamber, turned the cylinder to place a chamber beneath the rammer and with the lever forced the cartridge or bullet fully into the chamber. Once all chambers were loaded, placing caps on the nipples and bringing the hammer to full cock readied the weapon to fire.

One item Colt had begun making and selling during the Paterson era was what he advertised as "combustible envelope cartridges," more convenient than loading with loose powder and ball. These reached a high level of development in the 1850s and were made not only by the Colt Cartridge Works but by such other makers as Eley of London, England, and D. C. Sage & Co. of Middletown, Connecticut. To the base of the conical bullet a charge of powder was attached in an envelope of either animal membrane ("skin") or nitrate-impregnated paper which was consumed completely almost instantly upon firing. These were

47. Houze, *Samuel Colt: Arms, Art, and Invention*, 55.

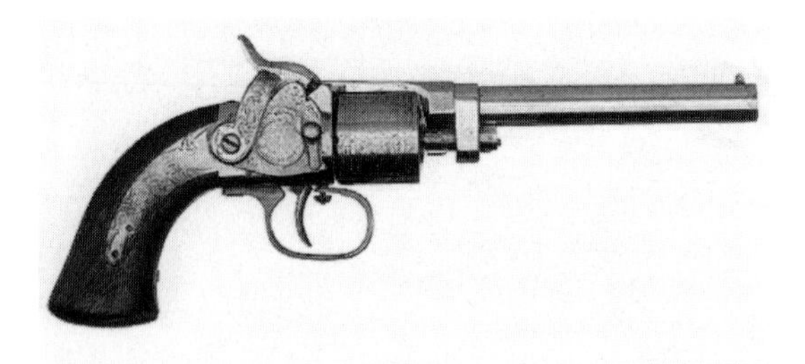

110. Massachusetts Arms .31 caliber revolver with Maynard tape primer. Following the successful Colt lawsuit against this firm, these arms were made with a hand-turned cylinder; the release button is located in front of the trigger.

packaged in a block of wood bored with the proper number of holes for the gun for which the cartridges were intended—five or six for most revolvers. These blocks were then wrapped in paper and sealed with varnish to keep out moisture. A string or wire circling the box beneath the paper was pulled to tear it open. The price for percussion cartridges for Colts ranged from $11.50 for .28 caliber to $18.50 per thousand for .44s. Percussion caps were packaged and sold separately from cartridges, often in small tin containers holding one hundred. Eley caps imported from England were quite popular; American brands included those by Hicks and Goldmark.[48]

An inconvenience sometimes encountered with most percussion revolvers occurred if a fragment of exploded percussion cap should become lodged behind the cylinder preventing or hindering its rotation. Such jams happen occasionally as shooters of percussion revolvers today will testify. In addition, carrying a loaded Colt or other revolver with the hammer on a capped nipple was a potentially hazardous undertaking, for dropping the weapon or some other unexpected blow to the hammer could cause an accidental discharge. One could leave one chamber uncapped or could carry a Colt with the hammer resting on one of the small safety pins located between each nipple, but the former was by far the safest method for these pins were rather easily worn down. World traveler Richard Burton later wrote of his 1860 journey to California and spoke of such a danger.

> As a precaution, especially when mounted upon a kicking horse, it is wise to place the cock [hammer] upon a capless nipple rather than trust to the intermediate pins. In dangerous places the revolver should be discharged and reloaded every morning, both for the purpose of keeping the hand in and, and to do the weapon justice. A revolver is an admirable tool when properly used; those, however, who are too idle or careless to attend to it, had better carry a pair of "Derringers."[49]

A Colt dealer price list of 1855 offered the Dragoon at $35 less 25 percent, and Model 1851 for $26 less 20 percent, and pocket models between $17.30 and $19.30 depending on barrel length, also less 20 percent to dealers. Engraving on the Model 1851 cost $6 more and $5 on pocket revolvers. A bullet mold, metal powder flask, and nipple wrench accompanied each revolver at no extra cost. In the late 1850s, a typical skilled factory worker was putting in six ten-hour days a week and earning about $2.00 a day.[50]

Another price list of about 1863 showed prices of $9.75 to $11 for a .31 pocket model, $14 for the .36 Navy or Belt model, and $1 more for the .44 Model 1860. If a customer wanted a more decorative piece, engraving added $4 to $5 and ivory grips $5 or $6. A detachable wooden shoulder stock or "carbine breech" was available for the belt or holster models for $7.50, or the even less common hollow stock with a canteen inside for $9. Sales of stocks were rare, however. The bullet molds were of iron or brass and had two cavities,

48. James E. Serven, *Colt Firearms from 1836* (La Habra, Calif., 1954), 149.

49. Louis A. Garavaglia and Charles G. Worman, *Firearms of the American West, 1803-1865* (Albuquerque, N.M., 1984), 312.

50. Ray Riling, *The Powder Flask Book* (New York, 1953), 48.

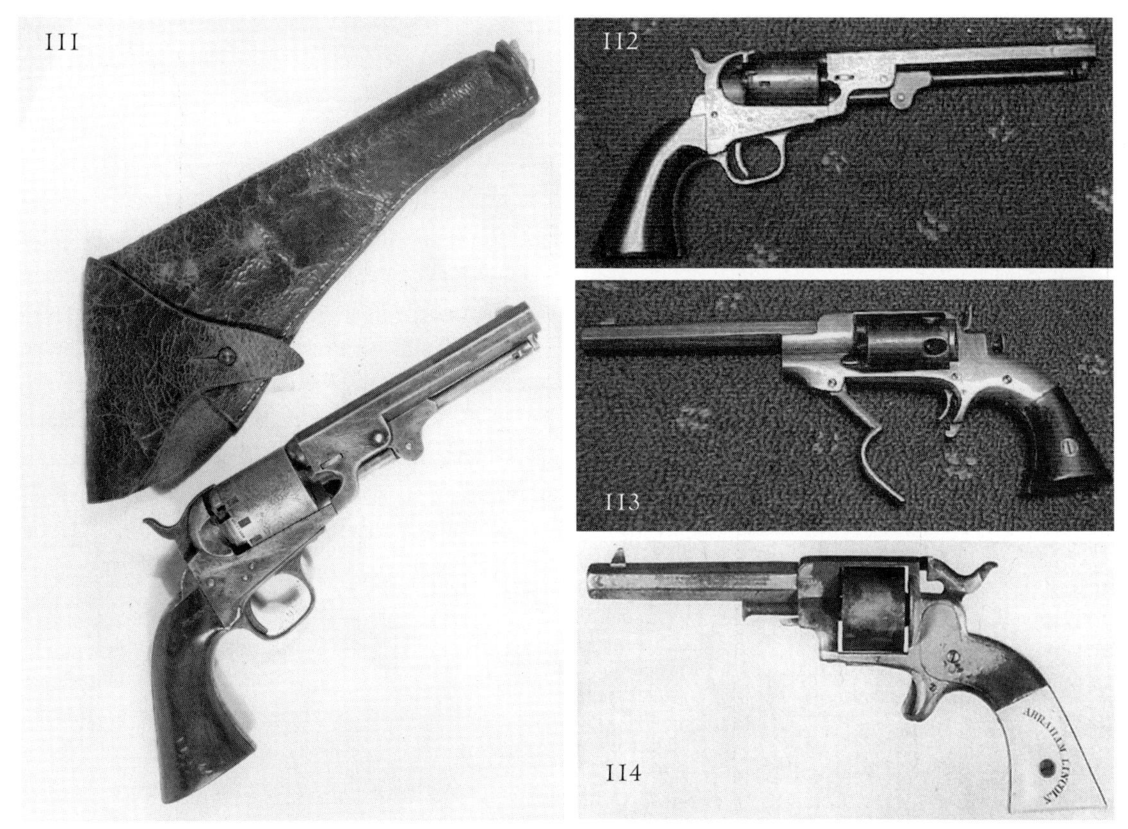

111. The .31 caliber Manhattan pocket revolver, a major competitor to Colt's 1849 pocket model which it resembles closely. (*Courtesy: Union Pacific Railroad Museum*) **112.** Manhattan .36 "navy" with six and one-half inch barrel. Closely resembling the Colt Model 1851 Navy, it is five- rather than six-shot and the frame is slightly smaller than that of the Colt. This specimen employs a spring steel plate between the frame and back of the cylinder intended to deflect any flame from a percussion cap which might set off more than one charge. (*Courtesy: Neil and Julia Gutterman*) **113.** Allen & Wheelock .31 percussion side-hammer revolver showing operation of the combination trigger guard and loading lever. (*Courtesy: Neil and Julia Gutterman*) **114.** Gold plated side hammer .32 rimfire Allen & Wheelock with "ABRAHAM LINCOLN" inscribed on the ivory grips. The revolver was passed down through the Lincoln family. (*Courtesy: Paul Henry*)

one to cast a round ball and the other a conical bullet. One might also purchase a wooden case of walnut, mahogany, rosewood, or oak with partitions inside to keep one or two revolvers in place along with appropriate accessories.

The usual finish on a percussion Colt handgun, and frequently on those by other makers, was a deep rich blue for barrel and cylinder. Frame, hammer, and loading lever often were case hardened giving them a pleasing mixture of shades of brown and purple. Standard grips were walnut, varnished on civilian guns and oil finished on military purchases. Brass fittings such as trigger guard often were silver plated on guns for the civilian trade and left plain on those intended to fill military orders.

Sam Colt in 1851 had successfully sued the Massachusetts Arms Company for patent infringement and forced that firm to redesign their revolvers to incorporate a hand-rotated cylinder rather than one which revolved with the cocking of the hammer. But he failed in his efforts (sometimes involving questionable ethics) to obtain a renewal

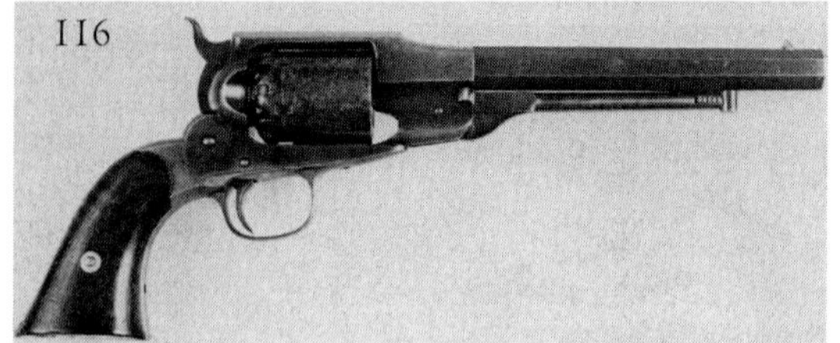

115. Double-action .31 Remington-Rider pocket revolver. Sliding the cylinder pin beneath the barrel forward allows removal of the cylinder to load it. **116.** A .36 caliber six-shot Remington-Beals.

of his 1837 patents when they expired in October 1856. Previously, except for pepperboxes, no contemporary revolvers had been produced in large numbers in this country. But now other American gun manufacturers were legally able to compete for a share of the revolver market, although the Colt name and reputation by then were well established. In fact, some of the better-selling competing handguns which came onto the market were obvious copies of Colt products (Figure 110).

The Manhattan Fire Arms Manufacturing Company of New York, for example, soon was offering both .31 and .36 caliber revolvers closely patterned after Colt's 1849 Pocket and 1851 Navy models. These arms achieved rather substantial sales successes, along with Manhattan pepperboxes and single-shot percussion pistols. During the Civil War years, J. M. Cooper & Company of Pittsburgh (and later Philadelphia) also produced .31 and .36 caliber close copies of Colt revolvers but in double-action rather than single-action form. Eli Whitney, Jr., who had manufactured the Colt Walker, became another Colt competitor with both .31 pocket and larger .36 caliber belt models, notable because they employed a solid frame in which there was a topstrap over the cylinder. Ethan Allen in partnership with brother-in-law Thomas Wheelock between 1857 and 1864, in addition to pepperboxes, produced various pocket and larger size revolvers based on Allen's patents, including one in which the trigger guard and loading lever were combined. Smith & Wesson's metallic cartridge revolvers too offered competition, but these were available only as a diminutive .22 or after June 1861 as a somewhat more powerful .32 (Figures 111, 112, 113, 114).

During the percussion era, another gun maker provided a major challenge, the Remington plant in Ilion, New York, beside the Erie Canal. It's sometimes claimed that Eliphalet Remington II made his first rifle, a flintlock, at his father's forge between 1816 and 1828. If he did, there is no evidence he continued such manufacture. Rather, the Remington forge after about 1828 was turning out gun barrels (and cowbells!) which other makers incorporated in their own rifles. Such barrels often bore the Remington name on the underside, hidden by the stock, rather than prominently displayed. Production of complete firearms appears to have begun with long guns in 1848, Jenks breech-loading carbines for the navy and Model 1841 "Mississippi" rifles for the government. Handgun manufacture got under way in 1857 with several series of compact .31 caliber five-shot pocket models. Most did not have an attached loading lever, which meant one had to remove the cylinder to reload (Figure 115).

In the spring of 1861, larger .36 six-shot Remington Beals revolvers were available followed by a .44 Beals the next year and many of these and improved versions saw Civil War service. All were rugged solid frame weapons with a topstrap over the cylinder. Unlike the Colt with its easily removed barrel, the barrel on these Remingtons screwed into the frame. To remove the cylinder, one lowered the loading lever mounted beneath the barrel, put the hammer on half cock, then slid the

117. English double-action "wedge frame" Webley. Some English revolvers of various make were imported into the United States in the 1850s and early 1860s, often to take advantage of their double-action mechanism. (*Courtesy: Donald Mitchell*)

cylinder pin forward to free it from the cylinder. The topstrap over the cylinder provided a stronger frame, but when one wanted to remove the cylinder to clean it, powder fouling could make the cylinder pin difficult to remove without hammering on the pin's "wings" and a fired percussion cap lodged between cylinder and frame could jam the gun. Scaled down five-shot versions in .31 and .36 caliber competed with comparable Colt models. Somewhat unusual among American revolvers for that period was a .36 double-action belt model Remington of which perhaps no more than 5,000 were produced (Figure 116).[51]

Double-action or "self-cocking" revolvers in which a long pull on the trigger cocked the hammer, rotated the cylinder, and then fired the weapon were faster firing than those of single action design in which the hammer had to be cocked with the thumb and then the trigger squeezed to fire the weapon. A single action did have the advantage that most people could fire one more accurately. "Self-cockers" were common among English makers, but such wasn't the case in the United States during the percussion era except for a few produced here plus those imported from abroad and many of the American pepperboxes. A fast firing double-action definitely had merit in some circumstances, however (Figure 117).

Between 1853 and 1857 Colt maintained a factory in London, England, and among the models produced there was the popular Model 1851 Navy. The Royal Navy purchased some in 1854 and thousands were procured in 1855 for the British army during the last months of the Crimean War. On one occasion during that conflict, a British officer reportedly was surrounded by a half dozen Russians stabbing at him with bayonets. He was armed with an English double-action five-shot Adams revolver with which he dispatched five of his attackers and used his sword to fight off the sixth. He later reportedly stated that a single-action six-shot Colt would not have saved him because of its slower rate of fire.

Dueling, Affairs of Honor

Dueling perhaps began when two cavemen argued over some trivial matter and settled their personal conflict with rocks or clubs. Swords later became common weapons of choice for more formal such affairs, and in the mid-1600s firearms were coming into more frequent use, putting the contestants on closer to an equal footing where physical strength was not a determining factor in the outcome. In 1777 in Ireland, gentlemen delegates to a legislative assembly drew up the Code Duello, rules of dueling which established it as a carefully choreographed affair. The code was only formulated for use in Ireland, but its twenty-six rules usually were followed by gentlemen in England and on the European continent, although there were some deviations. Pistols by the 1790s were the frequent weapons of choice throughout Britain and much of Europe even though the sword remained popular in France.

51. The design of some early large frame Remingtons allowed the cylinder pin to be slid forward without lowering the loading lever, but this was altered later to preclude the pin from inadvertently sliding forward during use and allowing the cylinder to jam.

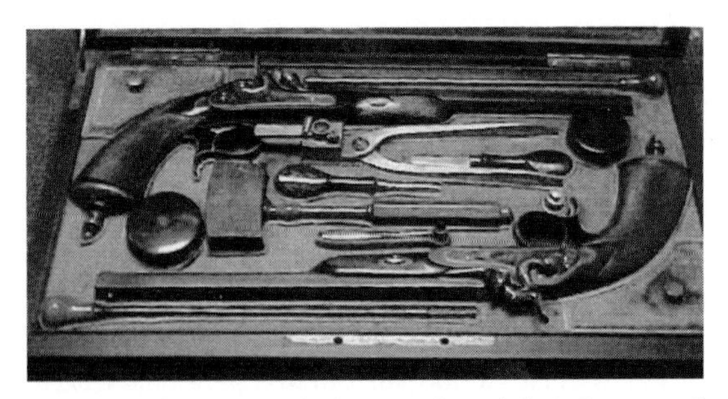

118. Pair of percussion dueling pistols made by J. Boussart of Liège, Belgium, and cased with accessories in the French contour style. The pistols are about .48 caliber with 11-inch barrels and measure 17 inches overall. (*Courtesy: Neil and Julia Gutterman*)

Foreign duels were a means of settling a grievance between gentlemen rather than commoners. But in America's more democratic environment, duels came to range from the tragically serious in which perhaps two politicians, newspaper editors, or even doctors who disagreed on medical procedures followed all the formal rules resulting in one or more deaths to those duels that could only be called comical. A story is told about the brash Israel Putnam who as a militia officer almost was burned alive by Indians during the French and Indian War after he was captured. During that war he was challenged to a duel by a British officer and Putnam is said to have selected as weapons two kegs of gunpowder, each with a long fuse. With each duelist seated on his keg, the British officer finally lost his nerve and ran as the fuse burned down. Putnam remained seated until his fuse sputtered out and then revealed the contents—onions rather than powder. It's a good story, whether true or not.

Several generations later, young Abe Lincoln was challenged over his criticism of the tax policies of an Illinois auditor and reportedly chose horse manure as weapons to illustrate the foolishness of a duel. In what perhaps is a different version of the same incident, Lincoln is reputed to have named cavalry sabers as weapons. With his long arms and greater reach he would have had the advantage if friends hadn't arranged a peaceful settlement, which undoubtedly is what the "rail splitter" wanted.

Some advocates of the duel felt it was a means of ensuring good manners but others disagreed. Both Benjamin Franklin and George Washington were among those in colonial America who condemned the practice, but numerous duelists ignored or evaded those laws which some states passed prohibiting it. For example, North Carolina in 1802 made it a crime to send a challenge or fight a duel within its borders. Duels could result from seemingly trivial matters, as when Henry Clay and Humphrey Marshall, then both in the Kentucky legislature, fought over Clay's proposal to require all Kentucky legislators to wear domestic homespun rather than British broadcloth, although perhaps this was the culmination of other disagreements between these two politicians. Each man was slightly wounded in the fray. Although dueling began to decline in popularity in the north in the early 1800s, it lingered on in the south until after the Civil War despite growing public opinion against it as a form of legalized murder. Under a Georgia law of 1873, conviction of sending or accepting a challenge carried a fine of $500 and imprisonment of six months or longer plus punishment for those acting as seconds.[52]

One of the most famous—or notorious—of Southern duelists was Alexander K. McClung, an attorney and West Point graduate also known as the Black Knight. In an 1834 duel he killed his opponent with a pistol shot at the unusual distance of 100 feet as the two men walked toward each other as previously arranged. In another encounter, he took the life of John Menifee of Kentucky with a Mississippi rifle at 60 yards, a contest which history claims resulted in later duels with six other members of the Menifee family, although this may be an exaggeration. Despite his reputation, McClung served in the Mexican War as second in

52. Brown, *Firearms in Colonial America,* 378-79. Barbara Holland, *Gentlemen's Blood: A History of Dueling from Swords at Dawn to Pistols at Dusk* (New York, 2003), 277.

command of the 1st Mississippi Infantry under Col. Jefferson Davis and in 1849-51 was this country's charge d'affaires in Bolivia. He committed suicide in 1855 (Figure 118).

Within the U.S. Army, dueling officially was prohibited by an 1806 Congressional act stipulating that any officer who should send or accept a challenge or fight a duel could be cashiered. A noncommissioned officer or soldier who did so could suffer corporal punishment. The navy imposed no such restriction until 1862. Among army officers, dueling most commonly was conducted among the lower ranking officers, but the naval service saw occasions of duels between ships' captains and even between naval officers and foreign diplomats, the former quick to defend the honor of the new republic. Several duels were averted after an American midshipman failed to stand during the playing of England's national anthem on the island of Malta. Perhaps the lengthy confinement in close quarters on board a ship was a reason tempers were prone to flare between brother naval officers. One of the most famous naval duels was that fought by captains Stephen Decatur and James Barron in 1820 in which Decatur lost his life.[53]

The site of the Decatur-Barron duel has come down through history as the Bladensburg dueling ground, located off Route 450 in Maryland not far from Washington, D.C. It was a secluded and convenient meeting place for a number of such encounters. One of the most unfortunate took place in June 1836 when a single shot from John Sheburne's pistol killed his friend Daniel Key, son of the author of "The Star Spangled Banner," Francis Scott Key. The cause of the disagreement reportedly was a festering argument over the speed of two steamboats. Not until two years later were duels outlawed in the District of Columbia. Other sites have become famous or infamous in history as a common meeting place for duelists. In New Orleans, City Park surrounds what was known as the Dueling Oaks, where on just one Sunday in 1839 ten duels reportedly were fought beneath the overhanging boughs. Belle Isle in the James River outside Richmond, Virginia, and Vidalia, Louisiana, across the Mississippi River from Memphis, are but two others.

By about 1800 certain features characterized those pairs of matched single-shot pistols often selected as weapons in a duel. One was that such a pistol "pointed" naturally as it was raised at arm's length, for the first man to align his pistol accurately and fire after the command was given had the advantage. Other common features were a barrel which was browned or blued to eliminate glare and checkering of the wooden stock to improve one's grip. Often the pistol was built so the frizzen operated against a lighter spring to speed ignition. Among Britain's better known makers of such pistols were Henry Nock, Durs Egg, Joseph Manton, and John Twigg. It was not unusual for dueling pistols by one of these gunsmiths to be imported into the United States, and when Alexander Hamilton and Aaron Burr met at Weehawken Heights in 1804, they fought with pistols made by England's Robert Wogdon.

American-made dueling pistols usually followed European styling, particularly English. Today it's sometimes difficult to distinguish between what could have been a single-shot dueling pistol or one sold for target shooting or other sporting purpose. Dueling pistols were sold as exact duplicates in pairs, nearly always in a wooden case complete with appropriate accessories. They generally were about .50 to .60 caliber with a barrel length of eight or more inches, often smoothbore as was most common among English duelers, even though rifling was found among some French- and German-made specimens. Since duels with pistols often were fought at a distance of a dozen paces, a well-bored smoothbore had sufficient accuracy. Single or double "set triggers" are most often found on percus-

53. *Military Laws and Rules and Regulations for the Armies of the United States,* Adjutant and Inspector General's Office, Washington, D.C., May 1, 1806.

119. French dueling pistols by Le Page of Paris used in the formal duel between former California supreme court justice David S. Terry and U.S. senator from California David C. Broderick on September 13, 1859. Each has a plate on the left side inscribed "Hon. David S. Terry SAN FRANCISCO." The pistols had been used in another California duel two years earlier. (*Courtesy: Mrs. W. H. Wood and Wells Fargo Bank History Room*)

sion rather than flintlock duelers, and could be adjusted to fire the pistol at the slightest touch.

Not many dueling pistols were made in this country, but William Haslett of Baltimore, who worked from about 1803 to approximately 1824, in 1806 advertised dueling pistols both of his own make and London manufacture. A characteristic embellishment of Haslett's duelers was a gold ribbon inlaid in the top of the barrel stamped "Haslett Baltimore." Simeon North of Berlin and later Middletown, Connecticut, producer of numerous long and short guns on military contracts, also fabricated a number of high-quality dueling-style pistols. One pair was presented to naval hero Stephen Decatur. The prolific Henry Deringer, Jr., and Richard Constable, both of Philadelphia, and John Happoldt of Charleston, South Carolina, also are known makers of pistols suitable for use as duelers.

A Georgia newspaper in 1817 described an unusual duel near the South Carolina line—between two women.

> The object of the rival affections of these fair champions was present on the field as the mutual arbiter in the dreadful combat, and he had the grief of beholding one of the suitors for his favor fall dangerously wounded before his eyes. The whole business was managed with all the decorum and inflexibility usually practiced on such occasions, and the conqueror was immediately married to the innocent second, conformably to the previous conditions of the duel.[54]

Not all duels ended in death or injury and it was the duty of those parties acting as seconds to the combatants to try and settle the disagreement without bloodshed. Sometimes they were successful. On other occasions, each dueler merely discharged his pistol harmlessly in the air, an act sufficient to satisfy honor. One such affair between Andrew Jackson and another attorney, Waightsmill Avery, ended in this manner. But in 1806 when Jackson met Charles Dickinson on the "field of honor," the latter fired first, striking the future president in the breast. True to the standards of conduct, Jackson gave no indication he had been wounded as he walked from the field while his opponent lay dying. In his lifetime, Jackson is sometimes said to have fought more than a dozen duels, some ending in fatalities. Jackson proclaimed that he would not interfere in differences between military officers whose profession was fighting.

54. Holland, *Gentlemen's Blood*, 81.

When California joined the Union as a state in 1850, its constitution denied duelists or their seconds suffrage or the right to hold political office, but this restriction usually was ignored. During California's gold rush era, duels were fairly frequent and often were less formal affairs than in the east and south. Newspapers sometimes published notices of such events and duelists occasionally sent out formal engraved invitations. Rifles, pocket derringers across a poker table, Colt revolvers, and even shotguns were sometimes employed. On one occasion a politician and a judge shot it out with Colt Navy revolvers at a distance of twelve paces. After each had fired five times, one finally drew blood with a minor wound whereupon the participants bowed to each other and departed, their individual honor upheld. Earlier in 1847 in California Captain John C. Frémont almost found himself at the wrong end of a shotgun when he challenged Colonel Richard Mason. The no-nonsense colonel as the challenged party chose shotguns loaded with buckshot, but a higher ranking officer forbade the encounter. Such an engagement could have been devastating to one or both participants (Figure 119).

A. J. "Natchez" Taylor of San Francisco, a dealer in Henry Deringer pocket pistols and other arms, operated a shooting gallery where duelists could practice. In an 1856 ad he announced that he carried in stock genuine Henry Deringer pistols as well as a large assortment of English dueling pistols and Colt revolvers. One bloodless duel in California's gold rush country occurred when a fiddler and a vocalist fought, but their seconds conspired to load one pistol with only powder and the other with a capsule of currant jelly rather than a lead ball. Thus the "blood" shed proved to be harmless. A most unusual duel was conducted in 1861 between two miners in Nevada City, California, who used 25-foot water hoses as weapons, "from whose unerring aim there was no escape." The contest thoroughly soaked the participants and probably spectators as well and lasted until one of the hoses burst![55]

Christian Sharps and His Breech-Loading Rifle (1848–1880)

Loading a shoulder arm at the breech rather than from the muzzle offered several advantages, primarily an increase in fire power. It facilitated reloading and, particularly significant for a soldier or hunter, made it possible to reload while in a prone or kneeling position behind cover. But a persistent problem in developing a truly effective breech-loading system was sealing the joint at the breech to prevent the escape of propellant gas. Breechloaders in very limited numbers were available as early as the 1530s. In England, Colonel Patrick Ferguson improved upon an earlier screw design and as mentioned before a few of his rifles were used in the American Revolution. English makers Henry Nock and Durs Egg made a few breechloaders in commercial form, and in America John H. Hall in 1811 obtained a patent which he first applied to a few civilian guns and involved a tip-up breechblock.

In 1848, Christian Sharps patented what would be the best known of American percussion breechloaders, employing a breechblock which slid vertically (slanted rather than vertically in early models) in a channel cut into the receiver. This block was linked to the trigger guard which acted as a lever to raise or lower the block. Lowering the guard dropped the block to allow the insertion of a paper or linen cartridge into the chamber and in raising it the sharp edge of the block cut off the end of the paper cartridge to expose the powder to the flame from the separate percussion primer. The Sharps rifle's popularity in both sporting and military form and its dependability led in 1876 to the company's adoption of the trademark "Old Reliable," which then was stamped on the barrel of many of its sporting rifles. Although the inventor's name continued to be associated with these breechloaders, he divested himself from the Sharps Rifle Manufacturing Company of Hartford in 1853. Instead he concentrated on the design and

55. Worman, *Gunsmoke and Saddle Leather,* 93, 99.

120. Model 1849 Sharps sporting rifle with the primer cover open to show the primer wheel holding eighteen regular percussion caps. It's thought that fewer than one hundred such rifles were produced and this unusual priming system was dropped in subsequent models.

manufacture of single-shot breech-loading pistols, a very few revolvers, and later a very popular series of four-barrel pepperbox-style cartridge derringers. In his later years, he raised trout in Connecticut.[56]

Sharps' first guns reached the market in the summer of 1849 but these early pieces were few in number and involved much hand fitting in their production. After the fall of 1852, Sharps rifles and carbines (and a very modest number of shotguns) became available in greater numbers, made and finished almost entirely by machine. Eventually the Sharps rifle became a standard against which other arms often were measured for reliability, power, range, and accuracy. H. E. Dimick & Co., a major St. Louis arms dealer, in the late 1850s advertised the Sharps as a gun which could be "loaded on horseback at a smart gallop, or lying down in the grass. . . . We will back our judgment with the 'filthy lucre'" (Figure 120).

The popularity of the Sharps grew steadily in the late 1850s, spurred on by complimentary remarks by users and in part by publicity sprouting from its role in the slavery dispute taking place in "Bleeding Kansas." That term is somewhat of a misnomer, it seems, for despite the unrest and the atrocities committed by both sides, the actual number of violent deaths in the territory during that period over the issue of slavery may have totaled not many more than half a hundred. The acquisition of Sharps was welcome by anti-slavery forces and was bitterly opposed by pro-slavers, yet many of the firearms carried by those on both sides of the issue were muzzle-loading rifles and shotguns, a sizeable number of surplus military muskets and rifles, and handguns as well.[57]

Abolitionists in New England felt that the immigration of anti-slavery voters was vital to ensuring that the Territory of Kansas would enter the Union as a state free of slavery. One of the first anti-slavery groups of immigrants settled in 1856 at Wabaunsee, where the colony's stone church building dedicated in 1862 still stands today as the Beecher Bible and Rifle Church. At that group's organizational meeting in New Haven, Connecticut, a Yale professor pledged $25 to buy a Sharps rifle for the company. The Brooklyn minister Henry Ward Beecher pledged that his congregation would buy twenty-five more Sharps if the audience would pledge the same number. The challenge was met, and twenty-five Bibles also were donated by one of Beecher's parishioners to accompany the Sharps. Reportedly the guns were packed in crates marked as Bibles to avoid suspicion on their journey to Kansas, but the association between Sharps and Bibles may well have evolved from the mixed shipment of guns and books rather than any subterfuge. Even though that shipment arrived in Wabaunsee, those Sharps were employed generally for target practice and when Topeka citizens requested them for their defense, the Wabaunsee residents refused to give them up.[58]

56. The term "sharpshooter" was not coined as a tribute to Christian Sharps or his guns but was in use at least several decades before the first Sharps rifle appeared on the market.

57. Dale E. Watts, "How Bloody Was Bleeding Kansas?" Kansas History 18, no. 2, Summer 1995, 129.

58 Dale E. Watts, "Plows and Bibles, Rifles and Revolvers--Guns in Kansas Territory," Kansas History 21, no. 1, Spring 1998, 34; e-mail from Dale E. Watts to the author, April 10, 2007.

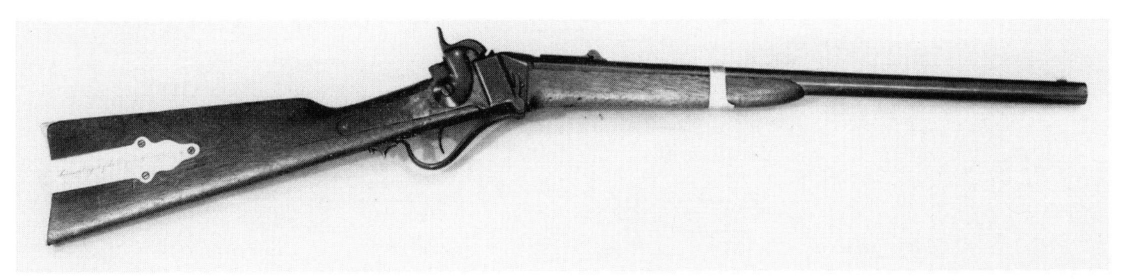

121. A Sharps .52 Model 1853 "slant breech" carbine, one of those sent to anti-slavery forces in Kansas. (*Courtesy: Kansas State Historical Society*)

The New England Emigrant Aid Company was a major anti-slavery organization whose members as private individuals funneled Sharps and ammunition to the free-staters. Pro-slavery Missourians, one free-stater observed, "frequently break open heavy trunks or boxes to search for Sharps rifles of which they stand in great fear." On one occasion in March 1856, a man to whom a shipment of one hundred Sharps was entrusted suspected such an attempt by pro-slavers to steal the guns so he removed the breechblocks and sent those essential components by another route. Missourians did seize the guns onboard the steamer *Arabia* but they were useless without the blocks (Figure 121).[59]

The fear of violence was prevalent in mid-1850s Kansas regardless of what side of the slavery issue you favored or if you were neutral. The radical abolitionist John Brown, Sr., in May 1856 oversaw the murder of five pro-slavers. In the same year, Axalla Hoole of Lecompton wrote: "I have my rifle, revolver, and old home-stocked pistol where I can lay my hand on them in an instant, besides a hatchet and axe. I take this precaution to guard against the midnight attacks of the Abolitionists."[60]

In all, it's thought that close to one thousand Sharps .52 caliber carbines (often mistakenly referred to as rifles) were shipped to Kansas by anti-slavery advocates, proportionately not a large number when compared with a territorial population which numbered in the tens of thousands. John Brown, Sr., and his interracial "army" of twenty-one men did include Sharps Model 1853 carbines as well as pikes among their arms when they attempted to seize the federal armory at Harpers Ferry in October 1859 and promote a slave uprising. Two Sharps carbines documented as having been taken from Brown himself and Aaron D. Stevens are exhibited today at the Harpers Ferry National Historical Park. Among those present either at the capture or hanging of "Old Brown" were Col. Robert. E. Lee, commanding the U.S. Marines; Maj. Thomas J. Jackson with Virginia Military Academy cadets; John Wilkes Booth; and elderly Edmund Ruffin who in April 1861 would fire the first cannon to begin the bombardment of Fort Sumter in Charleston harbor.

E. E. Cross journeyed over the western plains to Arizona in 1858 but once there his Sharps rifle was stolen. When he wrote the factory to order a replacement, he called the Sharps "the greatest weapon of the age. Three hundred would sell in this territory at from $75 to $85 each." A similar letter from the diminutive western explorer Captain John C. Frémont noted that on his journey to California in 1853 he had taken along both a

59. The steamboat *Arabia* sank later in 1856 after striking a snag in the Missouri River. The wreckage was discovered 132 years later buried beneath what had become a corn field as the river changed course over the years. Portions of the side wheel steamer and some of its cargo recovered from beneath the layers of silt can be seen today at the Steamboat Arabia Museum in Kansas City, Missouri.

60. Exhibit text, Kansas State Historical Society.

122. Grip, the dog, and his companions. The long guns from left are a muzzle-loading rifle, a Model 1873 Winchester, and a Model 1874 Sharps. (*Courtesy: Herb Peck, Jr.*)

Sharps carbine and rifle. "I found it [them] the most convenient gun[s] I ever used."

The Pony Express provided mail service between St. Joseph, Missouri, and Sacramento, California, for only eighteen months beginning in April 1860 before completion of the trans-continental telegraph line put it out of business. There are no precise records as to what guns the slight youthful riders carried, although evidence exists that Model 1849 and Model 1851 Colts were common choices. Even though weight was kept to a minimum, rider J. G. Kelley recalled carrying a Sharps, presumably a carbine, on several rides. On one occasion he was almost shot by an emigrant who mistook him for an Indian as he galloped past a wagon train. In July 1860, an army colonel at Camp Floyd in Utah Territory loaned the Pony Express 106 Colt .44 Dragoon revolvers and 60 Mississippi rifles to defend their stations from Indian attacks, some of which the Mormons were suspected of inciting.[61]

During the Civil War, the federal government purchased more than 85,000 Sharps from the factory, more than 75,000 of these as carbines for cavalry use. Most considered these the best of the single-shot percussion carbines used during that conflict. The availability of large numbers of these and other breechloaders auctioned off as postwar surplus hindered the sale of new Sharps, but as late as the early 1880s percussion Sharps were being converted by the manufacturer or dealers into sporting arms firing metallic cartridges. In the fall of 1867, the government contracted with the manufacturer to convert some of the army's percussion Sharps carbines to fire a .50-70 metallic cartridge. Almost 30,000 Sharps were thus altered over the next several years for continued military service. Sharps .50-70 carbines also equipped the Texas Rangers in the early and mid-1870s with $17.50 deducted from a ranger's pay of $30 to $40 a month for a private; the state provided ammunition. Later some rangers purchased Model 1873 .44 Winchester repeaters as replacements for their Sharps, paying $50 for a rifle and $10 less for a carbine.

By early 1871, production was under way for what the Sharps firm would later designate as its Model 1874. It would be marketed in many variations in .40, .44, .45, and .50 caliber, in different barrel weights and lengths, and with other features, depending on the purchaser's intended use and pocket book. It was the choice of many of the professional hunters who in just over a dozen years exterminated virtually all of what once was thought of as endless herds of bison in the west. In shooting contests at distances of up to a thousand yards, Sharps shooters made impressive scores on the target ranges at Creedmoor in New York, Walnut Hill in Massachusetts, and elsewhere. When the British rifle team visited the Sharps factory in 1877, team member Sergeant Gilder obtained a new Sharps and before departing for home made sixteen con-

61. Worman, *Gunsmoke and Saddle Leather,* 32.

123. Remington single-shot Model No. 1 rolling-block sporting rifle in .45-70 caliber and fitted with a telescopic sight. (Courtesy: Roy Marcot)

secutive bulls eyes at 1,000 yards with his acquisition (Figure 122).

Reliable repeating rifles were available in the mid-1860s, specifically the Spencer and Henry followed by the first Winchester in 1866. The number of shots they could discharge in a few seconds was impressive, but all of these rifles were of moderate range and their ammunition didn't equal those new centerfire cartridges which were coming on the market, the .50-70 to name just one. Until the late 1870s the hunter, target shooter, or other buyer who needed maximum stopping power, dependability, and long-range accuracy in a rifle was forced to rely on a single shot.

Among available single shots, a major Sharps competitor in the post-Civil War years was Remington, producing various rifles using a mechanism patented by Leonard Geiger in 1863 and improved by the prolific inventor Joseph Rider of the Remington firm, the "rolling block" action. After the hammer was cocked, the breechblock was rolled backward and downward to allow loading and then was closed. Closure activated a locking lever at the base of the hammer and as the hammer fell forward in firing, a bearing surface on its underside blocked the breechblock from opening at the moment of discharge giving additional support. It was a sturdy action and much of Remington's production capability from the late 1860s through the next two decades was devoted to producing what's been estimated as more than 1.5 million military rolling block rifles and carbines for more than fifteen different foreign governments such as Denmark, Norway, Argentina, and China as well as for some states' militia (now national guard) organizations in this country including New York's. Remington's survival during this country's great depression of the 1870s can be attributed in large measure to the success of Rider's design (Figure 123).[62]

Like the Sharps series of sporting rifles, rolling block Remingtons marketed to hunters and others in this country were available in a variety of calibers up to .50, in different weights and frame sizes. George A. Custer was an enthusiastic owner of a heavy No. 1 size Remington sporter and is seen with one on his knee in an often published photo with an elk he killed during an 1873 expedition into the Yellowstone country. Earlier in the summer of 1866, Nelson Story and twenty-odd drovers pushed a cattle herd from Texas to Montana to feed beef-hungry gold miners despite warnings Story's men wouldn't arrive with their scalps in place. Although attacked by Sioux, Story's cowboys were

62. Some factory records from such firearms manufacturers as Smith & Wesson, Winchester, Sharps, and Colt still exist and in some instances it's possible to determine such details as date of shipment, price, original purchaser, and features of a specific gun by serial number. However, such records unfortunately no longer exist for Remingtons, or at least have never been located.

equipped with an early version of the Remington rolling block which they used with sufficient skill to complete the historic drive. Later during the concentrated slaughter of the bison herds, one professional hunter, Frank Mayer, estimated that 80 percent of the bison killed fell to Sharps and Remington rifles.

Competition between advocates of Sharps target rifles and those by Remington using the rolling block action was stiff and undoubtedly sometimes vocal, but guns by these two often dominated the field among competitive shooters. There were claims of accidental discharges with Remington rifles as the action was closed, however. Other competing makes favored by some competitive shooters during the last quarter of the century were models by Marlin-Ballard, Maynard, and J. Stevens. Each of these firms also produced sporting models intended for hunters.

The Sharps company catalog in 1878 offered the Model 1874 as a nine-pound Hunters Rifle at $25 or for $13 more their Sporting Rifle of between nine and twelve pounds (or up to sixteen pounds at $1 per pound more). For $65 one could purchase a Mid-Range No. 1 with target sights. That year also marked the introduction of a new model designed by Hugo Borchardt, who in later years became famous for his developmental work with auto-loading pistols including the Luger. The Borchardt Sharps was an advanced design with a hammerless action which found favor with some sportsmen and competitive target shooters although others clung to the more familiar Model 1874 with its outside hammer. But by 1880, sales of new guns and custom work converting percussion guns to fire a metallic cartridge or rebarreling later guns couldn't keep the company afloat. In October, production of new Sharps ceased after a thirty-one-year run.

Initial Handguns by Smith & Wesson (1854–1861)

At the same time Colt's revolver patents expired there appeared a diminutive seven-shot pocket revolver that fired a metallic .22 rimfire cartridge in which the powder, projectile, and primer were contained in a waterproof copper case. It was a cartridge developed by Daniel B. Wesson and Horace Smith, with substantially less power than today's modern .22 Short and although it certainly was no "man stopper," the revolver nonetheless marked a milestone in firearms history in the United States.

In 1853 Wesson and Smith had applied for a patent on a lever action repeating pistol with a toggle joint action and a tubular magazine beneath the barrel. Once in production, their pistols came in two sizes, an eight-shot .41 caliber and a smaller frame .31 six shot, each using the hollow base "rocket ball" bullet designed by Walter Hunt which included a primer in the base as well as a charge of black powder. But more on Hunt's contributions later. The .41 ammunition was loaded with only 6.5 grains of powder with a 100-grain bullet producing a mere 260 feet per second muzzle velocity. The ammunition was unreliable, gun sales were sluggish, and production of both size pistols didn't reach 2,000 total. Walter Hunt had called his first rifle the "Volition Repeater" and because of their rapid fire, these S&W pistols were nicknamed "Volcanic," a term apparently first used in *Scientific American* magazine in 1854 when it reviewed the guns. Whether the magazine misapplied the earlier nickname or the gun was intentionally likened to a volcanic eruption isn't known, but the name stuck.[63]

In 1855 the former partnership was reorganized as the Volcanic Repeating Arms Company, including as a stockholder a wealthy manufacturer

63. R. Bruce McDowell, *Evolution of the Winchester* (Tacoma, Wash., 1985), 58, 63, 76. Roy G. Jinks, *History of Smith & Wesson* (North Hollywood, Calif., 1977), 25.

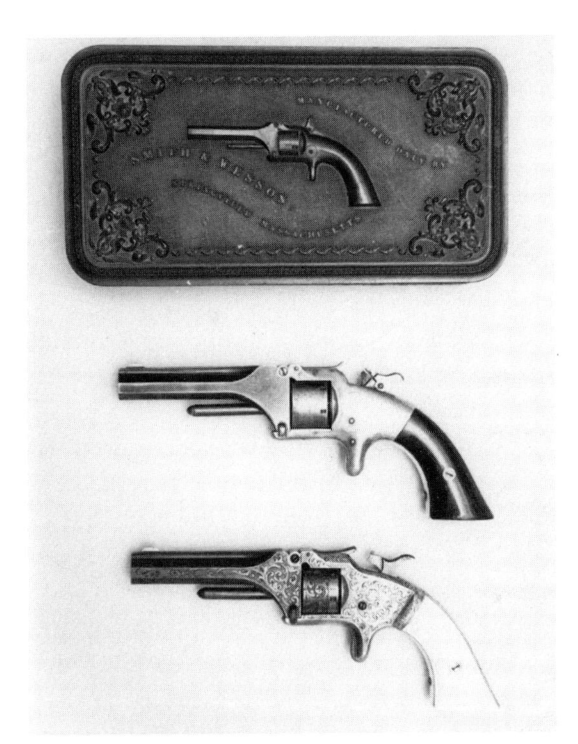

124. Smith & Wesson .22 First Model revolvers and a gutta percha case of the type which was available for these. Above is the First Issue with an engraved example of the improved Second Issue introduced in 1860 below.

of men's dress shirts, Oliver Winchester. Wesson remained on board as plant superintendent but Smith dropped out. The new firm would be short lived but for the moment continued with the production of an estimated 3,000 Volcanic .41 caliber pistols and pistol-carbines with a detachable shoulder stock. Depending upon barrel length, magazine capacity varied from eight to twenty shots. Anemic though its ammunition was, the Volcanic's basic design was sound and the weapon would be a significant predecessor to the famed lever action Winchester rifle (see below).

D. B. Wesson continued to experiment with his and Smith's earlier design of a thin rimfire copper cased metallic cartridge containing powder, ball and primer similar to the breech-loading system and low powered cartridge patented in France by Louis Flobert in 1849 for use in indoor target or "saloon" pistols. Wesson also was working on the design of a revolver to fire this rimfire cartridge. The key element in the story was Wesson's discovery that Rollin White, a former Colt employee, in 1855 had secured a patent on a cylinder with chambers bored straight through from end to end permitting rear loading. Ironically this was an unheralded element of White's design of what would have been an impractical revolver and which was never manufactured. Except for foreign-made pinfire revolvers, the chambers in a revolver cylinder of the early 1850s were sealed at the rear end by percussion nipples.[64]

Reunited with Horace Smith, the two partners in November 1856 signed an agreement with White giving them exclusive rights to manufacture revolvers under the patent. White would receive a royalty of twenty-five cents per revolver while the patent was in effect and, perhaps unwisely as time would reveal, he agreed to defend his patent against any infringers. The partners' control of this patent until 1869 prevented the legal production of any revolver in this country which loaded from the rear without violating that patent. Full-scale production of the first S&W .22 brass frame revolvers was under way in Springfield, Massachusetts, in early 1858. These revolvers were simple to operate. The barrel was swung upward to allow removal of the cylinder, empty cartridge cases were punched from the chambers upon a short rod mounted beneath the barrel, and the cylinder was replaced after reloading. The wholesale price was $12.75 in 1858 (Figure 124).[65]

Samuel Clemens (Mark Twain) when crossing the country in 1861 was armed with a .22 S&W. In his typical manner he wrote in *Roughing It*: "I was armed to the teeth with a pitiful little Smith & Wesson's seven-shooter, which carried a ball like a homeopathic pill, and it took the whole seven to make a dose for an adult."[66]

64. Jinks, *History of Smith & Wesson*, 34.

65. Ibid., 36-37, 41.

66. Twain, *Roughing It*, I:5.

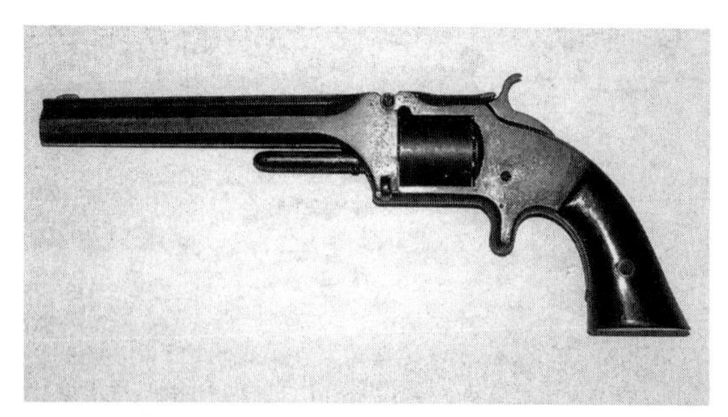

125. Smith & Wesson No. 2 .32 revolver carried by James H. McNeill of North Carolina during the Civil War. He originally entered Confederate service as a chaplain, but later became captain in a cavalry company and eventually commanded a cavalry regiment. He was killed at the battle of Five Forks near Petersburg, Virginia, little more than a week before Lee surrendered his army at Appomattox Courthouse.

Demand for these innovative revolvers was strong and for the man or woman who didn't need the shocking power of larger caliber percussion handguns, these No. 1 size S&Ws offered major advantages. Not only were they compact and easily concealed, they were easy to reload and the copper cartridge was virtually waterproof and not easily damaged. In June 1861, S&W began production of its larger and somewhat more powerful No. 2, a six-shot rimfire .32 with a wrought iron rather than brass frame. Dealer price in small quantities was $15.50 in the spring of 1865.

The No. 2 lacked the stopping power provided by percussion .36 and .44 revolvers and was not adopted by the Union army or navy. But substantial numbers of both the Nos. 1 and 2 were carried to war as personal property by Civil War officers and enlisted men, including a few Confederates. Major General John C. Frémont was presented with an engraved No. 2 with pearl grips in 1864 through a New York City fair held to aid the U.S. Sanitary Commission, a volunteer soldier relief agency. Another of numerous No. 2s with Civil War association is one inscribed on the frame "Capt. A. H. Bogardus/145th Regt. Ill. Vol." In later years Bogardus became a world renowned exhibition shooter. In August and September 1862, distributor B. Kittredge of Cincinnati sold the state of Kentucky 731 No. 2s, all of which are thought to have been issued to the 7th Kentucky Cavalry (Union). Wartime demand for these handguns and the backlog of orders was such that the manufacturer didn't catch up and begin advertising them until mid-1866 (Figure 125).[67]

To fulfill a need for a slightly smaller pocket size .32, S&W in the spring of 1865 added what the firm called the No. 1 1/2, a five-shot with a 3 1/2-inch barrel. There would be improvements and changes in the shape of barrel, grip, and cylinder in the three models along the way, but the Nos. 1, 1 1/2, and 2 S&Ws with the tip-up barrel design remained in production in .22 caliber until 1882 and .32 caliber until 1875.

There were numerous attempts by competing gun makers to cash in on the demand for these early S&W revolvers. Some of the revolvers resulting from such attempts were evasions and others outright infringements on the White patent. The Manhattan Fire Arms Company was one firm which ignored the risk of a lawsuit and by early 1860 was producing an almost identical copy of the No. 1 S&W. A court battle eventually did halt production but after thousands had been sold. Edwin A. Prescott by the beginning of 1862 was marketing both a .32 and a larger .38 caliber rimfire revolver. Because of the greater potency of the latter, it found favor with some Civil War soldiers for personal use—at least until a court injunction halted production in late 1863.

Other manufacturers through sometimes rather ingenious design successfully evaded patent infringement and produced metallic cartridge revolvers firing ammunition which loaded from the front or side of the cylinder. Such chambers were not bored completely through but had an opening

67. Charles W. Pate, *Smith & Wesson American Model in U.S. and Foreign Service* (Woonsocket, R.I., 2006), 27-29.

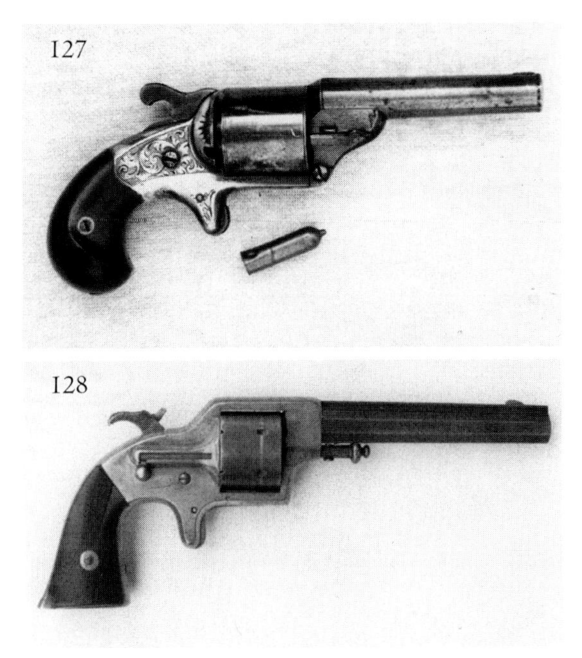

126. An ad for the .32 side-loading Slocum. **127.** Moore front-loading .32 revolver with one of its teat-fire cartridges. **128.** Plant .42 front-loading revolver. The rod mounted behind the cylinder is used to eject empty cartridge cases.

at the rear just large enough for the nose of the hammer to strike the base of the cartridge. Two popular models were the side-loading .32 Slocum made by the Brooklyn Arms Company and the seven-shot .32 Moore. The latter accepted a "teat fire" cartridge in which the priming compound was contained in a teat protruding from the base of the cartridge. Another of the more successful front-loading evasions was the Plant, made in .28 and .42 caliber sizes. A selling point in favor of the .42 Plant was that it could be purchased with two cylinders, one percussion and the other accepting metallic cartridges. All of these successful evasions of the White patent were available for civilian or unofficial military use by 1864. Occasionally an example of one of these revolvers can be found today in its original pasteboard box, often with a label on the lid illustrating the gun within(Figures 126, 127, 128).[68]

Tyler Henry's Rifle and the First Winchester (1860–1866)

An early and key player in the evolution of the Henry and Winchester rifles as well as Smith & Wesson arms was Walter Hunt, an individual with what certainly can be described as an inventive mind. Among those inventions attributed to him was the safety pin (to which he sold the rights for a mere $400), the fountain pen, an ice boat, a lock stitch sewing needle which led to Elias Howe's sewing machine, an improved nail making machine, and other diverse inventions. An initial step in the evolution of the Winchester was his invention in 1847 of his lever action repeating rifle and his "rocket ball," a bullet in which the powder was contained in the hollow base. The powder was held in

68. A cased, engraved, and gold-mounted Slocum was presented to General Phil Sheridan in 1865 after he was voted the most popular general at a Sanitary Commission Fair in Chicago.

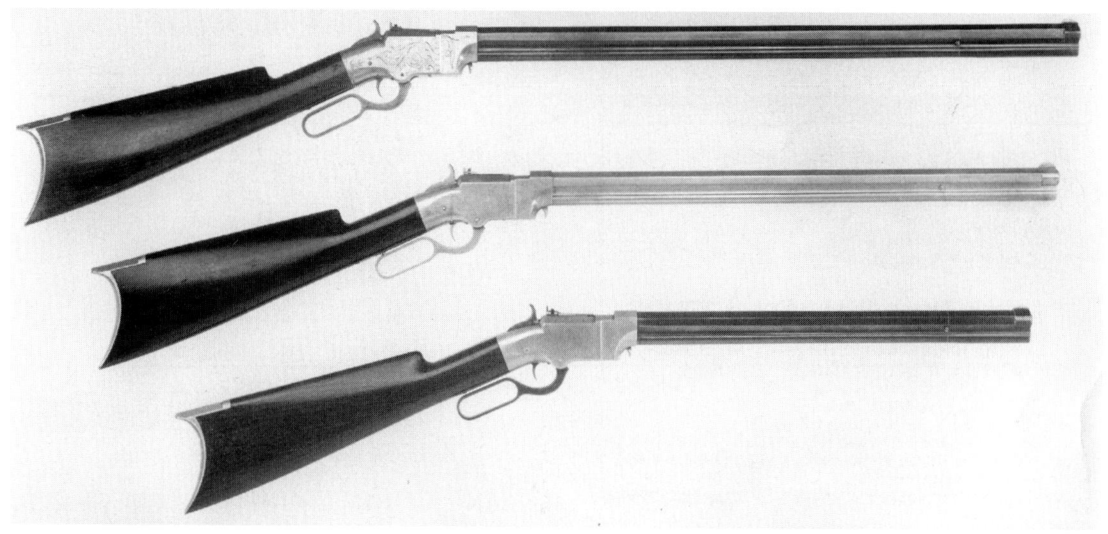

129. A trio of Volcanic shoulder arms.

place by a cork washer with a hole in the center to allow entry of flame from a separate priming pill. A significant feature of the Hunt repeating rifle was the tubular magazine located beneath the barrel. In the early 1850s, Lewis Jennings and then Horace Smith improved upon Hunt's rifle design, but with only modest production of hundreds rather than thousands of repeating rifles in several forms, all still using the "rocket ball."

In 1857, the Volcanic Repeating Arms Company was reorganized as the New Haven Arms Company. By year's end, .41 caliber carbines and .31 caliber small frame pistols had been added to the larger frame lever action repeating pistols already in production. Despite Winchester's abilities as a promoter and awards which the Volcanic arms won for design, by 1861 only about 3,000 guns had been manufactured and the company faced bankruptcy. Meanwhile, plant superintendent Benjamin Tyler Henry had perfected a rimfire metallic cartridge similar to but larger than the .22 of Daniel Wesson's design. On October 1860 he received a patent on a new firing pin and bolt mechanism designed to load, fire, and extract his new .44 rimfire metallic cartridge. Volcanic production ceased and the firm devoted its efforts to the new rifle, named the Henry (Figure 129).

A couple of problems plagued the use of the early copper rimfire cartridge, whether in a Smith & Wesson or other revolver or the new Henry. It couldn't be reloaded, but more important the rim had to be thin enough to allow the hammer nose or firing pin to crush it and detonate the fulminate located within around the rim, limiting the size of the powder charge and the power of the rimfire cartridge. Finally until ammunition manufacturing procedures improved, the first rimfire cartridges could be unreliable if the fulminate was spread unevenly around the rim. To counter this, the Henry firing pin had two prongs so the rim was struck in two places opposite each other. In the late 1860s two colonels, America's Hiram Berdan and Britain's Edward Boxer, developed centerfire cartridges with a small primer cap in the center of the base. Berdan's design was more difficult to reload than Boxer's so the latter became far more popular in the U.S. Berdan is best known in this country for his creation of a regiment of sharpshooters during the Civil War (Figure 130).

The Henry repeater operated like the Volcanic, but its .44 bullet left the muzzle at more than twice the velocity of the .41 Volcanic cartridge (1,250

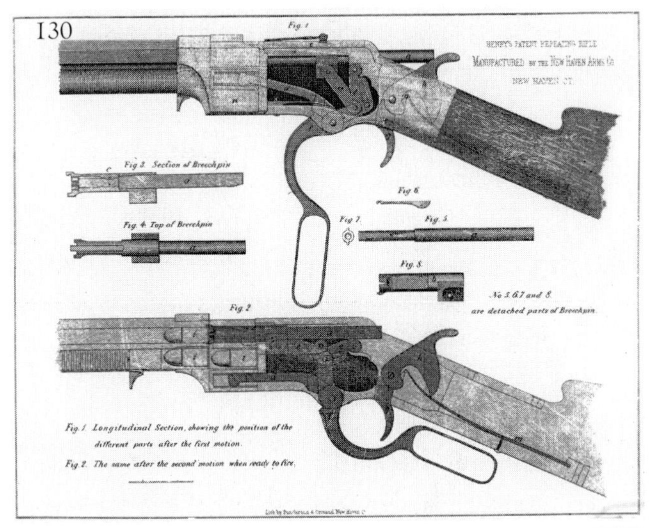

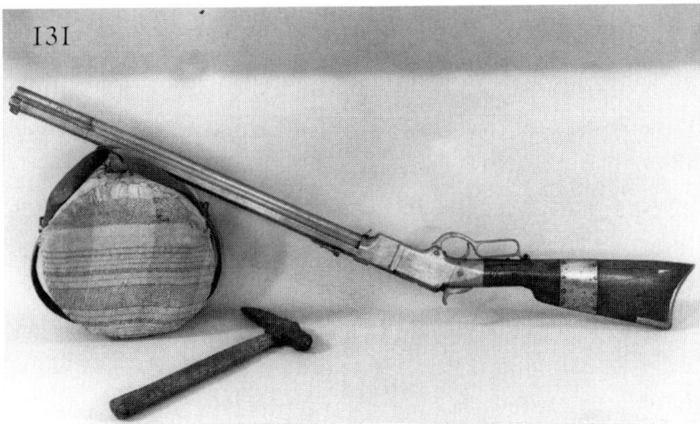

130. Illustration showing the interior mechanism of the Henry rifle. **131.** Prospector Ed Schieffelin was told all he'd find in Apache country was his tombstone. Nevertheless he persisted and from his discoveries of gold and silver the town of Tombstone, Arizona, rose. He relied on this Henry rifle for protection, exhibited with two other prospector's tools, a canteen and rock hammer. Repairs to the Henry's butt stock are evidence of the hard use to which it was subjected. (*Courtesy: Tombstone Courthouse State Historic Park, Arizona*) **132.** Henry rifle advertisement.

feet per second versus 500). Swinging the combination trigger guard and lever downward and then back again extracted the fired case from the chamber and ejected it, loaded a fresh cartridge into the chamber, and cocked the hammer. The tubular magazine beneath the barrel loaded from the front end by first pushing the cartridge follower to its forward limit which compressed its coil spring and allowed a collar at the muzzle to be turned aside. The magazine then could be loaded from the front end with fifteen copper cartridges. First inserting a round in the chamber made it a "sixteen shooter" (Figure 131).

Henry output began slowly with only about 1,500 completed by January 1863. Ultimately 14,000 Henrys were produced by 1867 including about 400 early ones with an iron rather than a brass frame. Oliver Winchester confirmed the new rifle's popularity in the west when in May 1863 he wrote that plain Henry rifles which retailed at $42 were selling for as much as $75 in California. Oliver Winchester as chief promoter for the New Haven Arms Company followed a sales promotion program similar to that of Sam Colt—presentations to individuals whom he felt could further the firm's cause. Hoping for military acceptance, Henry rifle with serial number 1 was a gift to Secretary of War Edwin Stanton, number 6 went to President Lincoln, and number 9 to Gideon Welles, Secretary of the Navy. In military tests the Henry had performed well, but its cost, the feeling that its rapidity of fire would cause a soldier to waste

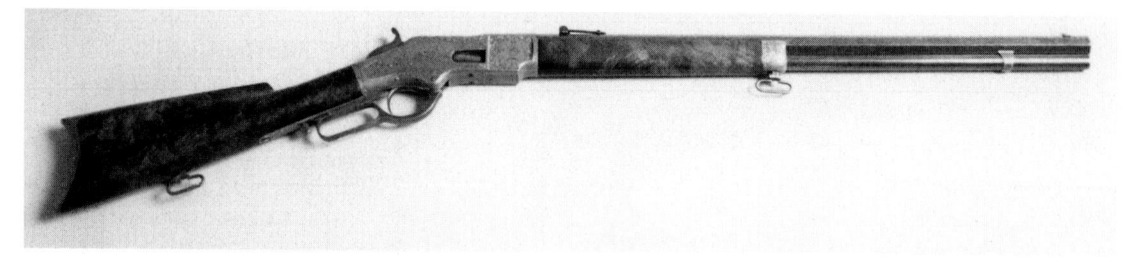

133. Winchester Model 1866 rifle, engraved and inscribed "Wichita Oct. 21, 1871 to Nehemiah Green from S. J. C. in Friendship and Gratitude." Governor Samuel Crawford resigned office in 1868 to assume command of the 19th Kansas Cavalry, raised to help control American Indians. Green as lieutenant governor succeeded Crawford. (*Courtesy: Kansas State Historical Society*)

ammunition, and the inability to use it as a muzzle loader if the magazine should become jammed were among reasons for rejection. Popular as the Henry was with those few Civil War soldiers who carried one in battle, its impact on the war's outcome was insignificant. The government only bought 1,731 although thousands more saw service as private purchases (Figure 132).[69]

A major Henry weakness was the open slot along the entire length of the magazine tube through which the thumb piece on the follower moved for it was susceptible to accumulating dirt or other debris which could hinder proper feeding. The solution to the problem was patented in May 1866 by Nelson King, Tyler Henry's replacement after Henry and Winchester had a falling out. King's improvement included a spring-tempered loading gate set in the right side of the frame. A tubular magazine with a coil spring similar to the magazine used in Walter Hunt's early repeater was mounted beneath the barrel and was closed to dirt for its entire length. The rifle was loaded simply by inserting cartridges one by one into the magazine through the gate. The new rifle was a little lighter than the Henry since the magazine now was merely a thin tube protected by a wooden forearm rather than being a part of the barrel. The improved Henry introduced another change for on the barrel it soon bore the reorganized firm's new name, the Winchester Repeating Arms Company.

The new Winchester fired the same cartridge as did the Henry and when all parts in stock had been used and production finally came to a halt in 1898, almost 160,000 had left the factory in New Haven, Connecticut. The Model 1866, as it later was designated, was sold in rifle form with a 24-inch barrel or carbine length with a barrel four inches shorter. With a round in the chamber and a fully loaded magazine they held eighteen and fourteen rounds, respectively. Somewhat less popular was a later Model 1866 "musket" with a 27-inch barrel. The new Winchester couldn't match the shocking power or longer range accuracy of such single shots as a Sharps or a good muzzle-loading "Rocky Mountain rifle," but it offered up to 18 quick shots without reloading. On the western frontier or elsewhere when one's life might depend upon a heavy volume of fire, this was a major consideration for some and both the Henry and the improved Henry, as the '66 sometimes was called, were popular civilian arms west of the Mississippi as well as with hunters and others in the east (Figure 133).[70]

69. Ibid., 129.

70. Thomas Addis served as Winchester's chief foreign sales representative for more than thirty years. A story is told of his delivery in late 1866 of 1,000 Model 1866s and ammunition for Benito Juarez's army. After delays, Addis finally had to take the guns to Monterrey, Mexico, where he was paid $57,000 in silver. Distrusting his guards and driver, Addis stayed awake throughout the three-day trip back to the U.S. border by periodically sticking himself with a pin.

The Model 1866 Winchester often is said to have been known on the western frontier as the "yellow boy" but in more than thirty-five years of research I have never found any use of this term in the hundreds of nineteenth-century accounts of western experiences which I've examined in the course of writing several other books. I wish I could recall the details more clearly, but years ago I did find the story of a particularly resourceful Winchester salesman who was making an overseas sales presentation of I believe the Model 1866 before a representative of a foreign country, perhaps Turkey. To make the demonstration a true test of the gun's qualities, the potential buyer loaded the gun's mechanism with dirt and sand making it virtually inoperable. With a star salesman's quick mind and ingenuity, the Winchester representative used the only liquid available to him and urinated on the rifle to clear the dirt and then continued with the demonstration. Is there a chance this was the origin of the term "yellow boy"?

Shotguns, Often Ignored but Often Present

As noted firearms authority Norm Flayderman has pointed out, "shotguns and fowling pieces were likely the most commonly owned and used of all antique American arms, the type of gun that just about every household had at one time or another." This was particularly true in the colonial period. Even later, throughout most of the nineteenth century, the country still had a large rural population and if a rural household had only one firearm, it probably was a shotgun. (There is no difference between a fowler or fowling piece and a shotgun although the latter was becoming the common term by the mid-1800s.)[71]

The use of a charge of small pellets or shot from a smoothbore gun when hunting birds and other small game for food began at least as early as the 1570s. Later wealthy sportsmen adopted the practice of shooting game as a form of target practice rather than solely for meat or to keep pests from eating one's crops. By the late 1700s in England a series of improvements in design allowed fowling piece barrels to be shortened from around three and a half feet to as little as thirty inches or so thus lightening them and making the double-barrel shotgun more practical and convenient than a longer single-barrel fowler at distances of up to around 60 yards. Few flintlock doubles were made or assembled in the United States although those which were often resembled English pieces and incorporated English parts. Eventually shot came in various sizes, depending upon the bird or animal being hunted, buckshot being the largest as the term implies.

In 1812, Johannes Samuel Pauly of Switzerland obtained a French patent on a breech-loading, centerfire design for a shotgun firing a self-contained cartridge with a soft brass head with a depression in the base to hold a primer of percussion compound. The gun caused no great stir among sportsmen, but twenty-three years later, Casimir LeFaucheux was granted a French patent for what became the first popular breech-loading double-barrel shotgun. Its barrels tipped down to load and it fired what came to be called a pinfire cartridge, in which a metal pin protruding at right angles from the side of the combination metal and paper case was driven into the primer when it was struck by the hammer. Later the case was improved by making it entirely of metal. No pinfire guns were manufactured in the U.S., although some were imported. In Europe and Latin America, revolvers and shotguns using the pinfire cartridge were quite popular.

71. Norm Flayderman, *Flayderman's Guide to Antique American Firearms . . . and Their Values* (Iola, Wis., 2001), 609. If someone interested in antique American firearms begins a reference library with one book, it should be the latest edition of Flayderman's 650-plus-page guide. It not only illustrates and describes in detail thousands of American long and short guns, but includes a wealth of historical data and advice to the collector while providing relative values.

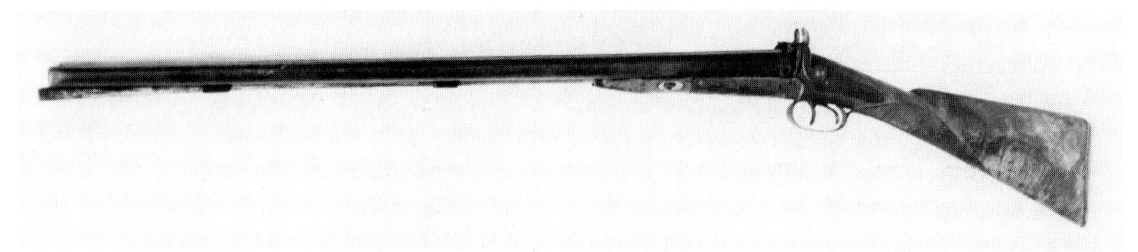

134. Muzzle-loading percussion double-barrel shotgun which started the "pig war." The U.S. and Britain disputed ownership of the San Juan Islands in Washington State's Puget Sound in the 1850s and citizens from both countries were settling there. In 1859, an American named Cutler used this shotgun to kill a Brit's pig which had gotten into his potato patch. An international dispute developed but ultimately they agreed to joint occupation of the islands until 1872 when arbitration decided the question in U.S. favor. (*Courtesy: Washington State Historical Society*)

A specialized form of shotgun present during much of the percussion era and even later was that used by market hunters along common flyways or Chesapeake Bay's waterways—anywhere ducks and geese abounded—and who shot large numbers for commercial sale. These men often hunted from small flat bottom boats or punts, using an oversize shotgun sometimes called a punt gun. These guns often were rather crudely made, with a barrel as much as seven feet long and a bore of up to perhaps two inches, too heavy a gun to be fired from the shoulder and so often one was mounted in the bow of the punt in a manner similar to a small cannon. At a single discharge from one of these mammoth fowling pieces, a hunter might kill (or unfortunately only wound) scores of birds. Prairie chickens and other fowl suffered the same fate at the hands of inland market hunters. The practice reached its peak in the 1880s but was followed by the deplorable trade in bird feathers in which "feather hunters" fulfilled a demand for plumage to adorn fashionable ladies' hats! This slaughter was largely curtailed by the Lacey Act of 1900 which prohibited the interstate shipment of illegally taken game. However it was not until 1918 that the practice of market hunting for meat was finally halted by federal law.

During at least the first half of the nineteenth century in this country, most American makers seemingly concentrated their efforts on rather plain utilitarian single-barrel muzzle-loading shotguns. Many of the finer quality muzzle-loading and later breech-loading double shotguns (and some single-barrel pieces) sold in this country before the 1870s and 1880s were manufactured overseas, particularly in England and Belgium, and bear such well-known names as W. W. Greener, who began business about 1829; Joseph Manton, who in 1806 patented the elevating sighting rib between the barrels; and James Purdey, maker of doubles in flintlock and then percussion form. Even prominent American makers such as the Tryon family of Philadelphia often used imported parts when assembling their single-barrel shotguns, particularly English barrels, even though their names may appear on such guns. One later example is the lot of several hundred double barrel 10- and 12-gauge shotguns which bear the famed Sharps Rifle Company name. Sharps in the 1850s had produced a few hundred percussion single-barrel, breech-loading shotguns, but announced in 1878 it was preparing to manufacture double-barrel guns. In reality, all the doubles were made entirely by Philip Webley & Company of England and imported even though they bear the Sharps name (Figure 134).

4 *A Very Uncivil War*

(1861–1865)

The Civil War marked a number of firsts. An income tax, railroad-mounted artillery, the large-scale movement of troops by rail, a battle between two ironclad ships, non-combat service for conscientious objectors, an organized ambulance corps, the use of fields of land mines, and the first successful rapid-fire machine gun were just a few of the innovations. Despite the glamour and pomp sometimes associated with that struggle, Captain William L. Nugent, a Mississippi cavalryman, described it in more realistic terms. He called it "an unmixed evil [of] . . . blood, butchery, death, desolation, robbery, rapine, selfishness, [and] violence."[1]

The Union army at the end of 1860 included only about 16,000 men, most of whom were serving in the west beyond the Mississippi River. The regular army was kept intact and was not dramatically increased in size during the war. Instead the war was fought primarily by volunteers in state regiments. Figures for the Union side show enlistments totaled about 1,500,000 men of which some 3,550 were American Indians and 186,000 African Americans. Those men serving in the navy and marines numbered about 106,000. About three-fourths of the Federal soldiers were native born; 65 percent of those of foreign birth were German or Irish. The age of privates ranged from eleven to eighty but twenty-five was typical. The average soldier stood about five feet seven inches tall and tipped the scales at about 135 pounds. A survey of 9,000 Union soldiers showed 62 percent listed their occupation as farmer. A private would earn about $13 a month in addition to whatever enlistment bonus he might receive. The Confederacy had no formal army when the war broke out—the call to arms was met by state militia and locally recruited units, and although figures are imprecise, probably between 800,000 and 900,000 southerners served. Many more men on both sides would die of disease than of battle-inflicted wounds.[2]

Arms Shortages

In terms of weapons, the war represented a microcosm of more than a century of firearms develop-

1. Exhibit text, National Civil War Museum, Harrisburg, Pennsylvania,.

2. About 109 African Americans would serve as commissioned officers within the U.S. Colored Troops, including eight as surgeons and fourteen as chaplains. One was Martin Delany, a major in the 104th U.S. Colored Troops, and who during his life was a physician, journalist, inventor of a railroad braking system, explorer, trial justice, and civil rights activist. Another 16,000 or so African Americans served as sailors in the U.S. Navy. The Confederate government did commission one woman officer during the war. Sally Tompkins of Richmond at her expense maintained a twenty-two-bed military hospital from 1862 until almost war's end and was given the rank of captain. Carol C. Green, *Chimborazo: The Confederacy's Largest Hospital* (Knoxville, Tenn., 2004), 16.

ment since small arms used extended from flintlock smoothbore muskets similar to those of Revolutionary War vintage to repeating rifles firing metallic cartridges. The variety of small arms employed by both Union and Confederate forces makes the collecting of Civil War guns a challenging field today. The U.S. Army's ordnance department between January 1, 1861, and June 30, 1866, purchased nineteen different makes of breech-loading carbines and eight of breech-loading rifles alone without regard to the variety of muzzleloaders and handguns it procured. Some of these purchases only involved a few hundred or were of guns which were delivered too late for service but nevertheless show the conglomeration of small arms with which Union soldiers and sailors were equipped. The Confederate story was little different. However, the north with its well-established industrial base eventually was able to adapt that to wartime production of suitable firearms, something the south was never able to match.[3]

Neither side was adequately prepared for war in manpower or materiel, and the Union was not eager to adopt advanced firearms which soon would become available. As late as February 1861, Colonel H. K. Craig of the federal army's ordnance department had written to the secretary of war: "It is not believed that what are called repeating arms are desirable for infantry of the line or riflemen." He complained they were complicated, more expensive, more liable to get out of order, and more difficult to repair than muzzle loading rifles and muskets. Excessive rapidity of fire, he pointed out, was not a critical consideration since a soldier could only carry so much ammunition. Of course he was talking about the Colt revolving rifle since no other repeater had been given serious consideration for adoption up to that time and the Colt did have its weaknesses, namely the occasional discharge of more than one shot at a time when improperly loaded. "The principle of the repeating arm is suitable for pistols, and should in my opinion be restricted to that weapon." A congressional act on June 23, 1860, prevented the army from purchasing patented firearms for such would have involved the payment of a royalty and the cost of arms produced at the federal armories generally was less than from commercial sources. The demands of war would change the picture quickly, however.[4]

Confederate forces seized the federal armory at Harpers Ferry on April 18, 1861, obtaining some arms stored there but even more important capturing up-to-date arms-producing machinery, mechanical drive belting and shafting, and parts including perhaps as many as 30,000 seasoned wood gunstocks. Rifle-producing machinery was shipped to Fayetteville, North Carolina, and rifle musket machinery to the Virginia State Armory in Richmond. The latter facility began turning out rifle muskets in October 1861, close copies of the U.S. Model 1855 but without the Maynard priming mechanism. With this loss, Springfield Armory alone couldn't keep up with the wartime demand for infantry arms. Between the initial contracts in July 1861 and war's end, the government contracted with almost two dozen private makers including Colt, Savage, Remington, and Whitney for one and a half million rifle muskets, most at $20 each. Only about 645,000 were delivered and it was not until early 1863 that significant quantities began to reach soldiers' hands. Meanwhile during the war, the Springfield Armory manufactured almost 800,000 rifle muskets in original Model 1861

3. For the author seeking to determine what make/model of long gun or revolver was issued to a particular Civil War regiment, there are several good sources. One is *An Introduction to Civil War Small Arms* by Earl J. Coates and Dean S. Thomas (Gettysburg, Pa., 1990), which includes a detailed listing of such information (primarily on Union regiments). Similar data can be found in several excellent books by John D. McAulay, *Civil War Pistols* (Lincoln, R.I., 1992), *U.S. Military Carbines* (Woonsocket, R.I., 2006), and *Rifles of the U.S. Army* (Lincoln, R.I., 2003).

4. Lewis, *Small Arms and Ammunition in the United States Service*, 63.

135. Springfield Model 1861 .58 rifle musket and its triangular bayonet. (*Courtesy: National Park Service, Fuller Gun Collection, Chickamauga and Chattanooga National Military Park*) **136.** The Springfield Armory museum's "Organ of Muskets," the only remaining one of thirty-six such storage racks constructed in the early 1830s to hold Model 1816 flintlock muskets. Later they were modified to accommodate the Model 1855 rifle musket. In September 1862 to meet critical arms needs of the Union army, rifle muskets were ordered to be boxed for shipment immediately upon their fabrication and inspection and the racks were set aside. Today it holds 645 Model 1861 Springfields.

form and later models with minor modifications adopted in 1863-64 (Figure 135).[5]

The arms making machinery Col. Turner Ashby's force seized at Harpers Ferry and sent south to Richmond represented some of the world's most advanced. The Confederate States Armory, as the Richmond facility became known officially, was the largest producer of arms for the Confederacy in the south but throughout its existence was beset by material shortages and production difficulties. When it ceased production in January 1865, it had completed an estimated total of 31,000 rifle muskets plus about 6,700 muzzle-loading carbines and short rifles. In contrast, Springfield Armory in 1863 was producing almost 20,000 rifle muskets per month (Figure 136).[6]

As another effort to meet the critical need for weapons, both sides sent agents abroad where in 1861-62 they competed vigorously between themselves as well as with representatives of some states to purchase whatever arms and other war materiel they could. Union agents as of June 30, 1862, had obtained almost 800,000 foreign arms. It was an ideal opportunity for European nations to dispose of their obsolete or worn-out weapons—sometimes at exorbitant prices—and they often did, including flintlock smoothbore muskets, some of which were first converted to percussion abroad or after they arrived in the North or South. These and other less desirable arms were replaced as soon as

5. In the face of urgent wartime demands, American contractor-made rifle muskets weren't held to the same rigid inspection requirements as were those made at Springfield. Inspectors in 1862 set up a classification system and contractors were to be paid $20 for Class 1 arms while those delivered which were determined to be serviceable but as low as Class 4 were paid only $16. Colt, with its well-established factory, in the fall of 1862 began deliveries which eventually totaled 75,000 rifle muskets, all accepted as Class 1. Claude E. Fuller, *The Rifled Musket* (Harrisburg, Pa., 1958), 151, 154.

6. Paul J. Davies, *C. S. Armory Richmond* (Ephrata, Pa., 2000), 141, 281, 349.

137. The Austrian Model 1842 Fruwirth carbine is an example of a foreign arm imported early in the war by the Union. Ten thousand of these .71 caliber rifled carbines arrived in early 1862 and were used in the western theater until better arms became available. Originally these had no provision for mounting a ramrod, which was carried separately. (*Courtesy: National Park Service, Fuller Gun Collection, Chickamauga and Chattanooga National Military Park*) **138.** British .577 Pattern 1853 rifle musket, imported by both the North and South in large numbers. (*Courtesy: National Park Service, Fuller Gun Collection, Chickamauga and Chattanooga National Military Park*)

possible with better weapons, but this was a slow process and many men in both blue and gray went off to war with outdated arms and in numerous instances retained them for many months (Figure 137).[7]

The primary foreign arm procured in large numbers that in quality equaled the .58 caliber rifle musket adopted less than a decade before by the U.S. was Britain's Pattern of 1853 (P53) rifle musket, often called the Enfield. The caliber (.577) was close enough to accept the .58 bullet if the bore wasn't badly fouled, and both sides contracted for the production of large quantities. The North imported about 428,000 of these brass-trimmed rifle muskets plus about 8,000 P56 rifles, similar and a little handier with a shorter 33-inch barrel and with a saber-type bayonet which could be used as a short sword, rather than the more common triangular bayonet. Of the rifles and muskets obtained abroad by Confederate purchasing agents in 1861-62, about sixty per cent (80,000) were Enfields and that government adopted .577 as its standard caliber (Figure 138).[8]

However, the quality of the Enfields varied rather widely. To maintain a show of neutrality, Britain refrained from selling P53s made at Enfield. The next highest quality P53s were produced by the London Armoury Company and were eagerly sought by both northern and southern agents. But other Enfields were assembled by smaller contractors using parts made by British and

7. The arms shortage in the south was such that Gen. "Stonewall" Jackson in March 1862 requested he be provided with 1,000 pikes, a request endorsed by Gen. R. E. Lee. Although thousands of pikes were produced in various southern state arsenals as last ditch weapons, their actual use in combat hasn't been documented. Generally they were around eight feet long including a blade about a foot or so in length on an ash or hickory pole.

8. Wiley Sword, *Firepower from Abroad: The Confederate Enfield and the LeMat Revolver* (Lincoln, R.I., 1986), 12.

Belgian subcontractors, often resulting in a lack of parts interchangeability and sometimes substandard quality. One federal commander complained that moisture swelled the unseasoned stocks of his regiment's Enfields making it impossible to remove the ramrod. In 1863 as deliveries of contractor-produced Model 1861 Springfield rifle muskets began to arrive in earnest, the federal need for imported Enfields diminished. Nevertheless, General U.S. Grant after the capture of Vicksburg in July 1863 ordered those regiments still equipped with inferior muskets to exchange them for captured Confederate Enfields.[9]

Northern purchasing agents also secured about a quarter million Austrian Model 1854 Lorenz rifles in .54 and .58 caliber, a very serviceable arm, along with more than 150,000 Prussian arms, primarily a .72 caliber musket often known as the "Potsdam." Other models of smoothbore and rifled arms from various countries including France, Britain, Germany, Belgium, and Spain served both Confederate and Union soldiers. Some were called "worthless" or "unfit to be placed in the hands of civilized troops" by those to whom they were issued. In some instances the criticism had merit, but at times these imports were purposely maligned or even intentionally damaged in the hope that their replacements might be the highly regarded Springfield. In defense of some of those condemned foreign arms, many once had been front line weapons in their country of origin and often they were subjected to improper care in the hands of inexperienced Americans. Yet during the first half of the war, they were essential to maintaining the conflict (Figure 139).

During the war, the Union army obtained by purchase or fabrication almost 471,000,000 .577 and .58 cartridges. The latter used a 500-grain bullet with three circumferential grooves, dipped in a hot mixture of one part beeswax and three parts tallow as a lubricant before being wrapped in paper with 60 grains of musket powder. Loading was done in the same manner as previous muzzle-loading muskets although the cartridge paper was now discarded—tear or bite the cartridge open with the teeth, pour the powder down the barrel, then ram the Minie ball down. Prime the gun by placing a percussion cap on the nipple (unless it employed a Maynard automatic tape priming device), bring the hammer to full cock, and the weapon was ready. The usual rate of fire was about three rounds per minute under good conditions. Reportedly a requirement for enlistment, at least early in the war, was four front teeth, necessary to bite open a cartridge, a more convenient method than tearing it open with fingers.

139. Civil War infantryman with an imported Belgian Model 1857 .58 rifle musket, one of the better-quality imports. (*Courtesy: Mississippi Department of Archives and History*)

One firm, Johnston & Dow of New York City, did offer a .574 caliber "waterproof and combustible cartridge" suitable for the .58 Springfield or .577 Enfield and which did not have to be torn open to expose the powder to the flame of the cap,

9. Ibid., 33; Claude E. Fuller and Richard D. Steuart, *Firearms of the Confederacy* (Huntington, W.V., 1944), 30.

140. Civil War infantry cartridge box with outer flap raised and tin liners removed from the interior. **141.** An unusual copper cartridge box with a hinged spring lid and belt loops on the reverse. It was patented in 1863 by B. Kittredge & Co. of Cincinnati and generally is thought to have been intended to contain .44 rimfire cartridges used in the Henry repeater and the single-shot .44 Wesson breechloader, two arms the Kittredge firm distributed. (Courtesy: Neil and Julia Gutterman)

thus speeding the loading process. The paper was made combustible by treating it with a solution of niter and waterproof by coating it with a form of lacquer, perhaps an early form of nitroglycerin. The government in 1862 ordered more than 1.3 million of these cartridges but despite strong endorsements from troops in the field to which they were issued, no further orders followed.[10]

A Union infantryman normally carried his ammunition in a black leather cartridge box suspended from a wide black leather shoulder belt. The box was fitted with two removable tin trays inside, each holding twenty cartridges in two packages of ten, a total of forty rounds. Caps were carried separately in a small leather cap box worn on the belt, lined with fleece and with a small wire nipple pick inside, useful to clear a fragment of cap or other blockage from the cone. Confederates sometimes weren't so well equipped and might have to carry ammunition in cartridge boxes made of cotton cloth stitched in three or four layers (Figures 140, 141).[11]

Ammunition for the rifle musket and musket was issued in paper-wrapped packets of ten plus a dozen percussion caps, the packages packed in wooden cases holding 1,000 rounds each and weighing about 100 pounds. The extra caps were to allow for misfires or loss. Caps also came in bulk, in cans of 100, 250, or 500. Cartridges intended for breech-loading rifles and carbines were packed in boxes or packages of seven to fifty rounds while the fragile skin or combustible cartridges for revolvers usually were packaged in bored-out blocks of wood containing the number required for one loading of the cylinder for the particular revolver intended.

Fouling caused by the residue left from burned black powder was a detriment to accuracy and loading speed throughout the black powder era. In extreme cases, it might be necessary to drive the end of the steel ramrod against a tree to force the bullet down the barrel. A successful means of countering this was the .58 Williams bullet, containing a zinc washer in its base which was flattened upon firing and scraped the bore relatively clean. Until late in the war, one or more of these cartridges was put in each package of .58 ammunition. One Union captain after the war wrote that he had found the .58 rifle musket often became difficult to load without cleaning after 20 or 25 shots. To

10. Elmer Woodward, "Mystery Ammunition of the 5th New York Volunteer Infantry, Duryee's Zouaves," *Military Collector and Historian*, Winter 2006, 222-23.

11. The design adopted late in the war for the Union 15th Corps' badge and flag was a cartridge box with the words "FORTY ROUNDS" above it.

relieve the situation, whenever possible he drew both .58 and .577 cartridges so his men could place the former in the top sections of their cartridge box tin liners to be fired first and the latter in the lower compartments to load more easily as the bore became fouled.[12]

On another occasion, some companies of the 101st Illinois Infantry in January 1863 learned that the enemy was near and prepared their .72 caliber Prussian muskets with fresh loads. "But some of the officers had reported that the charges were getting wet, and that they had no ball screws [to screw on the end of the ramrod and withdraw the ball], so I halted and tried to fire them, but after three snaps got but little over half the pieces discharged." Earlier in December 1862, an official report on the condition of the 110th Illinois Infantry stated: "The regiment is poorly armed. The arms are without bayonets, without ball screws, and until a few days ago were without shoulder straps for cartridge boxes. . . . The arms now in the possession of the regiment are of various caliber & manufacture ranging in the latter from 1812 to 1856."[13]

Despite the advantages of the new rifle musket, there still was nothing to prevent improper loading of any muzzleloader during the noise, confusion, and heightened emotion of battle. An excited soldier might load ball before powder or forget to cap his piece. He might fail to notice his gun had not fired yet continue to load one round on top of another. After the battle of Gettysburg, Union ordnance officers found that of the thousands of muzzle-loading guns retrieved from the field, many were improperly loaded with bullet before powder or with more than one load and as many as twenty rounds!

Union and Confederate recruits during the first two years of the war often found themselves issued a variety of obsolete foreign or American arms. The Confederacy's 19th Tennessee Infantry fought their first major battle at Fishing Creek, Kentucky, in January 1862. An officer wrote that "the rain poured down so they [flintlock muskets] would not fire at all. Several of the men after trying repeatedly to fire, just broke their guns over a fence or around a tree and went off in disgust." Another Tennessee regiment, the 20th, didn't replace their flintlocks until 1862. Farther north, some men of the 51st Illinois after testing their Belgian muskets and finding them unsatisfactory stacked them and refused orders to take them up. They were marched back to their barracks and the guns remained stacked outside all night in the rain. Their risky rebellion worked and next day they were issued Harpers Ferry rifled muskets which the men found more satisfactory. The 76th Ohio Infantry first carried "old second-hand Belgian rifles, a short, heavy, clumsy arm with a vicious recoil" before receiving Springfield rifle muskets at year's end in 1862. Smoothbore muskets still equipped at least in part roughly ten percent of the Union army regiments at Gettysburg half way through the war. Muskets in .69 caliber, both smoothbores and those which in the late 1850s had been updated with the addition of rifling, remained in frequent Confederate use in the east and within both armies in western theaters as late as 1864.[14]

When a regiment was only partially armed with first-class weapons such as the Enfield or .58 Springfield, commanders sometimes issued those to flank companies leaving the center companies with smoothbores or less effective arms. The 37th Illinois was known as the "Fremont Rifles" and carried a regimental banner with a painting of General Charles Frémont, the western explorer. When the regiment was issued former flintlock

12. Fuller, *The Rifled Musket,* 24.

13. Ken Baumann, *Arming the Suckers, 1861-1865* (Dayton, Ohio, 1989), 190, 201.

14. Earl J. Coates, Michael J. McAfee, and Don Troiani, *Don Troiani's Civil War Infantry* (Harrisburg, Pa., 2006), 11, 47, 60.

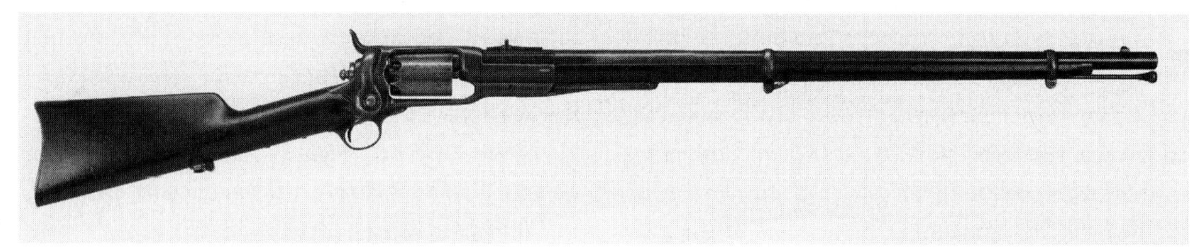

142. Colt Model 1855 revolving musket left with a family in New Mexico by a retreating Confederate soldier after the Union victory at Glorieta. He never returned and it remained in the family for a century. (*Courtesy: James D. Gordon, photo by Steven W. Walenta*)

muskets converted to percussion, Mrs. Frémont interceded and the flanking companies and non-commissioned officers received Colt revolving rifles and the other companies Enfields. In early 1862 the 33rd Illinois received only enough Springfield .58 rifle muskets to equip one of the two flank companies so the men conducted a shooting match to decide which company would receive them.[15]

Better Arms for the Infantry

With few exceptions, major battles of the Civil War were decided by infantry but these engagements frequently were fought using outmoded battle tactics, relying on massed formations and firefights often at ranges of 100 or so yards or even less, negating the longer range and greater accuracy offered by rifle muskets when such arms were available. Bayonet charges still occasionally determined the outcome of an engagement even if the enemy should flee before coming into contact with "cold steel." The "buck and ball" load was deadly at 50 yards and effective against a massed enemy at twice that range, the length of a football field. Even if buckshot wasn't always fatal at the longer range, a wound often was sufficient to take a man out of the fight and perhaps a comrade or two helping him off the field. This prompted some regiments to retain their .69 smoothbores by choice. However by the summer of 1863, particularly in the eastern theater of the war, first-class Springfields and Enfields were being distributed in growing numbers within both Union and Confederate armies. Union infantry and cavalry regiments operating far to the west often had lower priority and sometimes were slower in receiving improved arms.

Also with rather few exceptions, far more time was spent in drilling recruits in formation maneuvers than in actually giving them experience in firing their weapons on a target range. Many men in the Union ranks had never handled a gun before, yet unlike the situation in Britain's and France's armies, it was rare if they received any significant weapons instruction and practice. Men of the 13th New Jersey Infantry saw combat for the first time at Antietam in September 1862 without having fired their Enfield rifle muskets. Reportedly when the 8th U.S. Colored Infantry went into battle at Olustee, Florida, in 1864 they had not loaded their guns with live ammunition before. This was not as much of a problem within the Confederate army since many of its men were from rural backgrounds where hunting had been a more common pastime. Confederate Major General Patrick Cleburne, a veteran of the British army, was somewhat of a rarity among officers of both sides for he gave his men practice in estimating range and in firing at targets at various distances of up to 800 yards and his troops' success in battle reflected the wisdom of his efforts.[16]

15. Baumann, *Arming the Suckers, 1861-1865*, 107, 115.

16. Joseph G. Bilby, *Civil War Firearms* (Conshohoken, Pa., 1996), 77. Joseph G. Bilby, *A Revolution in Arms: A History of the First Repeating Rifles* (Yardley, Pa., 2006), 23.

While the muzzle-loading infantry musket or rifle musket was the primary weapon of foot soldiers throughout the war, repeating arms were available to a small proportion of the Union army. Five-shot .56 caliber Colt Model 1855 percussion revolving rifles, and a smaller number of six-shot .44s, were in service by late 1861 and some remained in use almost until war's end, particularly in the western theaters. Two prominent units, Berdan's sharp shooters and Wilder's "Lightning Brigade" of mounted infantry, were among the twenty-nine infantry regiments and twenty-six of cavalry which employed Colt rifles in their ranks at some time. Although the government paid between $30 and $50 apiece for these Colts, at war's end soldiers could take them home with them for $8. Later sales of Colts as surplus between 1866 and 1882 brought prices between $3 and 40 cents (Figure 142).[17]

Colts received varied marks for serviceability—some considered them fine weapons and others held them in lower esteem. There were complaints about splinters of lead striking the forward hand and wrist when fired, something that could happen if through wear the gun was slightly out of tune and the chamber wasn't in precise alignment with the barrel. Potentially more serious if one's supporting hand was positioned in front of the cylinder was the occasional and disconcerting discharge of more than one chamber at the same time. However such occurrences appear to be rare if the chambers are properly loaded with snug-fitting percussion caps and balls.

The sixteen-shot Henry rifle, firing a .44 rimfire metallic cartridge, was available in limited numbers by the summer of 1862, although wartime production couldn't keep up with demand. The federal government did buy almost 2,000 of these but far more were purchased by individuals, often in the later months of the war with soldiers' reenlistment bonuses, despite some criticism that the gun was too fragile for military service and its users would be tempted to waste ammunition. Nevertheless it merited the observation reportedly made by Confederates who faced it that it was a rifle which could be loaded on Sunday and fired all week.

Brigadier General James W. Ripley served as the army's chief ordnance officer from 1857 until replaced by George Ramsey in the fall of 1863. Ripley was ultra conservative when it came to considering the Henry and other new and unproven arms thrust upon him by numerous inventors and has been sharply criticized for this wartime stance. But in his defense, it must be pointed out that he was faced with the monumental task of arming thousands of infantry recruits as quickly as possible with arms which had been proven reliable in combat, in this case the .58 muzzle-loading rifle musket. Ramsey would be more accepting of new repeating rifles.

One account of Henry use later was seized upon and used by Oliver Winchester for promotional purposes in the company catalog. As the story goes, James M. Wilson, a Kentucky Unionist, purchased a Henry rifle for his own use. One day when he was recruiting men for the newly formed 12th Kentucky Cavalry, seven Confederate guerrillas burst into his home. He pleaded with them not to kill him in front of his family, and as he stepped outside made a dash for a log corn crib across the road where he'd left his Henry. In the course of the fierce gun battle which followed, he killed all of his attackers. Soon after the state of Kentucky purchased Henry rifles for Wilson's entire Company M. One can wonder why he didn't keep his Henry closer at hand and question why he didn't confirm the incident in his own words. The present owner of Wilson's Henry surmised that he might have remained silent knowing that after the war he might live among people who were his former enemies.[18]

17. John D. McAulay, *Civil War Breechloading Rifles* (Lincoln, R.I., 1987), 17-19.

18. Bilby, *A Revolution in Arms*, 97.

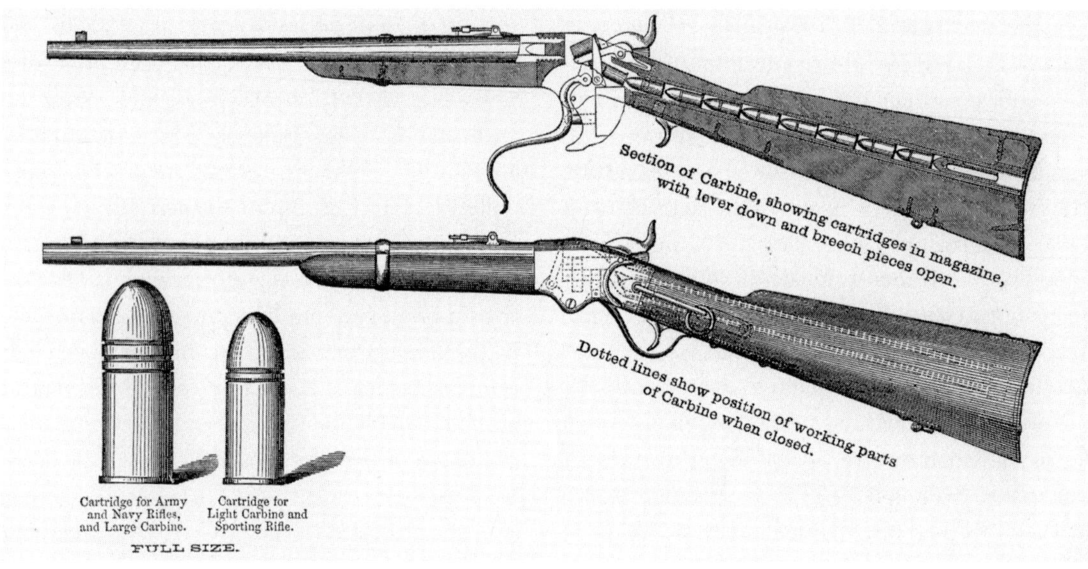

143. Magazine arrangement of the Spencer rifle and carbine.

The 7th Illinois Infantry obtained Henrys at the men's personal expense and Trueman Powell of the 64th Illinois in June 1864 wrote from Big Shanty, Georgia: "A part of the regiment are armed with these guns [Henry rifles] and all of the boys will be if the guns can be obtained. We have to pay for our own guns and they cost $41 each. The rest of the regiment have Whitney or Windsor [Mississippi] rifles and the Springfield rifle. These latter are good guns but half so good as the Henry rifle."[19]

An incident involving the fatal wounding of "Mountain Joe" Isenburg of the 73rd Illinois Infantry brought heavy retribution from a Henry user.

> At Kenesaw Mountain [Georgia] Joe was treacherously shot while carrying water on the picket-line—a duty which, by mutual consent of both pickets, had been performed unmolested [by either side]. Gil Harbison then and there, on his bended knees, over his wounded partner—who soon died—swore by the highest authority he called upon that he would be avenged. He sent for a Henry rifle, and from that time forward Gil Harbison could, at night or day, be found as often as elsewhere on or beyond the picket-line. During the night at Spring Hill he seemed to be in his glory, and during the short rest in the rear of the lines at Franklin he remarked to a comrade that he was abundantly satisfied with the way the account of Joe Isenburg stood. Very soon after this he was shot in the head and immediately killed.[20]

The army's first order for the famed Spencer rifle came in December 1861 over the objections of General Ripley, before Christopher Spencer had production facilities ready. He did secure space in a portion of the Chickering piano factory in Boston but quantity deliveries didn't begin until early 1863. Like the Henry, it needed no separate percussion cap but chambered a rimfire copper cartridge, designated as a .56-56 although it actually was .52 caliber. Seven of these cartridges were loaded through a trapdoor in the butt plate into the tubular magazine contained in the hollow butt stock. Working the combination trigger guard and operating lever

19. Baumann, *Arming the Suckers, 1861-1865*, 150.

20. Ibid., 156-57.

extracted and ejected the empty cartridge case and fed a fresh round into the chamber. One then manually cocked the hammer and was ready to fire. The Spencer rifle couldn't match the percussion rifle musket for range, but it did offer a heavy volume of fire. Although the Spencer often was called a "seven-shooter," with a round in the chamber seven more could be loaded in the magazine making it an eight-shot repeater (Figure 143).[21]

One of the first units to receive Spencer rifles was that of Colonel John T. Wilder, who in the spring of 1863 had sought to obtain Henry rifles for his mounted infantry. The factory's inability to fill the order prompted him to choose the Spencer instead. His men had been willing to buy Henrys with their own money, but the government provided the Spencer rifles without cost to them. Wilder's brigade gained fame for themselves and their repeaters on June 24, 1863, at Hoover's Gap in Tennessee when they outfought General Braxton Bragg's superior force of Confederates. Their designation as "Hatchet Brigade," so named because of their use of axes to prepare field fortifications, soon changed to that of Wilder's "Lightning Brigade."[22]

General George Ramsey, the army's new ordnance chief, in April 1864 wrote:

> Repeating arms are the greatest favorites with the Army, and could they be supplied in quantities to meet all requisitions, I am sure that no other arm would be used. Colt's and Henry's rifles and the Spencer carbines and rifles are the only arms of this class in the service. Colt's is both expensive and a dangerous weapon to the user. Henry's expensive and too delicate for service in its present form, while Spencer's is at the same time the cheapest, most durable, and most efficient of any of these arms.[23]

A myth connected with the Spencer claims that President Lincoln's test firing of the rifle and his intervention on behalf of the inventor were major factors in the government's adoption of the weapon. However between them, the army and navy had ordered more than 10,000 Spencer rifles by the end of 1861 and some already had been used in combat before Lincoln's introduction to the seven-shooter in August 1863. His initial impression was unfavorable for the two guns he examined malfunctioned. Christopher Spencer several weeks later was able to arrange for a personal demonstration and this time the rifle in Lincoln's hands performed properly. "The President made some pretty good shots," his secretary recalled following an hour's shooting near the present Washington Monument. However, there is no evidence that Lincoln exerted pressure to promote more government orders.[24]

Among the great variety of infantry arms used by or offered to the Union army by inventors, a most unusual example was a modest "repeater," the two-shot Lindsay rifle musket, designed by the same John Lindsay who produced two-shot civilian pistols. The weapon looked much like the typical Springfield rifle musket except it had two center-mounted hammers side by side, two nipples, but only one trigger. It was loaded with two superimposed rounds, one on top of the other, the rear Minie ball serving as a check to prevent ignition of the rear powder charge until the forward load had been fired. The initial trigger pull released the right hammer which fired the forward load while the second trigger squeeze fired the second (Figure 144).

The government ordered a thousand at $25 each and in the fall of 1864 some were issued to the 5th, 16th, and 26th Michigan Infantry and the 9th New Hampshire but they saw little actual combat use. Testing showed that the long channel from

21. In addition to his gun design, Spencer became wealthy with the invention of an automated screw-making machine and in 1901 invented a steam-powered automobile.

22. Roy M. Marcot, *Spencer Repeating Firearms* (Irvine, Calif., 1983), 51, 53. Bilby, *A Revolution in Arms*, 89.

23. Harold F. Williamson, *Winchester: The Gun That Won the West* (New York, 1963), 34.

24. Marcot, *Spencer Repeating Firearms*, 56-58.

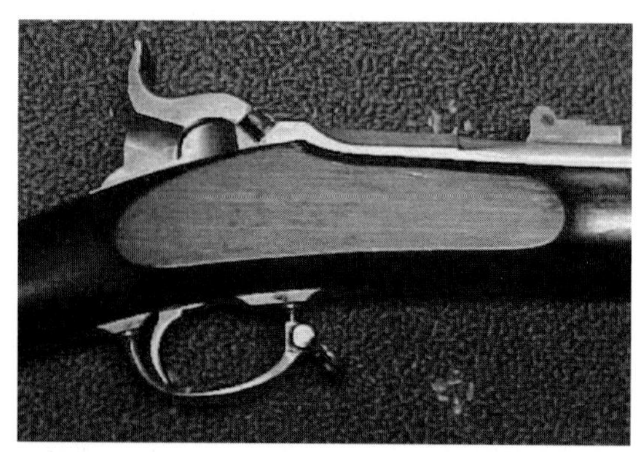

144. Breech area of the innovative but impractical two-shot Lindsay rifle musket with two hammers and nipples. (*Courtesy: Neil and Julia Gutterman*)

nipple to the forward charge apparently clogged easily and sometimes both loads went off at the same time. One private in the 16th Michigan was killed when a barrel burst and a lieutenant was wounded in another such accident. The historian for the 9th New Hampshire wrote: "In the afternoon, at battalion drill, the new double-shooting rifles were inspected and tested…and showed off their demerits to good advantage—flashed, fizzled, and failed famously, enough to secure their condemnation and a speedy exchange to the Springfield rifle."[25]

Among single-shot percussion breech-loading arms, undoubtedly the best was the .52 caliber Sharps, a weapon which could be loaded quickly and without error. It could accept either a paper or a linen self-consuming cartridge. The latter was preferred since when the rising breechblock sheared off the end of the longer paper cartridge, a little powder sometimes worked its way into the space between the frame and wooden forearm which could ignite. However not many Sharps rifles reached the infantry since the factory concentrated its production efforts on carbines, just as Spencer production too later was devoted primarily to carbines for the cavalry. The most famous regiments to employ the Sharps rifle were the 1st and 2nd U.S. Sharp Shooters, organized by Col. Hiram Berdan primarily for use as skirmishers. Berdan was a skilled marksman and inventor but capable as he was in civilian life, when the regiment was faced with battle he often managed to position himself well out of harm's way. Five of his officers were so disgusted with Berdan's timidity that in July 1862 they filed a complaint against him for cowardice, a charge which they later withdrew (Figure 145).[26]

These two regiments were made up of recruits who first had to prove their marksmanship ability. They had been promised Sharps rifles upon enlistment, but these were slow in coming and Colt revolving rifles were issued before the Sharps arrived, not without substantial complaining within the ranks of these regiments uniformed in green. Some of the recruits had brought their personal target rifles with them and a few retained these throughout their enlistment, cumbersome muzzle-loading affairs which sometimes weighed more than twenty pounds but which despite their weight and slowness in loading did prove useful for long-range use from fixed positions. In one engagement during the 1862 Peninsula Campaign in which McClellan's Union force threatened Richmond, Private George Case with his thirty-two-pound target rifle and telescopic sight prevented a Rebel gun crew from serving their artillery piece.[27]

A November 1862 poster recruiting qualified Michigan men for a company in Berdan's regiment of "Sharp-Shooters" noted they were equipped with the Sharps rifle, were not assigned to such tasks as building roads or fortifications, did not have to serve on night picket duty, and fought "Indian style" from behind rocks and other protective cover. Their mission was to down enemy officers and artillery horses. Only one man in the

25. John D. McAulay, "The Lindsay Rifle Musket," *Gun Report,* 27, no. 9, Feb. 1982, 18.

26. Roy M. Marcot, *Civil War Chief of Sharpshooters Hiram Berdan: Military Commander and Firearms Inventor* (Irvine, Calif., 1989), 52-53.

27. Bilby, *Civil War Firearms,* 111.

145. Model 1859 .52 Sharps military rifle, similar to those issued to the 1st and 2nd U.S. Sharp Shooters. (*Courtesy: National Park Service, Fuller Gun Collection, Chickamauga and Chattanooga National Military Park*)

company had been killed in battle, the poster trumpeted.

The Union army included other sharpshooter regiments such as the 66th Illinois Infantry, also known as Birge's Western Sharpshooters, outfitted with muzzle-loading sporting rifles secured through Horace Dimick, a St. Louis arms dealer and maker. Many of these men later secured Henry repeaters. Spencer repeating rifles were issued to some Ohio sharpshooters, but these rifles were inappropriate for long-range shooting.

Federal Cavalry Arms

Cavalrymen took longer to train than infantrymen and mounted regiments were more expensive to maintain so their numbers had been limited within the Union army before the war. It required several years before volunteer Union cavalrymen could compete on equal footing with Confederate troopers, many of whom already were skilled riders. The army in 1833 had adopted the Hall breech-loading carbine for its newly formed dragoon regiment and in the 1850s tested such other single-shot breech-loading carbines as the Burnside and Maynard. These two each used a metallic cartridge with a hole in the center of its base to admit flame from the separate percussion cap. The Sharps breechloader soon proved to be the preferred cavalry carbine of the 1850s, outperforming the Hall as well as the "worthless" muzzle-loading smoothbore .69 cavalry musketoon adopted in 1847 (Figure 146).

But Union cavalrymen would be faced with the same shortages and varied array of available weapons as was the infantry during the early months of the war. At first sometimes the only arms available were sabers and perhaps revolvers along with rifles or even longer rifle muskets, hardly appropriate for convenient use on horseback. The 1st Maine Cavalry throughout the winter of 1861-62 remained unarmed and men did guard duty with axe handles.

The primary handgun of the Union cavalryman was the .44 Model 1860 Colt, followed by the .44 Remington and lesser numbers of .36 caliber "navy" models by both makers and double- and single-action .36 and .44 Starrs. These revolvers all were six-shot percussion arms, preferably firing a self-consuming combustible cartridge or they could be loaded with loose powder and bullet, the conical slug preferred over the lighter round ball. Remington made a major contribution to establishing a fair market value for revolvers when in 1862 it offered its handguns at $12. It wasn't long before Colt dropped its $25 price to $14 to compete. These Remingtons were packed fifty to a box and came with fifty combination screwdrivers and nipple (cone) wrenches, fifty extra cones, and twenty-five bullet molds casting two balls each. Later the molds were eliminated as unnecessary. Spare nipples were included since the weapon could be unreliable if cones should become broken, worn, or deteriorated by the corrosive effect of the fulminate of mercury contained in percussion caps (Figures 147, 148).

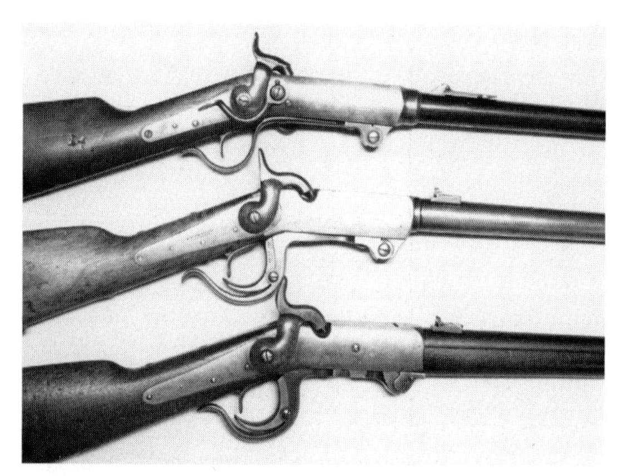

146. Three models of Burnside carbines (from top)—first, second, and third. The Burnside, first ordered for field trials in 1856, was the army's initial firearm to use a metallic cartridge although it was still fired with a separate percussion cap. The Civil War Burnside carbines had a wooden forearm as introduced in the third model.

The writer who wishes to depart from these most common handguns can equip a fictional volunteer cavalry trooper with a .36 Whitney or a French pinfire LeFaucheux or such pieces as a .36 Savage-North, .44 Pettengill, or .44 Joslyn. However by early 1864 the army had settled on .44 caliber as preferable to .36. When available to them, Colts and Remingtons were popular choices of southern cavalrymen, although shortages sometimes forced them to rely on such obsolete handguns as single-shot military pistols from a decade or two earlier (Figures 149, 150).[28]

The ordnance return for the 3rd Illinois Cavalry for the third quarter of 1863 is illustrative of the variety of revolvers and shoulder arms still found in some regiments. The list showed three makes of carbines (292 Burnsides, 65 Gallagers, and 39 Halls), six different models of handguns (385 .44 Colts, 31 .36 Colts, five .44 Pettingills, 34 .44 Remingtons, two .36 Savages, and 55 single-shot Deringer pistols), as well as 264 dragoon sabers (Model 1840) and 164 light cavalry sabers (adopted in 1857 but erroneously called the Model 1860 today). At the end of 1862, the regiment had had almost twice as many single-shot "horse pistols" as Colt revolvers. The great number of different arms used in the war by both cavalry and infantry sometimes made supplying spare parts and the appropriate ammunition to a unit a challenging task regardless of whether the troops wore blue or gray.[29]

It's been stated that the major fire which swept Colt's revolver factory in February 1864 was the cause of the firm's cessation of revolver deliveries to the army during the Civil War. Actually their last delivery of revolvers came in November 1863, and existing official government correspondence indicates it may have been Colt's unwillingness to reduce the price for their Model 1860 from $14 to $12 that precluded further wartime government contracts. During the remainder of the war, all of the Union's needs were met with .44 Remingtons and .44 single-action Starrs, a less expensive and less complicated mechanism than that of the double-action Starr. Deliveries of 5,000 .44 Rogers and Spencer revolvers contracted for in late 1864 came too late for wartime issue, and these revolvers quite often are found today in nearly new condition. At war's end enlisted men were authorized to purchase their revolvers if they chose to do so and of about 20,000 such sales, 9,047 were of Colts and 9,875 of Remingtons (Figure 151).[30]

28. In the past, the saber had been a favored cavalry arm before the advent of improved firearms and there were Civil War accounts of saber-wielding charges of up to regimental size and even occasional references to units sharpening the blades instead of employing the blades dull as originally issued. Colonel Philip St. George Cooke favored sharp sabers and wanted a scabbard that would not dull them, but Maj. Albert Brackett called the saber "a nuisance...and...of no earthly use." Despite some traditionalists' opposition, the revolver was fast becoming the favored weapon in a close-in fight, and some Confederate cavalry leaders eliminated the saber from their ranks entirely.

29. Baumann, *Arming the Suckers, 1861-1865,* 41-42.

30. Donald L. Ware, "Remington Army and Navy Revolvers, 1861-1888" (unpublished manuscript), 160-61. Pate, *Smith & Wesson American Model in U.S. and Foreign Service,* 42.

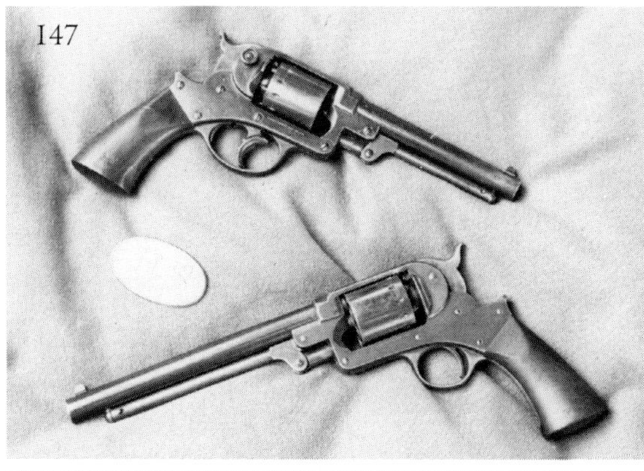

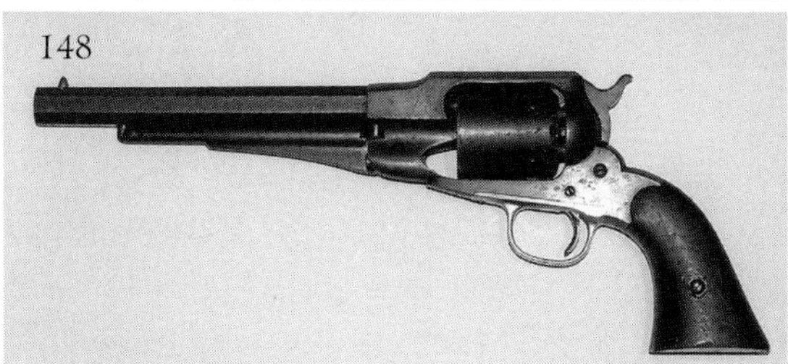

147. The third most widely used make of Civil War revolver was the Starr (about 2,000 .36s and 46,000 .44s). Shown here are the double-action .44 (top) with the single-action .44. The latter was less costly to produce. Removal of a thumbscrew in the frame allowed the barrel to tip downward to remove the cylinder for cleaning, but despite this convenient feature the Starr didn't receive a very high approval rating. **148.** A .44 Remington New Model inscribed on the backstrap "J. H. Moore/USV." Jesse Hale Moore, a Methodist clergyman and schoolteacher before the war, was colonel of the 115th Illinois Infantry. **149.** A Union cavalryman armed with saber and a 12-millimeter (.47 caliber) French LeFaucheux pinfire revolver. The federal government procured almost 13,000 of these imports during the war.

The revolver was strictly a close-range weapon and a veteran of the 1st Maine Cavalry recalled: "With revolver in hand, the trooper was more likely to shoot off his horse's ears, or kill his next comrade, than hit an enemy, however near." However Confederate partisans under Colonel John Mosby in the east and guerrilla leaders such as Charles Quantrill in Missouri did rely on handguns with deadly effect, although some of their shooting exploits have been exaggerated through the years. Mosby claimed that his men "did more than any other body of men to give the Colt pistol its great reputation."[31]

Breech-loading carbines were easier to load than a percussion revolver on horseback. But with a shorter barrel and sighting radius (distance between front and rear sights), a lighter powder charge and the bullet's looping trajectory, they lacked the accuracy and range of a rifle or rifle musket and were only effective against a massed group of men to about two hundred yards or so, perhaps half this distance against a single enemy. The fact that soldiers rarely received any instructions in estimating distance or such fundamentals as trigger squeeze or proper breathing further reduced the effectiveness of both carbine and rifle musket except at moderate range or against a massed formation. The opening scene at Gettysburg on July 1, 1863, found General John

31. Bilby, *Civil War Firearms*, 172.

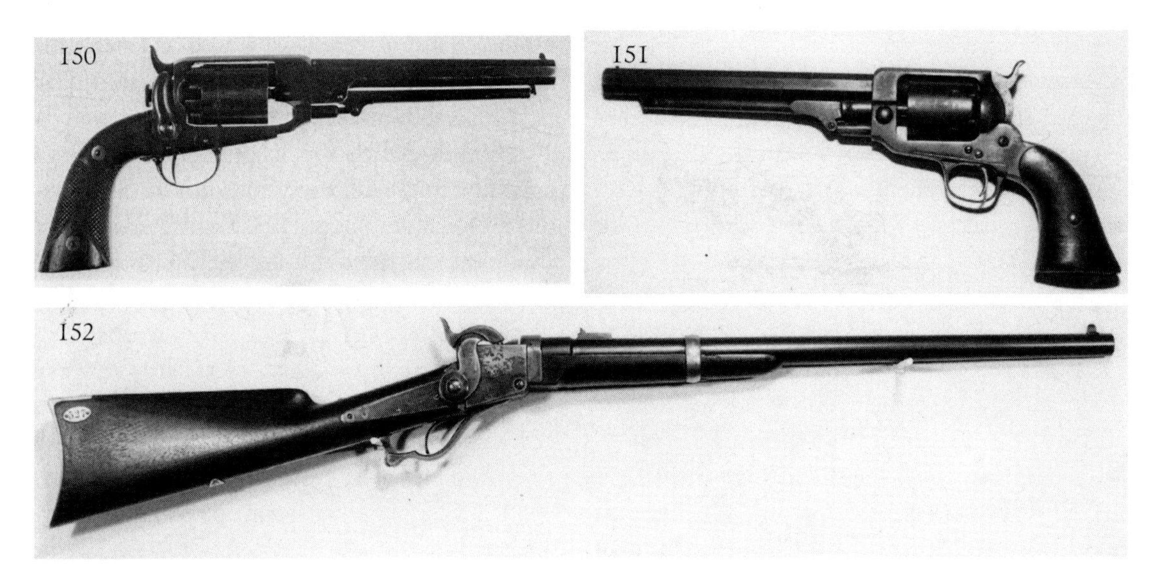

150. Five-shot .44 Joslyn, one of the least effective of Civil War revolvers. The federal army obtained 1,100 and the navy ordered 100 in 1861. **151.** The Whitney line of firearms covered almost a century and included a .36 caliber solid frame Civil War "navy" of which the army contracted for slightly more than 11,000 at prices from $10 to $17. The navy paid $10 to $12 for the 6,272 it ordered. **152.** Although similar in appearance to the Sharps, the Starr was considered by most who used it much less reliable even though 20,000 were purchased for the northern cavalry. (*Courtesy: National Park Service, Fuller Gun Collection, Chickamauga and Chattanooga National Military Park*)

Buford's cavalry troopers valiantly engaging Confederate infantrymen but armed with a varied array of carbines which were outranged by the gray infantry's arms. Nevertheless the cavalrymen were effective in forcing the southerners to take the time to deploy and delayed their advance until Union light artillery and then infantrymen arrived.[32]

During the war, the Union issued almost twenty different makes of breech-loading carbines using a variety of cartridges made of linen, paper, copper, brass, foil, and even rubber. These carbines were joined by a limited number of obsolete muzzle-loaders—musketoons and Springfield pistol-carbines. Many of these breechloaders required a separate percussion cap to ignite the cartridge, the Sharps generally being considered as the best of these, already in limited use at the time of the attack on Fort Sumter to open the war. Eventually wartime orders reached almost 80,000 Sharps .52 caliber carbines plus 8,000 rifles. The Ballard, most of which fired a .44 rimfire cartridge, was the best of the single shots firing a metallic cartridge and was the precursor to a long line of popular sporting arms. The Spencer repeater quickly earned a favored position. A properly outfitted trooper carried his carbine hung from the left shoulder suspended from a wide black leather sling with a swivel snap which secured it to a ring on the gun's left side. Thus when a trooper dismounted or was thrown from his horse, his carbine was attached to his body. When mounted, he thrust the muzzle into a short leather socket affixed to the saddle on the right side to keep it from bouncing wildly about. Cartridges were carried in leather belt pouches and percussion caps in infantry-style small fleece-lined cap boxes also worn on the belt (Figures 152, 153).

32. In an engagement by cavalrymen fighting on foot, their effective strength was reduced by one quarter since every fourth trooper was designated as a horse holder to restrain four animals.

Other Union-ordered carbines included the Burnside, patented in 1856 by Ambrose Burnside. He had sold his interest in the Bristol Firearms Company before the war and so made nothing on the wartime purchases of the carbines. He was a brave but rather ineffectual army commander during the war, remembered today primarily for his style of facial hair. The 53,000 percussion Burnsides ordered fired a brass cartridge shaped somewhat like an ice cream cone. The Smith carbine took a paper and foil or a rubber cartridge, as with the Burnside primed by a separate percussion cap. The hinged barrel swung downward 90 degrees to load. Thirty-one thousand Smiths were procured and like the Maynard, Joslyn, Cosmopolitan (also known as the Union or Gwyn & Campbell), and obsolete Hall received mixed marks from those to whom they was issued. The Starr and Gallager at the bottom end of the scale often were given a poor grade by those regiments which received them. Some troopers firing carbines using a metallic cartridge, like the Maynard or late model Joslyn, found that removing the empty case with their fingers from a chamber fouled with powder residue was easier if they wet the cartridge in their mouth before loading it. Of the army's breech-loading carbines, three began in percussion form but by war's end were being produced firing rimfire metallic cartridges—the Starr, Joslyn, and Gallager.[33]

153. Union cavalryman with Sharps carbine suspended from a shoulder sling and what appears to be a Colt Model 1851 Navy. (Courtesy: Herb Peck, Jr.)

Deliveries of the first Spencer repeating rifles ordered for the army began in December 1862. Colonel J. T. Copeland, a former Michigan state senator, from the summer of 1862 while recruiting for his 5th Michigan Cavalry fought hard to secure promises of Spencer rifles when they became available. They reached the 5th in Washington two days after Christmas with the overage going to partially equip the 6th Michigan. These Spencers were in the hands of the Michigan cavalry brigade at Gettysburg on the fateful July 3rd where under their newly appointed flamboyant and reckless commander George A. Custer they helped block Gen. J. E. B. Stuart's attempt to attack Union forces from the rear. The effectiveness of Spencer rifles on this day and by Wilder's brigade the previous month helped demonstrate the arm's worth. Although deliveries of carbines didn't begin until October 1863, from then forward, nearly all Spencers produced were carbines.[34]

Although Wilder initially had sought Henry rifles for his men, by January 1864 he had had many opportunities to witness the Spencer in action. In that month he wrote:

33. On a summer night in 1864 five men of the 11th New York Cavalry while reconnoitering a Confederate camp along a Louisiana river had to use the butts of their Burnsides and oars to fend off not rebels but a group of alligators which threatened them. McAulay, *U.S. Military Carbines*, 21-22.

34. Bilby, *A Revolution in Arms*, 83-84.

> I could enumerate at least thirty fights in which the "Spencer rifle" has triumphed over other arms in such overwhelming numbers so as to almost appear incredible. . . . I believe if the Government would arm ten thousand mounted infantry with these guns, and put them under a good enterprising officer, they could destroy all principal railroad lines in the South, and do more damage to the rebellion in three months than fifty thousand troops ordinarily armed could in a year.[35]

By war's end many Union troopers had received the highly coveted Spencer, and after the Sharps and Burnside it had become the most widely issued of the wartime carbines. The promise of arming them with Spencers was sometimes offered as an inducement to soldiers to reenlist in sufficient numbers to earn their unit's designation as a Veteran Volunteer regiment. Unfortunately production couldn't keep up with demand and these promises sometimes weren't kept or required many months to fulfill.

The Spencer carbine did lack the longer range accuracy of the rifle, due largely to the shorter distance between front and rear sights which magnified any sighting error. It was rather heavy but it was reliable and fast firing. Its rate of fire was increased but not until nearly war's end with the invention by Colonel Erastus Blakeslee, 1st Connecticut Cavalry, of his "quickloader," a leather cartridge box designed for use with the Spencer. A wooden block inside was drilled to hold six or ten tubes of seven cartridges each. To reload, a mounted trooper didn't have to handle seven individual rounds while holding reins, carbine, and magazine follower but merely emptied a tube into the magazine through the butt plate. An unknown number were issued for cavalry use, at least for trial purposes. One accidental discharge of a cartridge in a partially filled cartridge box was reported. The cause probably was a defective cartridge in which not all of the priming compound had been "spun" outward into the rim leaving some in the center of the base. When the round bounced against the bullet behind it, the concussion probably detonated it. A thirteen-tube box was intended for infantry use, although it's possible none were issued by war's end (Figure 154).[36]

Beside the faster rate of fire the Spencer and Henry offered, even taking into consideration the time required to reload the magazine, another major advantage was the durability of the copper cartridges which they fired. Not only were they waterproof, but they withstood rough handling. During the war August V. Kautz rose from lieutenant to brigadier general and noted: "Sharps carbine is a favorite arm, but the ammunition in a few days' marching deteriorates so much as to be a serious objection, as ammunition trains can seldom be taken on cavalry expeditions, and therefore only a limited supply can be carried by the men. The same objection exists against all paper cartridges."[37]

Not long after the 2nd Iowa Cavalry received their Spencer carbines in the summer of 1864, they engaged the enemy in a torrential downpour at Oxford, Mississippi. The regiment's historian later noted that their "Spencer carbines were impervious to rain. We now had it all our own way, for the rain had been as injurious to the rifles of the enemy . . . while our pieces emitted their deadly stream with as much certainty as if the day had been one of cloudless beauty."[38]

One unusual Union cavalry weapon worthy of mention wasn't a firearm but rather a nine-foot lance with a foot-long three-sided blade. Only one regiment carried it, the volunteer 6th Pennsylvania Cavalry also known as Rush's Lancers. A dozen car-

35. Ibid., 145.

36. Ibid., 212. Although issues of the Blakeslee box during the war were limited, some were in use by bandsmen of the 18th Infantry in Dakota Territory in 1866 and others with the 10th Cavalry as late as 1871. McAulay, *U.S. Military Carbines*, 157.

37. Bilby, *A Revolution in Arms*, 164.

38. McAulay, *U.S. Military Carbines*, 99.

154. Blakeslee "quickloader" Spencer cartridge box with one of seven tubes partially removed. (*Courtesy: National Park Service, Fuller Gun Collection, Chickamauga and Chattanooga National Military Park*)

bines were issued to each company in 1862 as supplementary weapons for use on picket duty. Colorful as the regiment might have been, this pole arm was ill suited and proved ineffectual in the largely wooded eastern theater.

On both sides, mounted troops sometimes were employed in the role of mounted infantry rather than in the better known cavalry tasks of scouting and raiding. Instead they used their mounts to achieve mobility but once in position relied on rifles rather than carbines and fought on foot as infantry. One well known Union example of mounted infantry was Wilder's "Lightning Brigade." Its success in the western theater of Tennessee and Georgia beginning in the summer of 1863 was in large measure achieved by its mobility and its fast-firing Spencer rifles.

The 1866 Spencer Repeating Rifle Company catalog included an 1863 letter to the firm from Wilder in which he confirmed the advantages of the Spencer's copper cartridges over those of fragile material such as paper or linen. "I believe that the ammunition used is the cheapest kind for the service, as it does not wear out in the cartridge boxes and has the quality of being water-proof—the men of my command carry 100 rounds. . . in their saddlebags, and in two instances went into a fight immediately after swimming their horses across streams twelve feet deep and it is very rare that a single cartridge fails to fire.

Artillery Small Arms

Federal artillery, rooted in the north's industrial supremacy, was a great asset to the Union army. Basically artillery regiments were divided into two types of units—heavy, which were equipped with siege guns too weighty to take into the field and so occupied fixed fortified positions, and light. In the latter regiments, artillerymen could travel on foot or riding depending upon whether they were supporting infantry or faster-moving cavalry units. Arms for light artillery batteries usually consisted of revolvers and either cavalry or artillery sabers and occasionally carbines. Revolvers were frequently issued only to drivers and non-commissioned officers and were used to destroy wounded battery horses as often as to defend the cannon positions against an enemy. Heavy artillery regiments, unable to withdraw their guns if the enemy threatened to overrun their position, often drilled as infantry and were issued infantry weapons.

A history of Battery M of the 1st Illinois Light Artillery contained the following:

> As [light] artillery men are supposed to be armed with sabers, and revolvers, on the first of September [1862] saber bayonets ("cheese knives" as we styled them) were issued to the privates, each of the "noms" [non-commissioned officers] receiving also a French [LeFaucheux?] revolver. On the 11th [January 1863] our cheese knives were turned over to our Uncle Samuel, and we doubt whether Bunyans pilgrim felt more relief at the falling off of his load of sin than did we when no longer burdened with those detested sword bayonets.[39]

39. Baumann, *Arming the Suckers 1861-1865*, 22.

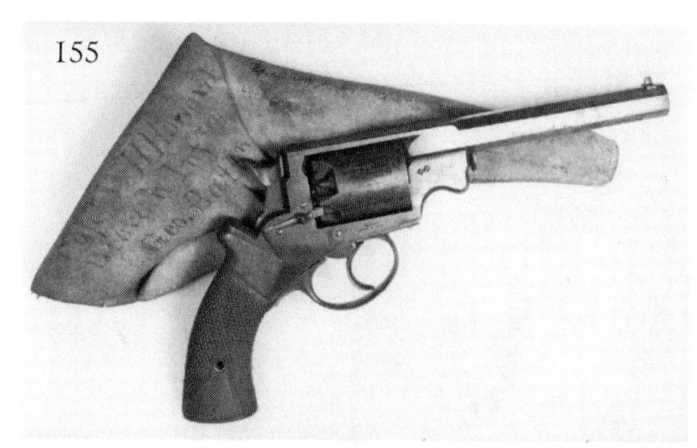

155. A Confederate officer's Massachusetts Arms .36 double-action revolver. Its original white buff leather holster is marked on the flap "Cap't. W. H. Howard/Etowah Inf'ty/Geo. Battn." This probably was the Etowah Invincibles, a company of Georgia infantry. (*Courtesy: Norm Flayderman*) **156.** "The accuracy and GREAT FORCE with which it shoots" was a grossly exaggerated advertising claim for this youth's Allen & Wheelock seven-shot .22, but its rimfire ammunition was waterproof and not easily damaged. (*Courtesy: Paul Henry*) **157.** Union soldier with a studio prop or his privately purchased Allen & Wheelock side-hammer percussion revolver. Percussion caps for his musket are contained in the cap box at the front of his waist belt, cartridges for his musket in the larger box at this side. (*Courtesy: Paul Henry*)

Gun collectors today sometimes encounter what appears to be a standard .58 Springfield or contract rifle musket but with a barrel length shorter than the usual 40 inches and with only two rather than three bands securing the barrel to the stock. These sometimes are described as an "artillery model." However it appears that no such weapon was officially adopted during the war and that most such guns are pieces which have been shortened by dealers who purchased surplus rifle muskets after war's end and cut them down for sale to military schools or to reunion groups for parades or some other unofficial use.

Private Handgun Purchases

Officers on both sides generally purchased their side arms—swords and revolvers—although their primary responsibility was directing and controlling their men rather than engaging in combat. Existing southern officers' handguns with documented history of wartime use include single-shot pistols, foreign imports, and southern-made revolvers in addition to numerous Colts. Some Union officers opted for the convenience of the Smith & Wesson .32 No. 2 revolver and its metallic cartridges, but the most popular seem to have been one of two models of Colts, the .36 1851 Navy or smaller frame .31 1849 Pocket. Some specimens found today bear the owner's name and perhaps regiment engraved on the backstrap or a presentation inscription indicating it was a gift from a family member, friend, or even a token of respect from the men he commanded. The Confederacy's Major General John B. Magruder, sometimes known as "Prince John" because of his flamboyant style, received a cased and handsomely engraved five-shot .44 English Beaumont-Adams revolver from a friend, an artifact that is preserved today (Figures 155, 156, 157).

Enlisted men on either side might add a handgun to their personal effects, as did some men of the 73rd Illinois Infantry after they received their enlistment bounty payment. "A good many of the boys in our company has got revolvers, they cost 18 and 20 dollars for good ones." But after a few long marches with their load of a ten-pound musket, bayonet, ammunition, blanket, canteen, and other issue equipment, a handgun became an encumbrance and often was disposed of rather quickly. Private Bill Nicholson of the 37th Mississippi Infantry was an exception and carried his revolver for two years before reportedly using it to dispatch five of his pursuers at the battle of Peachtree Creek in Georgia. Regimental orders sometimes prohibited enlisted men from possessing a handgun, as happened in 1862 in the 15th New Jersey Infantry after an accidentally fired pistol ball entered a major's tent and struck his boot! (Figures 158, 159).[40]

158. A private with the 2nd Rhode Island Infantry with what may be a personally owned James Warner pocket revolver in his belt. His regulation rifle musket is a Model 1855. (*Courtesy: National Archives, no. 111-B5325*) **159.** Rare .22 derringer by J. C. Terry of Springfield, Massachusetts. It's inscribed "John A. Lewis/Co. G. 34th Reg. Mass. Vs." Lewis was an eighteen-year-old farmer who enlisted as a private in July 1862 but, like many Civil War casualties, died of disease rather than a combat wound. (*Courtesy: Norm Flayderman*)

The Union Navy and Marines

The U.S. Navy and Marine Corps served during the war not just at sea but also on the Mississippi, Potomac, James, and other navigable rivers. Their personnel sometimes found themselves operating in small boats or on shore conducting reconnaissance or carrying out raids or major assaults on land, as in the hard fought capture of Fort Fisher,

40. Ibid., 156. Bilby, *Civil War Firearms,* 160, 162.

160. U.S. Marines photographed at the Washington Navy Yard in 1864, still armed with smoothbore Model 1842 .69 muskets with bayonets attached. (*Courtesy: Library of Congress*)

North Carolina. In a reverse exchange, during Grant's 1863 campaign to capture the river stronghold at Vicksburg, Enfield-armed infantrymen from several Illinois regiments were added to the Mississippi River squadron as sharpshooters to supplement the undermanned vessels' crews. By 1850 the navy had procured breech loading long guns—Hall rifles and carbines plus substantial numbers of percussion "mule ear" or side-hammer Jenks carbines and rifles. In the early 1840s it had toyed briefly with Colt Paterson revolving carbines, however the smoothbore musket became its primary shoulder arm.

In the years of peace before the Civil War, the navy relied rather heavily on the army for its small arms due to the modest number required. When in 1855 the army had adopted .58 caliber for its long arms, the navy continued its reliance on the Model 1842 musket, although many by 1861 would have been converted to rifled muskets with the addition of rifling and some an adjustable rear sight. On board a ship the sailor rarely carried ammunition on his person, so the greater weight of the .69 rounds over those of .58 caliber was not a problem and there was little need for the advantage of the rifle musket's long-range accuracy.

On the other hand, enlisted marines by 1861 were being issued regulation army long guns, the Model 1855 .58 rifle musket followed by its later versions. Deliveries of these .58 rifle muskets to the navy began late in 1861, but shortages forced continued reliance to some extent by both the navy and marines on .69 muskets and rifled muskets throughout the war. In the tradition of the corps' service on land and sea, marines occasionally found themselves participating in major land engagements beginning with the first battle of Bull Run. There and apparently after, marine Lieutenant Robert Huntington carried a Model 1855 Springfield pistol-carbine, a rather unusual choice for an officer's

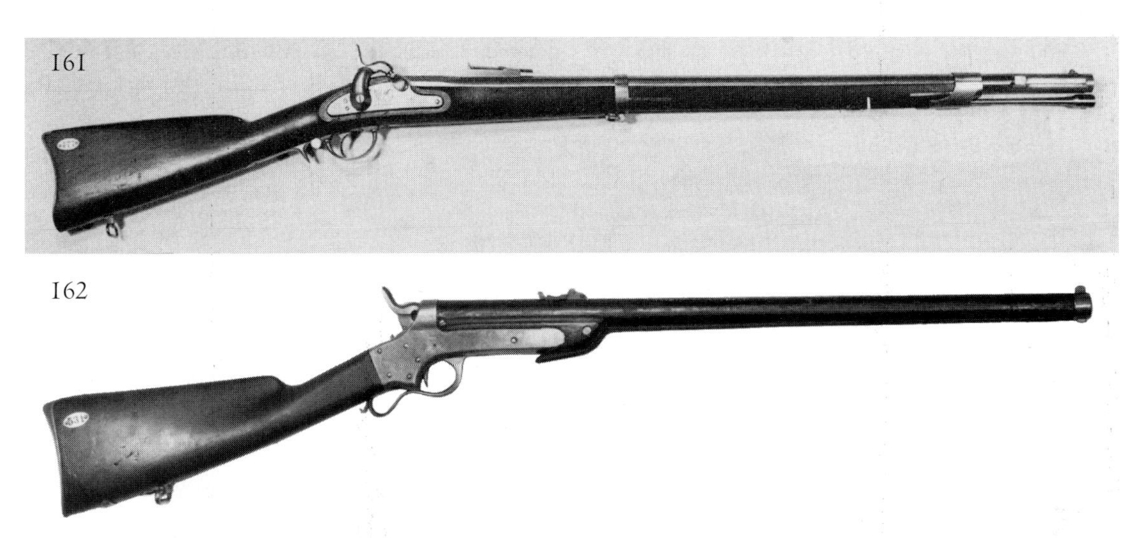

161. Whitney "Plymouth" navy rifle, seen with the finger spur behind the trigger guard, a feature found on most specimens. **162.** Navy issue Sharps & Hankins carbine with leather-covered barrel. (*Courtesy: National Park Service, Fuller Gun Collection, Chickamauga and Chattanooga National Military Park*)

sidearm, but an artifact that later became part of the Marine Corps' historical collection and exists today (Figure 160).

One gun used exclusively by the Federal navy was the two-band Plymouth rifle, the only rifled arm produced for the government in .69 caliber that wasn't a converted smoothbore. Captain John A. Dahlgren, a prominent navy ordnance officer, had designed it in 1856 and arranged for the production of a trial quantity, but had not been able to secure its adoption before the war. Its name was derived from the sloop of war USS *Plymouth,* a vessel commanded by Dahlgren and used primarily to test new naval ordnance. Ten thousand of these .69 caliber rifles eventually were delivered in 1863-64 from Eli Whitney, Jr.'s factory in Connecticut. It accepted either of two bayonets, a heavy knife-type weapon with a twelve-inch blade also designed by Dahlgren, or a sword bayonet with a twenty-two-inch blade (Figure 161).

The Plymouth was a muzzleloader, but navy purchases did include rimfire breech-loading rifles by Sharps & Hankins with bayonets in 1862 to arm marines on naval vessels at sea and with the Mississippi River squadron. Navy carbines by the same maker secured in substantial numbers featured an unusual 1/16th inch thick leather covering over the barrel to resist saltwater rusting. Other Civil War naval breechloaders included Sharps rifles and carbines. One of the best-known U.S. vessels of the war was the sloop *Kearsarge,* which in a famous battle in June 1864 off Cherbourg, France, sank the dreaded Confederate raider *Alabama.* When the former was outfitted for duty in the Atlantic two years earlier, she carried boarding pikes, Ames cutlasses, Colt 1861 revolvers, and Model 1859 Sharps rifles. Beginning in February 1863 Spencer repeating rifles joined the navy's arsenal, but not all looked on the Spencer favorably. Complaints that a sharp jolt as when jumping into a small boat could cause the gun to discharge accidentally caused several vessel commanders to exchange Spencers for Plymouth rifles (Figure 162).[41]

The navy's enlisted men's handguns included obsolete single-shot .54 Model 1843 pistols by Deringer and Ames and some of the larger 1842

41. McAulay, *Civil War Small Arms of the U.S. Navy and Marine Corps,* 135, 138-39.

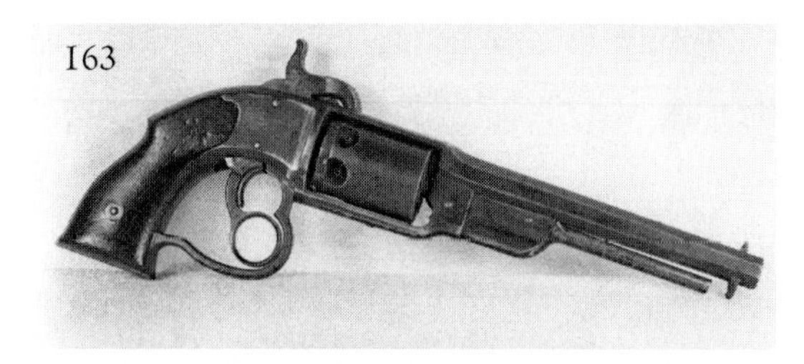

163. One of the more unusual Civil War revolvers in terms of design was the .36 Savage-North. The federal navy procured 1,126, the army 11,384. The ring trigger cocks the hammer and rotates the cylinder and the upper conventional trigger fires it. (The tip of the loading lever is broken off in this specimen.) **164.** Handsome cased English single-action "long spur" .44 five-shot Webley closely associated with two U.S. naval officers, each of whom eventually became an admiral. On the side of the barrel inscribed in gold is "Captn D. D. Porter, from Captn D. G. Farragut – 1862" and on the other side "Evans & Hall, 418 Arch Street, Philadelphia," well-known military goods dealers. The cylinder is engraved with figures of Union army and navy officers in combat poses and embellished in gold with the figure of a Union army officer with sword and revolver in hand standing over a fallen soldier. Each chamber is numbered in gold. (*Courtesy: Norm Flayderman*)

pistols by Aston and Johnson secured from the army. Such pistols were on hand to repel boarders and usually were kept loaded. Periodically they would be fired at target practice and then cleaned and reloaded. Revolvers included Colt .36 Model 1851s and 1861s and the .44 1860s. Five days after Lincoln called for 75,000 volunteers in April 1861, the Union Defense Committee of New York City donated fifty .31 caliber Colt 1849 pocket revolvers for issue to ships' captains. The UDC the following month purchased and donated one hundred iron mounted Model 1847 musketoons (either the artillery or the sappers and miners model). The initial deliveries of .36 Remingtons and .36 Savage-North revolvers began in the summer of 1861. The latter was unusual with its two-trigger design. A pull on the ring-like trigger moved the cylinder back away from the barrel, rotated it, and cocked the hammer. Releasing it allowed the cylinder to move forward to form a gas seal with the barrel, and a squeeze on the conventional trigger fired the weapon. Remingtons along with Whitneys and 1851 and 1861 Colts, all in .36 caliber, continued as the navy's principal wartime revolvers. However, the navy's last order for Colts from the factory was in 1862 and after that, purchases were of Remington and Whitney .36s (Figure 163).[42]

Some idea of the performance of .36 caliber Civil War era revolvers can be drawn from an 1862 navy test of a Remington. Five hundred shots were fired at 100 feet at a target consisting of inch and a half pine boards 30 inches square. All 195 shots which struck the target passed through the first board and penetrated the second, but only four penetrated the third. For comparison, thirty shots were fired from a .36 Colt and six of the ten which hit the target penetrated the third board. After the 150th shot it was necessary to clean the Remington, for the cylinder revolved with some difficulty as powder fouling built up at the junction of cylinder and barrel, and at the 582nd shot it wouldn't revolve at all until it was cleaned. Removal of the cylinder for cleaning was simple, so long as that operation wasn't hindered by fouling around the cylinder pin.[43]

42. Ibid., 95. McAulay, *U.S. Military Carbines,* 120. Ware, *Remington Army and Navy Revolvers, 1861-1888,* 416.

43. Ware, *Remington Army and Navy Revolvers, 1861-1888,* 223-24.

Today's collectors of Civil War naval arms have difficulty in identifying those received prior to August 29, 1864, for only a minority were stamped with an inspector's initials. But on that date a Navy Bureau of Ordnance circular directed that muskets, carbines, and pistols were to be stamped on the top of the barrel near the breech with an anchor and on the lock plate a "P" over the inspector's initials. Revolvers would bear the anchor on the top of the barrel near the cylinder while the face of the cylinder would be stamped "P" with the inspector's initials. The circular also stated that each inspector would be issued two initial stamps, the larger (.15-inch) for use on muskets, carbines, and cutlasses and a smaller one (.1-inch) for pistols and revolvers.[44]

Naval officers' personally owned sidearms could be almost any handguns of the period. A prized example is a cased English Webley percussion revolver inscribed in gold on the side of the barrel "Captn D. D. Porter, from Captn D. G. Farragut-1862." A convenient pocket revolver of the Civil War era was a .32 teat fire Moore, highly engraved and inscribed "A.H. Gilman, U.S.N." He joined as a lieutenant and rose to captain. Another Civil War naval officer, paymaster Eugene Littell, carried a privately owned .36 Manhattan revolver with his name engraved on the brass backstrap encircling the rear of the grip or handle—"Eugene Littell USN"—a convenient place for an inscription on almost any handgun (Figure 164).[45]

Firearms of the Confederacy

Gray-clad soldiers depended on various sources for their arms. Even though southern states began seceding in 1860, it wasn't until April 1861 that the U.S. placed an embargo on arms sales to the south. In one instance, southern purchasing agents were able to acquire more than 24,000 surplus U.S. .69 muskets converted from flintlock and priced at $2.50 each. In addition, there were substantial numbers of arms in federal and state arsenals in the south at the outbreak of war. Other wartime sources included weapons produced in the south, those captured from the enemy, imports from Europe (close to 300,000 long guns), and civilian guns pressed into military service.

Samuel Colt sometimes has been accused of disloyalty to the Union cause by continued sales to southern states up to the declaration of war in April 1861. However, in partial defense of Colt it should be noted that he was a staunch abolitionist and these state purchases were legal and authorized under the Militia Act of 1808. He did begin to delay such shipments south in 1861, and in April of that year organized and equipped a volunteer infantry regiment which in June was disbanded and became absorbed by the 1st Regiment Connecticut Rifles. E. Remington & Sons was another firm which sold guns south in the hectic months before war's outbreak, 1,000 of their .44 Beals model revolvers to the state of South Carolina (Figures 165, 166).[46]

Blockade runners began delivering their cargoes of arms from abroad in the fall of 1861. Such shipments often came by way of the Bahamas or Cuba, where the cargoes were off-loaded and transferred to fast blockade-running ships. The second ship to slip through from England was the *Fingal* and as cargo carried 11,340 Enfield rifles, 60 pistols, 12 tons of gunpowder, 550,000 percussion caps, 409,000 cartridges, 500 sabers, and other ordnance supplies. But once such ships reached southern ports, sometimes there would be bitter competition among Confederate and state forces for those weapons, particularly the highly coveted Enfields. The Union blockade became more effec-

44. Ibid., 128-29.

45. David D. Porter and David G. Farragut were foster brothers. Each had a highly successful Civil War career and achieved the rank of admiral.

46. Houze, *Samuel Colt: Arms, Art and Invention*, 81-82. John D. McAulay, *Civil War Pistols* (Lincoln, R.I., 1992), 67.

166

MEN WANTED FOR CAVALRY.

The undersigned having been appointed to raise a Company of Cavalry for the war, calls upon the young men to come forward and volunteer for that purpose.

$100 BOUNTY

will be paid to each man as soon as mustered into service; plenty to eat and good clothes will be furnished by the Government.

Men who come and bring their horses will be paid 24 dollars per month. Those who have good shot guns can get a good price for them.

JOHN A. WREN.

Volunteers will for the present address me at Greensboro, N. C.

165. Confederate trooper with a Starr double-action (right) and probably a Colt. (*Courtesy: Herb Peck, Jr.*) **166.** Second Model Colt Dragoon probably carried by Rutherd D. Beck, part Cherokee Indian and sergeant and later brevet 2nd lieutenant in the Confederacy's Indian Brigade. Composed of Cherokee Indians, portions of the brigade fought in numerous engagements in Arkansas and Missouri as well as Indian Territory. At war's end the brigade was commanded by Stand Watie, the only Indian to achieve general officer rank in the Confederate Army. The backstrap and butt are inscribed "May Victory Crown Our Banner" and "R. D. Beck." Surprisingly the gun was found in 1959 still loaded on the surface of the ground near Stillwater, Oklahoma, as land was being cleared for an oil well site. (*Courtesy: Bob Everhart*) **167.** Note the reference to "good shot guns" in this Confederate recruiting poster.

tive as more vessels joined the patrolling squadrons, but by early 1863 the south was still getting substantial numbers of Enfields, many of which were distributed to Confederate troops in the east. At Gettysburg in July of that year more than three-fourths of 3,000 shoulder arms captured by one federal corps were Enfield rifle muskets. Despite the improving quality of firearms in the hands of Confederate fighting men, superior Union resources eventually would prevail, but only after four years of bloody struggle.[47]

One Georgia captain in July 1861 wrote that his command was armed chiefly with "country rifles & bowie knives" and while recruiting men he was buying privately owned arms at from $7 to $15 each. A lieutenant at the Richmond Arsenal reported to his commander in February 1862 that he was paying $1 (perhaps a misprint) for muskets, many of which were trophies from the first battle at Manassas (Bull Run), Virginia, the preceding July. The use of pri-

47. McAulay, *Civil War Small Arms of the U.S. Navy and Marine Corps,* 60.

vately owned arms was still so common in April 1862 that the Confederate congress passed a measure under which any recruit who brought a musket, rifle, shotgun, or carbine could be paid for it, or if he didn't want to sell it he'd be paid $1 a month for its use. Colonel Josiah Gorgas, the able Confederate ordnance chief, gave an idea of the ratio of arms obtained from different sources when for the year ending September 30, 1864, he reported 30,000 small arms imported, 20,000 manufactured, and 45,000 captured (Figure 167).[48]

The 12th Virginia Cavalry was recruited in the Shenandoah Valley of Virginia, the region in which it operated. Ordnance returns listing ammunition issues for the regiment reflect the array of firearms in use in the spring of 1863—.577 Enfield rifles, .36 and .44 Colt revolvers, Colt revolving rifles, and breech-loading carbines by Sharps, Robinson (an inferior Richmond-made Sharps copy), Merrill, and Smith. Later they added English Kerr revolvers to the mix. Such an assortment was common within gray-clad cavalry regiments.[49]

Revolvers imported primarily from Britain and France constituted a greater number of handguns for the Confederate army and navy than came from manufacturing facilities in the south. In addition to those procured by government contracts, some were privately purchased. Except for those foreign handguns which were brought into this country before the conflict began, once Union ships started patrolling the coast, all had to slip through on steamers running the gradually tightening Union naval blockade.

A Confederate entry in the category of Civil War oddities was the "grapeshot" revolver designed and first patented in the United States by a French doctor living in New Orleans, Jean Alexandre François LeMat. Initially he was aided in his efforts by the U.S. army's Major Pierre T. G. Beauregard, who later became a prominent Confederate general. Efforts to interest an American firm in manufacturing these revolvers failed so production took place in Belgium, France, and eventually England. The LeMat featured a nine-shot percussion cylinder which revolved around a five-inch smoothbore 18-gauge (.63 caliber) barrel which acted as a cylinder pin. The nose of the hammer was adjustable to fire either .42 caliber bullets from the cylinder or the single shot cartridge. It's thought that about 1,500 Lemats made it safely through the Union naval blockade and saw service with the Confederate navy and cavalry. Some LeMats also were personal sidearms of such gray-clad generals as J. E. B. Stuart, P. G. T. Beauregard, and Thomas J. "Stonewall" Jackson. Such a weapon certainly would be an appropriate although uncommon sidearm for a southern officer in a novel of the "War Between the States" or whatever one chooses to call that tragedy (Figure 168).[50]

The English five-shot Kerr manufactured by the London Armoury Company, was available in both .36 and .44 caliber but the latter was more com-

48. Sword, *Firepower from Abroad: The Confederate Enfield and the LeMat Revolver,* 109. Fuller and Steuart, *Firearms of the Confederacy,* 210. The arms shortage in the south early in the war was such that in March 1862 General "Stonewall" Jackson asked that he be provided with 1,000 pikes, a request endorsed by General R. E. Lee. Thousands of pikes were manufactured at various southern state arsenals as last-ditch weapons, but their actual use in combat hasn't been documented. Most were about eight feet in length, including a blade about a foot or so long mounted on an ash or hickory shaft. Colonel Gorgas suffered from the effects of yellow fever contracted during the Mexican War. His son, Major General William C. Gorgas, later achieved fame for his efforts in controlling mosquito populations and halting the spread of yellow fever, facilitating the construction of the Panama Canal.

49. Earl J. Coates, Michael J. McAfee, and Don Troiani, *Don Troiani's Civil War Cavalry & Artillery* (Mechanicsburg, Pa., 2002), 27.

50. General J. E. B. Stuart, who commanded Lee's cavalry in the Army of Northern Virginia, also carried an imported English Terry bolt-action percussion breech-loading carbine, which is preserved in the Museum of the Confederacy in Richmond, Virginia.

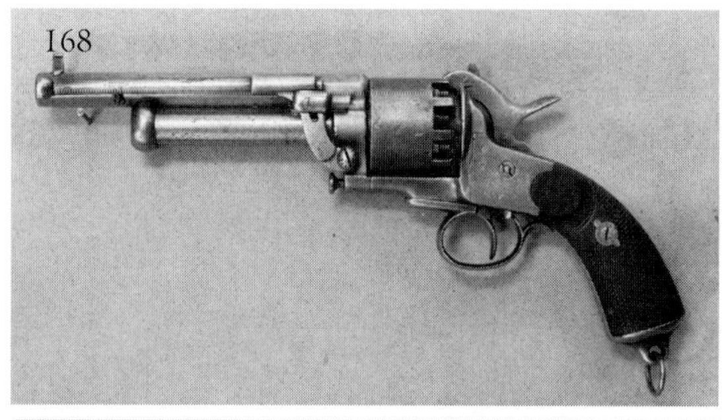

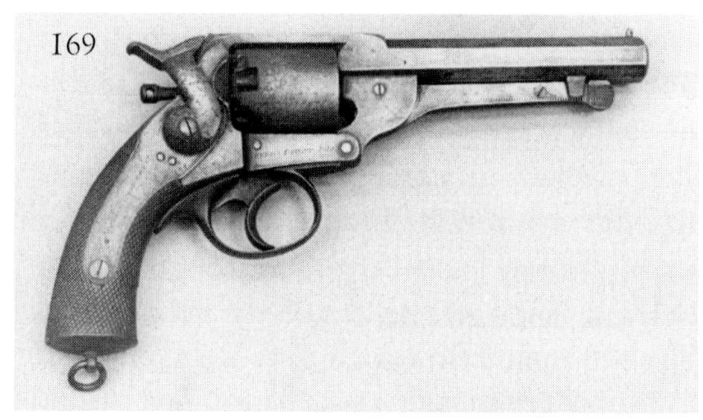

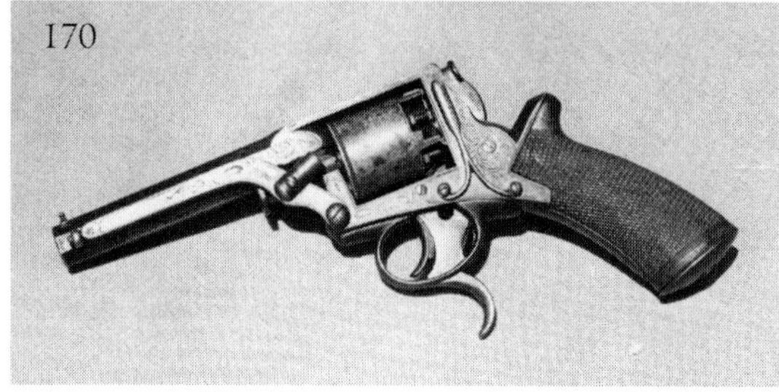

168. LeMat "grapeshot" revolver. A smoothbore barrel serves a dual purpose as the axis pin about which the cylinder rotates. **169.** English five-shot .44 Kerr imported in substantial numbers by the Confederate government. **170.** A "self-cocking" English Tranter imported by A. B. Griswold & Co. of New Orleans and so marked on the top strap over the cylinder.

mon. The Confederate government purchased around 8,000, all for the army except about 1,000 issued to the navy. English double-action five-shot handguns designed by Robert Adams, most common in .44 and .50 caliber, saw some Confederate use and a few were found in several Union cavalry regiments as well including the 8th Pennsylvania and 2nd Michigan. The 5th Virginia Cavalry was equipped with a mixture of Adams and Colt revolvers and single-shot flintlock and percussion pistols. Some English Webley revolvers and others by William Tranter were personal choices for sidearms. Two importers of Tranters, both operating in New Orleans, were Hyde & Goodrich and A. B. Griswold & Company. These firms sometimes marked the top strap over the cylinder of their imported Tranters or other revolvers with their firm name. In March 1863 H. E. Nichols of Columbia, South Carolina, proudly advertised in a Richmond, Virginia, newspaper that he had just received six Tranters which he was offering at $220 each, a price which perhaps was indicative of the value of Confederate currency at the time (Figures 169, 170).

From France came the war's most widely used foreign revolver, the pinfire 12mm (.44 caliber) Model 1853 LeFaucheux, although it was officially purchased only by the federal government (almost 13,000). Nevertheless some served Confederates as individually procured arms, including one handsomely engraved specimen presented to General "Stonewall" Jackson. The use of an internally primed metallic cartridge was an advanced design, but it wasn't popular with some who questioned the safety of the pinfire cartridge in the event a loaded revolver or its cartridges were dropped. Also perhaps imported into the south and obtained individually were the Perrin and Raphael revolvers, of which the federal government purchased fewer than a thousand of each along with the metallic cartridges which these French revolvers required.

Handgun manufacturers within the Confederate states during the war generally were small firms with a limited output, sometimes only a few hundred pieces. Their efforts were hampered by a lack of raw material, production equipment, and skilled workers as well as the later threat posed by invading northern troops. The quality of southern-made handguns without exception was inferior to that of those such as Colts made in the north. Sam Colt's Model 1851 .36 Navy was the pattern

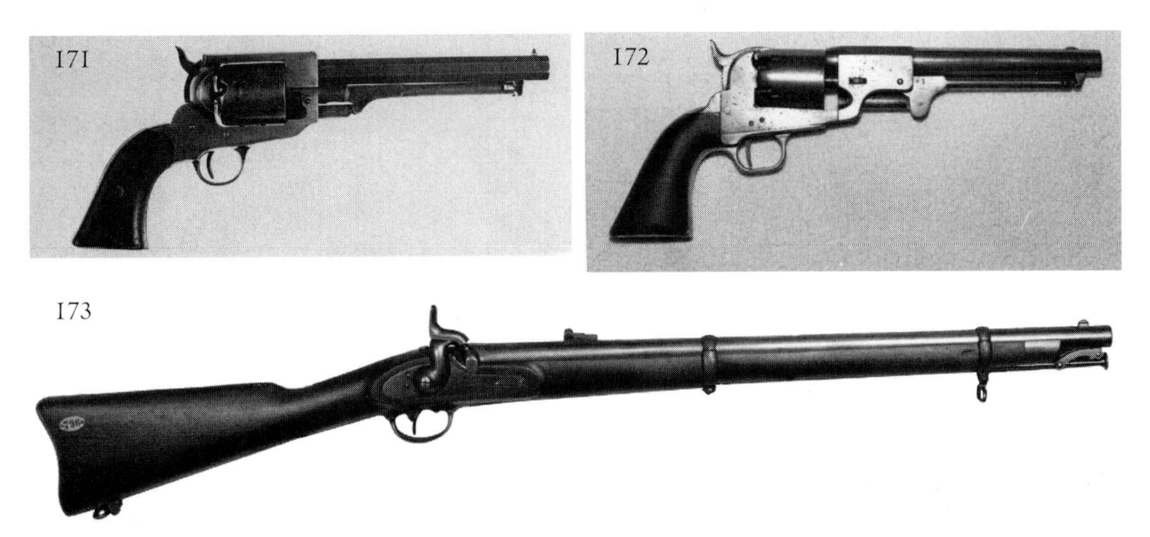

171. Confederate .36 Spiller and Burr, a close copy of the far more common .36 revolver made by E. Whitney of New Haven, Connecticut, although the former has a brass rather than iron frame. **172.** Dance & Brothers, Texas gun makers for the Confederacy, probably didn't produce more than 400 .36 and .44 revolvers including this .44. A characteristic feature of most Dances is a frame lacking a recoil shield. (*Courtesy: James Wertenberger*) **173.** Cavalry carbine made at the Confederate armory in Tallassee, Alabama, near the end of the war. It's patterned closely after Britain's two-band Enfield rifle and may have been made too late to see service during the war. (*Courtesy: National Park Service, Fuller Gun Collection, Chickamauga and Chattanooga National Military Park*)

for many southern revolvers including those by Leech & Rigdon; Augusta (Georgia) Machine Works; Columbus (Georgia) Fire Arms Mfg. Co.; Rigdon, Ansley & Co. of Augusta, Georgia; and Griswold and Gunnison of Griswoldville, Georgia. Revolvers by this last maker were the most numerous (close to 4,000) and probably the best in terms of quality. With a workforce of about two dozen workers, nearly all of whom were slaves, the factory turned out about five revolvers a day. Another Georgia maker was Spiller & Burr, which patterned its brass solid frame .36s after the Whitney revolver. In Texas where J. H. Dance & Brothers of Columbia made a few copies of the Colt Model 1851 Navy and .44 Dragoon, the need for arms was so great that the state's governor exempted Dance factory workers from military service (Figures 171, 172).[51]

Makers of shoulder arms for the southern states' militia or under Confederate contract faced the same problems as did handgun makers, not the least of which was a shortage of workers which persisted throughout the war. Diversity was a key word describing the Confederacy's supply of shoulder arms, and in the face of early severe shortages, virtually any usable arm was pressed into service whether it was a muzzle-loading "squirrel rifle," shotgun, or obsolete flintlock musket. The Asheville Armory in North Carolina received its first contract "for repairing and fitting up double barrel shotguns for service." M. A. Baker of Fayetteville, North Carolina, in 1861 is known to have had a contract with the state to cut down double-barrel shotguns and fit them with sling swivels for cavalry use. Later he altered flintlock U.S. muskets and rifles to percussion (Figure 173).

Gradually southern firms geared up for production and substantially more shoulder arms than handguns were manufactured in the south. The

51. Bilby, *A Revolution in Arms*, 193.

174. Early .35 caliber Maynard carbine. Lowering the trigger guard tips the barrel downward to load. The wooden tompion protruding from the muzzle was a common military accessory of the period intended to keep dirt and other debris out of the barrel of a musket or rifle when not in use. (*Courtesy: National Park Service, Fuller Gun Collection, Chickamauga and Chattanooga National Military Park*)

Fayetteville Armory in North Carolina used machinery and parts captured at the Harpers Ferry Armory to produce an estimated 8,000 to 9,000 very serviceable .58 rifles. From machinery relocated from Harpers Ferry to the Richmond, Virginia, armory came a substantial number of decent quality .58 muzzle-loading carbines, musketoons, and rifle muskets patterned after the U.S. Model 1855, although they lacked the provision for the Maynard tape primer.[52]

Georgia arms makers had been in the forefront of Confederate handgun production and the state offered its share of long arms as well. Cook & Brother of Athens made .58 carbines, musketoons, and rifles, while Dickson, Nelson & Co. despite having to move their operation several times provided carbines and rifles, as did J. P. Murray of Columbus. John C. Peck of Atlanta made rifles for the Confederacy but also offered a rampart gun of .93 caliber, one of which is exhibited at the Springfield Armory National Historic Site. Unnamed gunsmiths in Georgia and elsewhere did their share for the war effort by converting civilian muzzle-loading sporting rifles into serviceable military arms by boring them out to .58 caliber and rerifling them. In Danville, Virginia, N. T. Read and John T. Watson in an unusual reversal converted a number of breech-loading Hall rifles and carbines in that state's militia inventories into muzzleloaders. Samuel Sutherland, a leading gun dealer in Richmond, was among those who altered civilian guns to military use and repaired others. An existing example of Sutherland's work is a flintlock hunting rifle of about .50 caliber with the barrel cut down to only 17 inches for cavalry use with a stamping on the barrel "C.S.A. S.S. 1861."[53]

Several models of southern-made carbines are worthy of mention. Before the war, George W. Morse of Greenville, South Carolina, had invented a single-shot breechloader firing a .50 caliber forerunner of a centerfire metallic cartridge. Perhaps a thousand of these went to South Carolina militiamen and other southerners. In the Confederate capital in Richmond, the S. C. Robinson factory

52. One example of the material disadvantages often facing Confederate arms makers was the shortage of steel, which prompted the manufacture of socket bayonets made of iron with only the tip being steel. A shortage of leather also forced the manufacture of some musket and rifle slings of cotton cloth folded over several times. A Confederate general in 1863 wrote from Arkansas that it had been necessary to use public documents from the state library for cartridge paper. Fuller and Steuart, *Firearms of the Confederacy*, 1.

53. A Cook rifle illustrated the use to which many former military arms were put after the war, a gun for hunting small game. When it was disassembled in 1934, it was found to be loaded with a light charge of homemade bird shot, each having the tell-tale oblong shape with a small teat indicating they had been dropped only a short distance into water. The wadding appeared to have been newspaper from the 1870s. Ibid., 149, 208.

fabricated around 5,000 copies of the famed Sharps carbine, although their quality did not approach that of the genuine article—they were accused of "spitting fire" at the breech and sometimes bursting.

One of the most efficient of carbines was the Massachusetts-made Maynard, available in both .35 and .50 caliber. Some were seized from federal arsenals in the south and others had been purchased by some southern militia companies between Lincoln's election in 1860 and the outbreak of fighting in April 1861. The Confederate ordnance manual of 1862 included it in the list of carbines in their service along with the Hall, Burnside, Sharps, Merrill, and the Colt revolving carbine. Any arm seized from Union forces could be pressed into Confederate service. While the captured examples of the highly regarded Spencer and Henry repeaters were prizes, the south had no facilities for the mass production of the metallic cartridges they required, so their usefulness depended upon the capture of ammunition. Eventually by general orders southern soldiers were prohibited from exchanging their weapons for those requiring special cartridges. However during the siege of Petersburg, Confederate sharpshooter Barry Benson acquired a captured Spencer rifle but soon used all of the forty rounds of ammunition he had for it. He discarded it in exchange for an Enfield muzzleloader (Figure 174).[54]

175. George Maddox was a scout with Quantrill's Confederate guerrillas and participated in the infamous raid on Lawrence, Kansas. He's holding a pair of Remingtons in this view. (*Courtesy: Dr. and Mrs. Thomas Sweeney*)

To take advantage of their long-range effectiveness, some southern cavalry commanders partially equipped their men with rifles, particularly the Mississippi or a short-barrel version of the imported Enfield rifle musket. However, double-barrel shotguns found favor among some southern cavalrymen for close fighting. Then-colonel Nathan B. Forrest, former slave trader, farmer, and private, in early 1862 called the double-barrel shotgun the best gun with which cavalry could be armed. Some fast-hitting cavalry leaders like the Confederate John Mosby and the notorious southern guerrilla Charles Quantrill preferred that their men be outfitted with two, three, or more revolvers each rather than shoulder arms. Hard-drinking Major John Edwards fought with General Joe Shelby's gray cavalry in Missouri during the war and later became a propagandist for Jesse James. In 1877 he wrote a

54. Ibid., 194. The breech-loading Maynard had been available commercially in 1858. English author and world traveler Richard Burton visited the United States in 1860 and called the Maynard "the best of breechloading guns," pointing out several useful features. If a Maynard user ran out of metallic cartridges, he could use it as a muzzleloader by placing an empty case in the chamber and with a cleaning rod load the gun with loose powder and ball. He also mentioned the ease with which one could substitute a smoothbore barrel to fire small shot. Garavaglia and Worman, *Firearms of the American West 1803-1865,* 253. Susan W. Benson, ed., *Barry Benson's Civil War Book: Memoirs of a Confederate Scout and Sharpshooter* (Athens, Ga., 1992), 183.

176. Confederate guerrilla Sam Hildebrand with a pair of Colt 1851 Navies. He was active in Missouri and Arkansas and reportedly sometimes dressed in Union garb as a disguise as shown here. When identifying a revolver in an old photo as a Colt 1851 Navy, one must offer a caveat, for the Metropolitan Arms Co. of New York City beginning in 1864 manufactured perhaps 6,000 revolvers which resemble the Colt so closely that one has to look at the markings to tell one from the other. (*Courtesy: George Hart*)

glamorized account defending the excesses of the guerrilla bands which included such men as the James and Younger brothers, "Bloody Bill" Anderson, and George Todd. In it he made frequent reference to their use of "navy revolvers," probably Colt Model 1851s, which as seen in some photos seem to have been popular with the raiders. It's sometimes said that guerrillas and others often reloaded their Colt revolvers on horseback while in a fight. Managing one's horse while holding barrel, frame, and two cylinders as separate pieces would have been a difficult task for anyone. Replacing an empty revolver with one of several spares was much easier (Figures 175, 176).[55]

Confederates like their northern counterparts as early as 1862 sometimes made use of sharpshooters, particularly within the Army of Northern Virginia and in the western theater, even though the use of such men in either army was considered "unsportsmanlike" by some. In fact, such men on each side sometimes felt their lives were at risk if captured. (The term "sniping" was used in the 1770s in England but "sharpshooter" was a more common Civil War term for what we'd today call a sniper.) Even though the two well-known Union regiments organized by Berdan were allowed to decline in numbers and significance, both sides as the war progressed did organize and use units of sharpshooters, often smaller than regimental size, to act as skirmishers in advance of the main lines or as rear guards in a retreat. The Enfield rifle musket or shorter rifle was preferred by some southern marksmen, but some long-range English .45 Kerr and Whitworth muzzle-loading rifles were imported in very limited numbers apparently beginning in the winter of 1862-63. The Kerr resembled the Enfield, and one veteran of the Kentucky Orphan Brigade recalled that it had to be cleaned after every fourth or fifth shot if regular powder was used but it could kill at a distance of a mile or more.[56]

The rather expensive Whitworth (up to $100 or so) had a hexagonal bore firing a six-sided bullet, and when tested in 1857 in England against the Enfield rifle musket its accuracy at 800 yards was four times better than that of the Enfield. Probably no more than several hundred were purchased for the Confederate army, and it isn't known how many of those made it through the blockade. The Union

55. Fuller and Steuart, *Firearms of the Confederacy*, 209-10.

56. The Confederate 1st Kentucky Brigade was organized in a state which remained loyal to the Union. It eventually acquired the nickname Orphan Brigade since its men couldn't return to their homes during the war to recruit or visit families. In 1864 it was equipped with horses and mules and served as mounted infantry, primarily fighting on foot with Enfield rifle muskets.

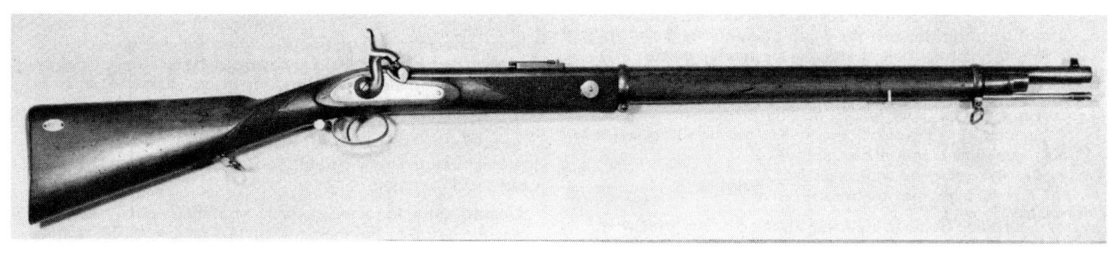

177. English Whitworth rifle with a hexagonal bore. A limited number reached the hands of Confederate sharpshooters and sometimes were fitted with telescopic sights. (*Courtesy: National Park Service, Fuller Gun Collection, Chickamauga and Chattanooga National Military Park*)

army took no particular interest in importing Whitworths, even though Confederate General Patrick Cleburne noted that at the battle of Liberty Gap in Tennessee his sharpshooters' Whitworths struck mounted federals at ranges of between 700 and 1300 yards! Sergeant Grace of the 4th Georgia Infantry has been credited as the sharpshooter who killed the Union's General John Sedgwick at Spotsylvania, Virginia, in 1864 with a shot from a Whitworth at 800 yards (Figure 177).[57]

A few Confederate sharpshooters' rifles, some fitted with telescopic sights, were produced at the Macon, Georgia, arsenal using barrels taken from "old country rifles." In August 1863, the commander at the arsenal wrote that production was halted since his skilled workers had gone to work for higher wages at the state armory. The effectiveness of Confederate rifle fire against Union officers, probably in part fire by southern sharpshooters, prompted Major General William Rosecrans in July 1863 to authorize officers to replace their conspicuous shoulder straps with less prominent badges of rank.[58]

Little detailed information exists on arms procured specifically for the Confederate navy or marines. LeMat "grapeshot" revolvers were among the navy's sidearms, and one model of shoulder gun imported from England for that service's use was a percussion breech-loading rifle designed by Thomas Wilson. An unusual feature was its long cutlass bayonet with a 26?-inch blade. It's a safe presumption that Enfield rifle muskets and shorter Enfield rifles were among the arms used by both sailors and marines. Officers' personal sidearms varied as within the army. One known example is a copy of a Colt Model 1851 by the Metropolitan Arms Co. of New York, on which on the backstrap is inscribed the name of Lieutenant W. A. Webb, captain of the Confederate ship *Atlanta.*

The frequent difficulties encountered in providing appropriate war materials to the southern navy and army as the conflict progressed were frustrating. General Robert E. Lee revealed his own aggravation in a comment he made. "I have been up to see the [Confederate] Congress and they do not seem to be able to do anything except to eat peanuts and chew tobacco while my army is starving."

57. David Noe, Larry W. Yantz, and James B. Whisker, *Firearms from Europe* (Rochester, N.Y., 1999), 28-30. Bilby, *Civil War Firearms,* 122. Fuller and Steuart, *Firearms of the Confederacy,* 229.

58. Sword, *Firepower from Abroad,* 32, 108.

5 Post-Civil War Military Arms (1866–1900)

Service in the postwar army often was difficult for both enlisted men and officers. It might involve occupation duty in the former Confederate states, but it frequently required service at isolated posts in the west far removed from family and society. Army life offered low pay, a rigid caste system between officers and enlisted men, extremes of weather in the west, slow promotion, and sometimes unreasonably harsh discipline. Drunkenness among both officers and men and desertion were common problems throughout much of the late 1800s, particularly on the frontier. Within the 4th Cavalry alone, between April 1871 and year's end 171 men attempted to desert, some selling their Spencer carbines to civilians to finance their effort. But the army still drew enlistees and for a variety of reasons—escape from debts or the law, a chance for adventure, or in the case of new immigrants to this country economic opportunity and a chance to learn their new country's language and customs. These postwar army enlistees would witness an evolution in the arms with which they fought.[1]

New Rifles and Carbines

Even before the end of the Civil War in April 1865, the army had begun efforts to adopt a breechloader for its infantry regiments. The cavalry already was making extensive use of the Sharps and other single-shot breech-loading carbines and the Spencer repeater. But the infantry's primary weapon still was the .58 caliber muzzle-loading rifle musket, awkward to load in any position other than standing and easy to load improperly. In December 1864 the army invited the submission of designs for converting their thousands of rifle muskets into breechloaders. The intent was to adopt a breechloader, but not necessarily a repeater because of the moderate range and power offered by ammunition used in such repeaters of the day as the Henry and Spencer despite their successes on the battlefield.

Meanwhile at Springfield Armory, master armorer Erskine S. Allin was directed to develop his own conversion system. This he did, removing a section from the top of the barrel at the breech end and replacing it with a hinged breechblock. His design was simple to operate—put the hammer on half cock, release the thumb-operated latch, and with a continued motion tip the breechblock upward and forward as the extractor flipped out the empty .58 rimfire cartridge case, insert a fresh cartridge, close the block, and bring the hammer to full cock. When the hammer was down in fired position, it locked the latch in place. The long firing pin passed entirely through the breechblock giving rise to a common nineteenth-century nickname of "needle gun." Today the more common nickname is "trapdoor." Late in 1865, 5,000 rifle muskets were ordered converted to the Allin system (Figure 178).[2]

1. McAulay, *U.S. Military Carbines,* 161.

2. Allin worked in various capacities at Springfield for 50 years beginning in 1829 until his death in 1879. His father Obediah joined the armory staff in 1810 and served as foreman for 25 years.

While the Model 1865 (or 1st Model) Allin conversion was undergoing field testing, in the spring of 1866 more than forty designs for breechloaders underwent scrutiny by a board of officers. These included entries by Remington, Sharps, Maynard, Peabody, and others, as well as both a .44 and a .50 caliber Henry repeater. However the ultimate choice was decided on in part for reasons of economy by avoiding payment of a royalty and taking the opportunity to use up large quantities of available rifle musket parts. It was a modified Allin design incorporating an improved extractor and a rifled barrel liner which reduced the caliber to .50, creating the Model 1866 or 2nd Model Allin conversion. On the firing range, the new .50 caliber arm proved to be twice as accurate at 500 yards as was the .58 Model 1865 Allin (Figure 179).[3]

Deliveries of the new .50 Model 1866 Springfields took several years to accomplish throughout the army. The .50-70 cartridges were issued in cardboard packages of twenty rounds each, and with the tin liners removed from the Civil War era leather infantry cartridge boxes, two packets of the new ammunition fit easily in place. From the mid-1870s forward, however, a number of different belt pouches were adopted to replace the earlier over-the-shoulder box along with various styles of cartridge belts.

Eventually Springfield Armory turned out 52,000 of the new M1866s. Deliveries came just in time to prevent tragedy among soldiers along the Bozeman Trail in present-day Wyoming near the Montana border. The arrival of these new rifle muskets by ox train occurred in July and on the first day of August 1867, when twenty soldiers of the 27th Infantry and six civilians cutting hay were attacked by Sioux Indians. Several of the civilians present had Henry repeating rifles, but an officer credited the new Springfield M1866 breechloaders with their survival during the day-long siege. The following day some miles away, several hundred other Sioux warriors fell upon another small party

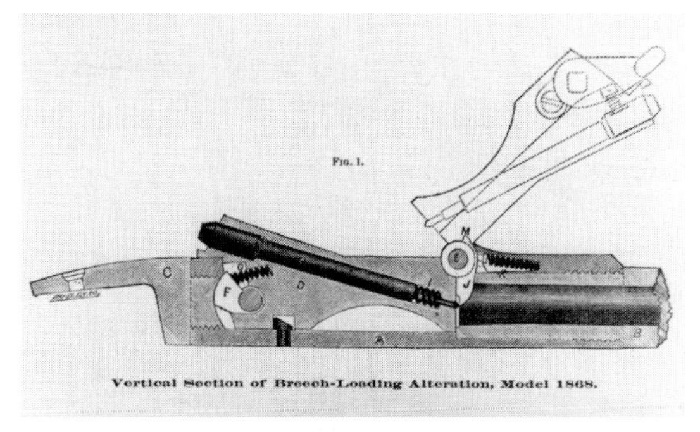

178. Drawing of the Allin "trapdoor" design showing the breechblock in both open and closed position and the long firing pin.

of soldiers and civilians cutting wood. The whites had little hope of survival, and one sergeant later admitted that he and others hadn't waited long before tying shoe laces together to loop over the trigger of their new Springfields to facilitate suicide when all hope was lost. However the volume of firepower they faced from the Springfields surprised the attackers, who presumed the soldiers were still equipped with slower loading muzzleloaders. They eventually withdrew. These two dramatic defensive engagements have persisted in history as the famed Hayfield and Wagon Box fights.

There would be improved .50 caliber "trapdoor" models in 1868 and 1870, and at this time the army also was conducting trials of other single-shot rifles including the Remington "rolling block," Sharps, and the bolt-action Ward-Burton. However in 1873 a new model Springfield "trapdoor" was adopted, a modified Model 1870 but now in .45 rather than .50 caliber. It's cartridge offered a flatter trajectory than the .50-70 and England's Joseph Whitworth in the 1850s had demonstrated the superior ballistics of a reduced caliber coupled with the same weight elongated

3. Even though it was decided that .45 was the optimum caliber, there were a number of .50 caliber arms in service and for the sake of uniformity, .50 was selected for the Model 1866 Allin conversion.

179. Springfield .50-70 Model 1866 rifle musket. (*Courtesy: Richard K. Halter*) **180.** A .50 caliber bolt-action Ward-Burton carbine. (*Courtesy: National Park Service, Fuller Gun Collection, Chickamauga and Chattanooga National Military Park*) **181.** Springfield Model 1873 .45 carbine recovered from the 1876 Custer battlefield along the Little Bighorn River or the "Greasy Grass" as the Indians called it. (*Courtesy: Smithsonian Institution, neg. no. 77826*)

bullet and a fast twist to the rifling. In this move, the U.S. joined other nations in going to smaller caliber military rifles. One other departure from its predecessor "trapdoor" models was the external blued finish given to the '73's barrel and most other metal parts rather than leaving these parts in "National Armory bright" (Figure 180).[4]

The board's 1873 rifle selection was not unanimous and their report stated that: "the adoption of magazine-guns [repeaters] for the military service, by all nations, is only a question of time; that whenever an arm shall be devised which shall be as effective [when used as a single-shot rifle] as the best of the existing single breechloading arms, and at the same time shall possess a safe and easily manipulated magazine, every consideration of public policy will require its adoption."

It would be two decades before the army found such a weapon.[5]

Issues of the new arms in both rifle and shorter carbine length began in early 1874, and by mid-1875 all regular army units were at least partially equipped with them. George Custer delayed the departure of his 7th Cavalry on its 1874 expedition into the Black Hills until the new carbines arrived. With numerous modifications and experimental variations along the way, the basic .45

4. Even though the army didn't adopt the Remington system, Springfield Armory did manufacture 10,000 Model 1871 .50-70 rolling block rifles for issue by various states to their militia.

5. Louis A. Garavaglia and Charles G. Worman, *Firearms of the American West, 1866-1894* (Albuquerque, N.M., 1985), 31.

"trapdoor" would remain the regular army's standard infantry and cavalry weapon until the mid-1890s. The regulation cartridge for the new rifle was the .45-70, a round for which many popular civilian single shot and later repeating sporting rifles of the period would be chambered including Sharps, Remingtons, Marlins, Winchesters, and Ballards. Its ability to penetrate seventeen inches of white pine at a distance of 100 yards made it a formidable cartridge. To lessen the recoil in the lighter Springfield carbine, the powder charge was reduced to 55 grains rather than 70. Among cavalrymen, a practical joke sometimes played was to substitute a rifle round for a carbine cartridge on the practice range to surprise the shooter with the unexpected greater recoil (Figure 181).

Before all the cavalry regiments received the new .45 Springfield carbine, they had relied primarily on the Spencer and the Sharps plus just after the war a smattering of other carbines such as Smiths, Maynards, and rimfire Joslyns and Starrs. In 1871 modest numbers of .50 caliber Remington rolling block and Springfield Model 1870 carbines undergoing field trials were added to the mix and in early 1872 .50 Ward-Burton carbines were also being tried in the field. The Model 1860 Spencer already had been tested thoroughly in combat and it had proven to be rugged and dependable. In an improved model of 1865, the caliber was reduced to .50 and a magazine cut-off switch designed by Edward Stabler was added just in front of the trigger. The device could be positioned so it limited the downward movement of the lever, allowing the gun to be used as a single shot while holding the rounds in the magazine in reserve.[6]

But by the end of 1872 the Spencer had largely disappeared from cavalry hands. Before its departure its firepower had saved some troopers of the 4th Cavalry at Blanco Canyon in Texas in 1871 when they were cut off by Comanches. Captain Robert Carter directed an orderly retreat with their Spencer magazines held in reserve. As they neared a sheltering ravine, Carter ordered his men to

182. An American Indian scout for the army aims a Spencer carbine during the war against the Modoc Indians in the jumble of jagged rocks known as the Lava Beds in northern California (1872-73). (*Courtesy: National Archives, photo no. 111-SC-82303*)

"unlock your magazines" then pump their shots into the enemy as they made a dash for safety. "Thank God for those Spencers," he later wrote (Figures 182, 183).[7]

A retired 6th Cavalry sergeant recalled the Spencer with some fondness. "It had many good features, among which its strength and durability

6. The Ward-Burton was a good example of the bolt-action design, a system that eventually would be used by many armies throughout the world including the U.S. Rotating the bolt handle upward and to the left 90 degrees allowed the longitudinally sliding breechbolt to slide straight back to expose the chamber for loading. When the shooter slid the bolt forward, it automatically cocked the weapon. While this self-cocking feature was a prerequisite for nearly all later bolt actions, it led to accidental discharges when the Ward-Burton was issued to men unfamiliar with its operation, a fault not of the system's design but rather of its users.

7. For detailed information on what specific cavalry units carried what firearms beginning in 1866, see the massive volume *Arming and Equipping the U.S. Cavalry, 1865-1902* by Dusan P. Farrington (Lincoln, R.I., 2004).

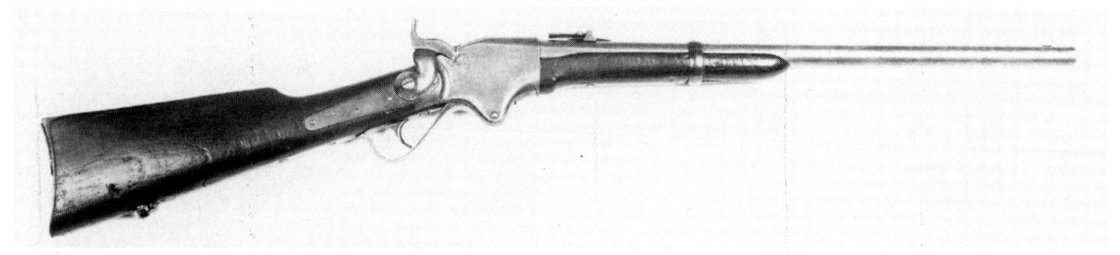

183. Gen. Edward S. Godfrey's distinguished army career began as a private in the Civil War. Later after graduation from West Point he was assigned to Custer's 7th Cavalry and fought at the battle on the Washita River in 1868 and survived the Little Bighorn fight. He received the Medal of Honor for gallantry against the Nez Perces at Bear Paw Mountain in Montana where he was wounded severely. He probably had this Spencer carbine with him on the Washita and perhaps on the later Yellowstone and Black Hills expeditions with the 7th. (*Courtesy: Dale C. Anderson*)

were prominent." Its rate of fire was a factor in the survival of one group of about fifty army scouts commanded by Major George Forsyth and equipped with Spencers, Colt Model 1860 revolvers, several longer range Springfield rifles, plus a Henry repeater or two. In the fall of 1868 while trailing a large war party of Sioux and Cheyenne warriors, Forsyth's men were attacked in eastern Colorado and forced to take cover on an island in the shallow Arikaree River. Relying on the firepower of their repeaters, they withstood a number of charges but they were kept under siege for almost a week as Indians watched from nearby bluffs in what's known today as the battle of Beecher's Island. Before they were rescued by men of the African American 10th Cavalry, they had been reduced to eating the flesh of their dead horses and pack mules, meat liberally sprinkled with gunpowder to mask the taste and odor of the rotting flesh.[8]

The 10th Cavalry was created as part of the army's reorganization of 1866. Two cavalry regiments (9th and 10th) and four of infantry (later reduced to two regiments, the 24th and 25th) were designated to be filled with African Americans led by white officers. American Indians sometimes referred to these troops as "buffalo soldiers," a reference to the color and texture of their hair. Despite racial prejudice, these regiments served with distinction on the frontier with lower rates of drunkenness and desertion than other units. George A. Custer once was offered command of the 9th but declined it. In contrast, Colonel Benjamin H. Grierson, commanding the 10th, once argued vehemently when the colonel of the white 3rd Infantry at Fort Leavenworth ordered the 10th's troops not to parade close to his men.[9]

In late 1868 in preparation for its participation in a campaign against hostile Indians in what is now Oklahoma, the 7th Cavalry under Custer began daily target practice with its Spencer carbines. This was an uncommon move, for the army generally provided little incentive or opportunity for practice. For example, in the face of severe postwar cutback in funds, army General Order No. 50 in May 1869 would allow only ten metallic cartridges per man per month for target practice.

8. At the time of this writing, the site of this historic battle has been preserved only through local volunteer efforts. It's adjacent to a small church, reached by a well-maintained dirt and blacktop road north off Rt. 36 east of Idalia, Colorado, close to the Kansas border.

9. Grierson, a former music teacher, was something of an enigma as a cavalry officer, for he had a distrust of horses resulting from a youthful accident. The daring 600-mile diversionary raid he led deep into Confederate territory during the 1863 Vicksburg campaign was the subject of the fictionalized movie *The Horse Soldiers* starring John Wayne and William Holden.

184. Sharps .50-70 carbine issued to the 3rd Cavalry. In 1871, most of the 3rd Cavalry's companies were stationed in Arizona Territory then in Wyoming Territory and Nebraska in 1872-73. (*Courtesy: Kenneth L. McPheeters*) **185.** Officer's Model Springfield .45-70 owned by Lieutenant John Murphy, 14th U.S. Infantry, with its custom leather carrying case. Between 1875 and 1885, Springfield Armory produced 477 such rifles rather than continuing the annoyance of receiving officers' requests for customized rifles. Special features found on Murphy's and other examples include a half-length stock with checkered wrist and forearm, wooden cleaning rod mounted beneath the barrel, and engraved hammer, lockplate, breechblock, and receiver. Among the officers who purchased one of these handsome rifles were Lieutenant General Phil Sheridan and Colonel Benjamin Grierson. (*Courtesy: Roy Kinzie*)

Custer's men fired at ranges of 100, 200, and 300 yards and individual trooper's scores were recorded. The forty best marksmen were organized into an elite company of sharpshooters, exempt from guard and picket duty. Late in November 1868 as the regimental band played "Garry Owen," the shivering 7th struck Black Kettle's Cheyenne camp on the Washita River, but it wasn't long before it found itself in a struggle against a larger force than was anticipated. The firepower the repeating Spencers offered undoubtedly was a factor in the regiment's successful withdrawal from the field as Indian opposition swelled. The Spencer's service life would not be long after such incidents, however. Without a major redesign it was not able to use the longer and more powerful cartridges being developed for both civilian and military use.

Serving beside the Spencer in the postwar cavalry was the Sharps carbine, superior in range to the seven-shooter, but a single shot. In early 1868, the army began receiving deliveries of carbines converted from percussion by Sharps under government contract to fire the .50-70 metallic cartridge. In January 1872 the army adopted and by summer was issuing a .50-55 carbine load, reducing the powder charge as well as the recoil and making the Sharps carbine a little more comfortable to shoot. The Sharps had its critics, complaining largely of extraction problems and the lack of firepower as was offered by the Spencer. Issues of the new Springfield .45 carbine began in early 1874, but it was not until mid-1875 before most units had traded all their Sharps for the new arms. Some .50 Sharps carbines continued to equip some civilian employees such as teamsters and packers who handled the pack animals. Meanwhile infantry regiments continued to rely on .50-70 Springfield rifle muskets as rearming with the new .45 rifles progressed. After receiving the new arms, each infantry regiment of "doughboys" was allowed to retain a few of its .50 trapdoor rifles for hunting and target practice. Distribution of the new .45 Springfields to state and territorial militia units had a lower priority and took longer to accomplish (Figure 184).[10]

Although the .45 "trapdoor" was not without its critics, particularly in carbine form, in general these guns gained a favorable reputation for accuracy, range, and killing power. The carbine was accurate out to about 500 yards and to a greater range when fired with the 70-grain rifle cartridge. A general lack of marksmanship training prevented the average soldier from taking advantage of its potential, however. In some cavalry regiments a limited number of rifles were issued for use at ranges beyond the carbine's capacity. After the Custer fight at the Little Bighorn, the 7th Cavalry's carbines were sharply criticized when fired cartridge cases stuck in the chamber and had to be pried out with a knife. However, other regiments weren't complaining of such carbine failures and the primary cause of the difficulty appears not to have been due to a faulty design of the Springfield but rather dirty ammunition combined with the limited elasticity of the copper cartridge case, a problem rectified in the 1880s with a switch to brass cases. But in recognition of the problem, in 1877 the carbine butt stock was modified with the addition of a three-hole compartment accessed through a trap in the butt plate to hold a three-piece jointed steel cleaning rod and a broken shell extractor tool (Figure 185).

Troops in the field often found that green verdigris formed on the copper cases when carried in cartridge belts with leather loops. This problem was largely corrected when soldiers cleaned their cartridges and when belts were replaced with those with cotton rather than leather cartridge loops. Nevertheless the "trapdoor" was still only a single shot, and in the summer of 1876 energetic Colonel Ranald Mackenzie unsuccessfully sought to have a portion of his 4th Cavalry armed with faster firing but shorter range and less accurate Winchester repeaters.

Two months after the Custer battle of June 1876 the commander at Springfield Armory in a letter defended the Springfield with these words:

> The most powerful Winchester rifle for frontier service carries a cartridge of 40 grains of powder [the .44-40 Model 1873] and 200 grains of lead. The extreme range for which this arm is sighted is 300 yards. The cartridge of the Springfield carbine contains 55 grains of powder and 405 grains of lead and is sighted for an extreme range of 1,300 yards. The penetration of the Winchester rifle, in pine, at a distance of 100 yards, is less than one half of that of the Springfield carbine at the same distance, and not so much as the penetration of the latter arm at the distance of one half a mile.[11]

As within the civilian community, firearms accidents were an occasional occurrence within the army. During the 1876 campaign against the Sioux after Custer's defeat, a trooper in the 5th Cavalry placed his carbine in a wagon after coming off guard duty. When he removed the weapon from the wagon, it accidentally discharged and killed a companion seated nearby. For his neglect in not emptying his carbine, he was court-martialed, served a term in a military prison, and was dishonorably discharged. Corporal E. A. Bode never fired his Springfield in combat while serving with the 16th Infantry in the west (1877-82). But he described

10. The term "doughboy," usually thought of as a reference to a World War I U.S. infantryman, actually was in use at least as early as the 1870s.

11. Dr. Albert J. Frasca and Robert H. Hill, *The .45-70 Springfield* (Northridge, Calif., 1980), 2.

186. Remington-Lee bolt-action repeater. Five .45-70 cartridges were contained in the detachable box magazine in front of the trigger guard. Although the army conducted field tests and the U.S. Navy issued them in limited numbers, most of the military-style Remington-Lees were exported to other countries. Some also were sold on the commercial market configured as sporting rifles. (Courtesy: National Park Service, Fuller Gun Collection, Chickamauga and Chattanooga National Military Park)

one near accident in which another corporal playfully "dry fired" what he supposed to be his unloaded rifle at a horse grazing nearby. Fortunately his aim was bad and the uninjured horse continued grazing unconcernedly. Bode himself had a near accident as he prepared to fire at a wolf and retrieved his rifle which had been lying on some sacks of corn. The wolf disappeared before he could fire, but upon unloading his rifle Bode noticed some kernels of corn fall from the breech and found the barrel full to the muzzle. "Had I fired the barrel would have exploded and blown my head off, and made a cornfield of a wolf's hide."[12]

During the twenty or so years the .45 "trapdoor" was the army standard, the army did experiment both with efforts to increase its rate of fire and with designs of repeating rifles. In 1875 Springfield Armory was ordered to alter 1,000 rifles to accept what's known as the Metcalfe attachment, a rectangular wooden block which held eight cartridges and could be attached to the right side of the stock just in front of the lock plate. Although Lieutenant Henry Metcalfe's device placed the rounds close at hand to facilitate reloading, field tests found too many disadvantages.

In 1878 a board of ordnance officers convened to review available repeating rifles suitable for military service. Late that year the board recommended field trials of the Winchester-produced Hotchkiss, a bolt-action weapon designed by Benjamin Hotchkiss and holding five .45-70 cartridges in its tubular magazine in the butt stock, similar to the Spencer magazine. Various difficulties with the Hotchkiss rifles and carbines surfaced during the trials, and in 1880-81 those in the field were replaced with the regulation Springfield. An improved version of the Hotchkiss was tested in 1884-85, but again the Springfield prevailed, in part because of a continued problem with stock breakage. Ordnance boards approved field testing of other .45-70 bolt-action repeating rifles, the Chaffee-Reece, the Remington-Keene, and Remington-Lee in the mid-1880s, but none of these replaced the "trapdoor" despite many other nations' adoption of bolt-action magazine rifles in the mid- and late 1880s (Figures 186, 187).

Not until 1892 did the army adopt a repeater as a replacement for the "trapdoor" as the army's standard rifle and carbine, the smooth operating bolt-action Norwegian-designed Krag-Jorgensen. Five cartridges, rather than being contained in the butt or in a tubular magazine beneath the barrel, were dropped into the horizontal box magazine mounted on the right side of the receiver. By means of a magazine cutoff switch, these cartridges could be held in reserve and the weapon fired as a single loader, a feature which was present on Spencers of the post-Civil War years. Actual issue of the new Krags required several years to inaugurate, but with

12. McAulay, *U.S. Military Carbines*, 188. Worman, *Gunsmoke and Saddle Leather*, 352-53.

187. In the mid-1870s, the government contracted to have the socket of Civil War bayonets cold pressed to fit the reduced diameter of the "trapdoor" .45-70 rifle's barrel. When this supply became diminished by 1880, some rifles were equipped with a combination cleaning rod-bayonet as was used on Hall carbines in the 1830s. Here an enlisted man, Robert Strachan, is photographed at Camp Meade, Pennsylvania, with a Springfield with the rod bayonet extended. (Courtesy: Erich Baumann) **188.** An 1880s trooper with the 6th Cavalry, his Springfield carbine in a saddle scabbard and a Colt Single Action revolver and saber at his waist. The flap covering the holster apparently is folded back behind his cartridge belt. During the Indian campaigns of the 1870s and 1880s, the saber often was left behind in the barracks as being a mere encumbrance on the march. By the 1880s, the shoulder sling for the carbine sometimes was dispensed with as a possible hazard since the rider could be dragged if he was unhorsed with the carbine securely held in the scabbard while still slung about his shoulder. (*Courtesy: Fort Davis National Historic Site*)

the Krag, the army not only had a repeater but one which fired a .30 caliber cartridge loaded with smokeless powder which no longer produced a disconcerting and position-revealing cloud of smoke as well as powder residue (Figure 188).

Although the .30-40 Krag equipped many regular army troops sent to Cuba during the Spanish-American War and to the Philippine Islands in 1898-1902, some volunteer forces and national guard units still relied on the now obsolescent "trapdoor" rifle and would do so for the next several years. Successor to the Krag would be another bolt action, the famed 1903 Springfield rifle with a 24-inch barrel and intended for both cavalry and infantry use. One of the most accurate of the world's mass produced military rifles, it saw service as late as World War II (Figures 189, 190, 191).

No regiment which served in the Spanish-American War of 1898 received more publicity or is better remembered today than that organized by the Assistant Secretary of the Navy Theodore Roosevelt, who served as its flamboyant lieutenant colonel. Better known as the Rough Riders, the 1st U.S. Volunteer Cavalry existed for only about four months. Recruits came largely from the southwestern United States but included a true cross section of society—cowboys, miners, farmers, football and baseball players, polo players, sailors, teachers, lawyers, and even a judge. Just before leaving San Antonio for Tampa, Florida, they received Krag

189. A .30 caliber Model 1896 Krag-Jorgensen carbine. (*Courtesy: National Park Service, Fuller Gun Collection, Chickamauga and Chattanooga National Military Park*) **190.** Palmer LaPlante, probably a private in the Wisconsin National Guard, with the standard bayonet mounted on his Krag rifle. Three other styles of Krag bayonets were adopted or at least tested—a shorter cadet model, a combination entrenching tool-bayonet (known by today's collectors as the Bowie knife bayonet), and the bolo bayonet intended for use in hacking one's way through the jungle as in the Philippines. The last two styles are rare. (*Courtesy: Erich Baumann*) **191.** U.S. cavalrymen armed with Krag carbines entrenched in the Philippines in 1899. (*Courtesy: Wade Lucas*)

carbines and Colt Single Actions with the barrel cut to what today is known as the artillery length of five and a half inches. In Cuba, they participated with other cavalry regiments in a dismounted attack on Kettle Hill, part of the successful attempt to seize the San Juan Heights outside Santiago. To help prevent their Krags from rusting in the damp climate, they rubbed them with bacon grease.[13]

Improved Revolvers for the Postwar Cavalry

For the first five years after the Civil War ended, the regular cavalry continued to rely heavily on .44 Remingtons and Model 1860 Colts plus an increasingly smaller number of Colt and Whitney .36s and Starr .44 single actions, all percussion arms. Difficult as it was to reload a percussion handgun on a moving horse and as easily damaged as percussion revolver cartridges were, Smith & Wesson's control of the Rollin White patent until

13. McAulay, *U.S. Military Carbines*, 235. It was during the fight for San Juan Heights that Lieutenant John Parker moved his detachment of rapid-fire Gatling guns forward with the infantry to provide fire support, an early offensive use of these early machine guns which previously had been considered best suited for the defense of fixed positions.

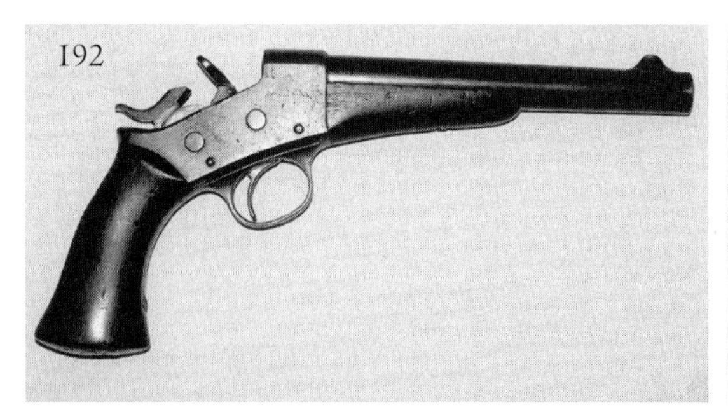

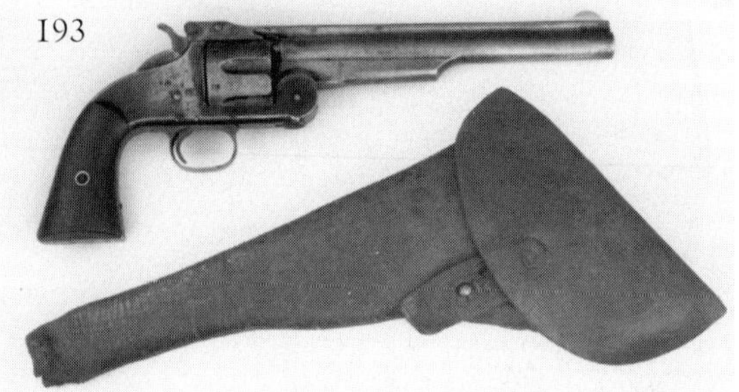

192. Model 1871 .50 caliber Remington rolling-block pistol, one of 1,000 procured for field trials by cavalry regiments. It is shown with the hammer cocked and the breechblock rolled partially back exposing the breech. **193.** Captain Robert H. Young's .44 Smith & Wesson American. During the Civil War he rose from private to captain before being mustered out at war's end. Two years later re reenlisted as an infantry lieutenant and served at various posts in the west. (*Courtesy: Lloyd Jackson*)

1869 restricted any significant move to adopt a metallic cartridge revolver.[14]

However in 1870 an army ordnance board ordered 1,000 of each of two handguns for distribution among the cavalry regiments for testing—a .50 single-shot Remington pistol using the rolling block action and a new model .44 Smith & Wesson six-shooter, the No. 3 (later designated the American) with an automatic self-extracting feature for eliminating the spent cartridge cases. Two hundred of the No. 3s were to be nickel plated rather than blued in a test of the two finishes. By the end of 1871, there were blued or nickeled No. 3s scattered among various companies of all cavalry regiments up to and including Custer's 7th. Not long after the board's selection, a third entry was incorporated in the field trials, 1,200 Colt Model 1860 .44s converted to centerfire using a system designed by C. B. Richards. Issues of trial quantities began in early 1872 and these Colts seemed to have been well liked.

The single-shot Remington pistol was an enigma. One may wonder why the army considered it, but at the time it was a centerfire breechloader and no metallic cartridge revolver had proven itself. Also, there was some enthusiasm for a standardized caliber and .50 caliber rifles and carbines were then in widespread use with the infantry and cavalry. Issue of the .50 pistol got under way in the spring of 1872 but hadn't progressed very far by year's end. The frequency with which specimens today can be found showing little or no use indicates the limited service these by then obsolete handguns saw. All .36 percussion Colts had been recalled from service in the spring of 1871 so in 1873 and early 1874, a cavalryman could have carried either a .44 percussion Colt or Remington or any of the three cartridge guns undergoing testing (Figures 192, 193).[15]

In December 1872 a newly designed solid-frame Colt centerfire .44 completed initial army tests. It soon would become the standard-issue service revolver. In comparison with the S&W, Colt's new Single Action couldn't provide the ease of extracting the fired cases by merely releasing the barrel latch and tipping the barrel downward. But it offered the advantages of a solid frame, fewer parts, better balance, easier disassembly for cleaning, and

14. Efforts by Smith & Wesson to gain an extension of the White patent failed when President Grant vetoed the bill. "It is believed the government suffered inconvenience and embarrassment enough during the war in consequence of the inability of manufacturers to use this patent." Farrington, *Arming and Equipping the U.S. Cavalry*, 1865-1902, 80.

15. In some official documents the Remington pistol was listed as "single barrel" rather than "single shot."

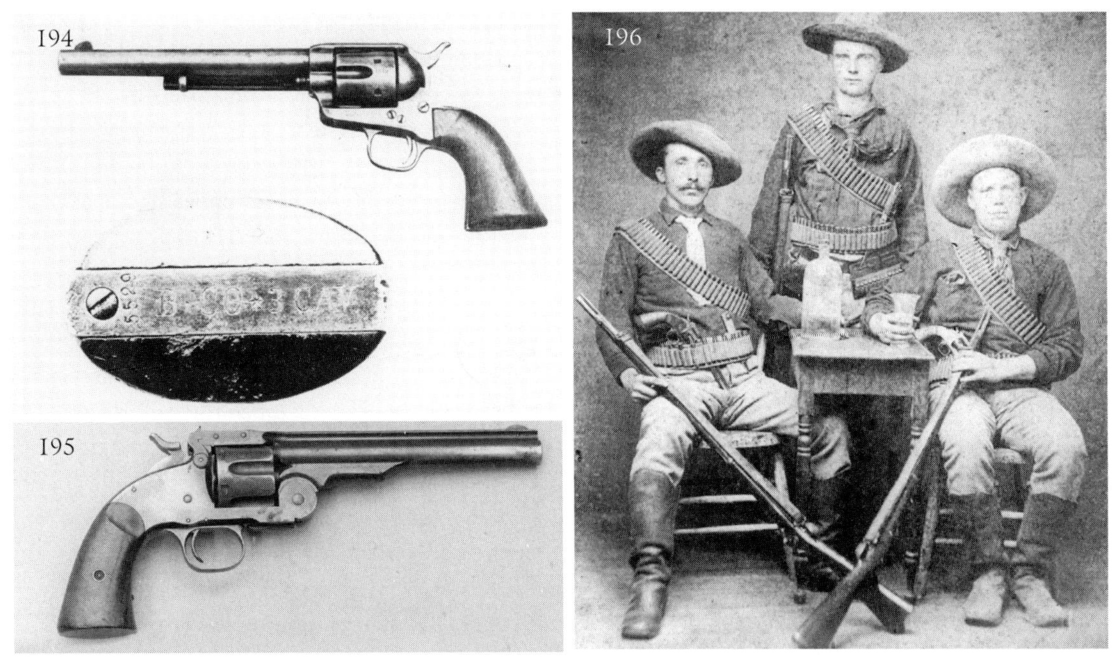

194. Colt .45 Single Action issued to Company B, 3rd Cavalry and so stamped on the bottom of the butt. Regimentally marked guns of the period are rare, for the practice, intended to reduce theft or illegal sale, was soon halted as disfigurement of government property. **195.** Martially marked Smith & Wesson .45 Schofield. **196.** Casually uniformed members of the 3rd Cavalry in the early 1880s equipped as unofficial mounted infantry with Springfield .45-70 rifles and Smith & Wesson Schofield revolvers. (*Courtesy: Thomas N. Trevor*)

greater durability under the rigors of field service. The following July Colt received an initial order for 8,000 of the new six-shooters, but in .45 caliber. When in 1890 the last contract for Colt Single Action Armies was issued, the government had procured 37,000 .45s, for use by the regular army, militia, post office department, and other federal agencies. Most of the army's Single Actions went to equip cavalrymen, but in May 1882 revolver issues were expanded to include each company sergeant of all branches of the army (Figure 194).

Durable and reliable as the Single Action proved to be, the army tested other handguns not long after it adopted the Colt. The army already had conducted field trials of the .44 S&W No. 3 American, but this model had lost out to the Single Action, although it was a favorite among some officers as a personally owned weapon. Major George Schofield of the African American 9th Cavalry had a reputation among some as a miser, but he made several improvements in the S&W No. 3 including a redesigned barrel latch. He was successful in arranging tests of his modified S&W design and in September 1874 the government ordered 3,000 but in .45 rather than .44 caliber. The cylinder was too short to handle the .45 Colt cartridge and ultimately the army adopted a slightly less powerful round that could be used in either revolver. Eventually the government ordered about 8,000 Schofields, but when in the early 1880s it attempted to obtain more, S&W said none were available (Figures 195, 196).[16]

16. As a corporal in the 16th Infantry described Schofield, he was "a perfect gentleman, but the penny was his god." Schofield transferred to the 6th Cavalry in 1881 and rose to the rank of lieutenant colonel before committing suicide with one of his S&W revolvers in December 1882.

197. Colt .38 Model 1894 army revolver issued to Troop H, 4th U.S. Cavalry. The 4th served in the Philippines and sought to capture the insurgent leader Emilio Aguinaldo. (*Courtesy: Neil and Julia Gutterman*)

Remington, Whitney, Forehand & Wadsworth, and Hopkins & Allen (makers of the Merwin & Hulbert revolver) sought unsuccessfully to interest the army in their own military size cartridge revolvers in the later 1870s. Colt's new 1878 double action model also underwent testing but was rejected. Some British and American revolvers did see service as nonstandard army officers' personal property, including George Custer's British Webleys. Harry W. Miller (rank unknown) of Troop L 5th Cavalry presumably owned a Model 1878 Colt for on December 8, 1882, he wrote to B. Kittredge from Fort Sidney, Nebraska, asking for the cost of pearl grips for that model. He included a sketch showing the scene he wanted engraved on the grips, a cavalry trooper on horseback with "L 5" clearly visible branded on the mount's flank. The Colt factory had presented one of the first Model 1878s produced to one of the army's ordnance officers, Capt. O. E. Michaelis, who praised the model. Although an officer was expected to obtain his own sidearm, if he had none he could draw one from company supply. If he chose to purchase a military issue revolver or long gun from the army he could do so or he could obtain whatever he wanted from a commercial source. Enlisted men had the latter option only if they wanted to purchase a personal gun, for they could not buy new guns from the government.[17]

The Colt Single Action remained the primary army issue revolver until replaced by Colt .38 double-action handguns with a swing-out cylinder beginning in 1893. It wasn't long before use in the field demonstrated the inadequacy of the .38's lighter 150-grain bullet and prompted the return to service of some .45 Single Actions but with a barrel shortened to five and one-half inches. These are known by today's collectors as the "artillery model," a modern and unofficial designation derived from the issue of several hundred early examples to artillery units during the Spanish-American War. It's usually said that the army reverted to using the .45 Colt when the recently adopted .38 revolvers failed to stop charging natives during the Philippine Insurrection of 1898-1901. Actually, the army as early as January 1898 if not before had recognized the inadequacy of the .38 Colt for cavalry service and was working to determine proper ammunition and ballistic qualities of an appropriate .45 revolver. But combat in the Philippines did dramatically illustrate the need to return to a larger caliber (Figures 197, 198).

In 1911 after fifteen years of development work, the army adopted the Colt .45 auto-loading pistol, which would serve American armed forces for three-quarters of a century. Its design was the work of the inventive genius, John M. Browning, from whose fertile mind sprang the lever-action Winchester model rifles of 1886, 1892, 1894, and 1895 as well as Browning auto-loading shotguns, the army's Model 1918 Browning automatic rifle, .30 and .50 caliber Browning machine guns, and other firearms for both military and civilian use.

Arms for the Sea Services

The navy continued its postwar reliance on many of its Civil War arms—imported Enfield rifle muskets and rifles, .69 rifled muskets, Plymouth rifles, Sharps & Hankins carbines, and others for several years. In 1868-69 it took delivery of 5,000 .50

17. Don Wilkerson, *Colt's Double-Action Revolver Model of 1878* (Marceline, Mo., 1998), 185.

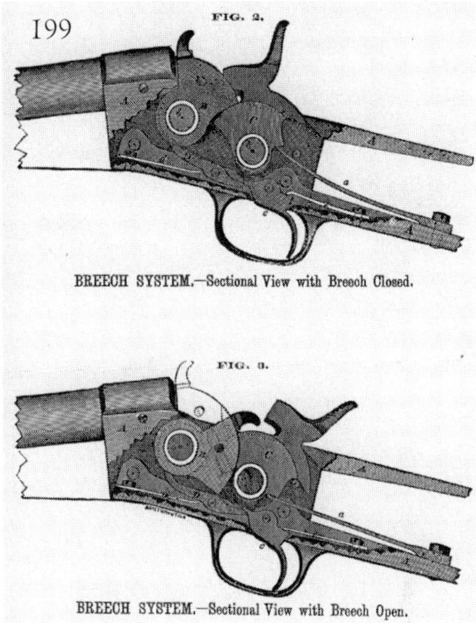

198. Interior of enlisted cavalry barracks at Fort Robinson, Nebraska, in the mid-1890s. The gun racks hold old and new arms—obsolescent Springfield "trapdoor" carbines and newly issued .38 Colt double-action revolvers. (*Courtesy: National Archives, neg. no. 92-F-54-8*) **199.** Page from the Springfield Armory's 1871 Description and Rules for the Management of the Remington Navy Rifle, Model 1870 illustrating the "rolling block" action.

Remington rolling block breech-loading carbines, unusual in that they were fitted with swivels to allow use of a shoulder sling rather than the normal carbine fixture of a sling bar and ring. Some of these carbines saw service with the fleet during the 1871 Korean expedition, but they remained in service only until the late 1870s, since their lack of a bayonet was considered a handicap. In 1871 the navy began receiving 12,000 Remington .50-70 Model 1870 rolling block rifles, which by the end of 1873 had replaced most of the Remington and Sharps & Hankins carbines. Among the rarest of these Navy Model 1870 rifles were those 100 or fewer which in 1889 were converted by Winchester to .22 rimfire by inserting a barrel liner, for use for shipboard target practice. This marked the first known use by the U.S. government of the .22 rimfire cartridge (Figure 199).[18]

Remington in the early 1880s also provided for trials a small number of its bolt-action Keenes, a repeater with a tubular magazine mounted beneath the barrel. The navy went so far as to issue for service some of the Remington-Lee rifles, another bolt-action .45-70, but one which held its reserve cartridges in a detachable box magazine on the underside of the stock just in front of the trigger guard. In the early 1880s, the navy also turned to Winchester for some of its long guns—several thousand Winchester-Hotchkiss bolt-action .45-70 repeaters. In line with the army's move to a small caliber repeater, the .30 caliber Krag-Jorgenson in 1892, the navy took a similar course in 1895. However its choice at first wasn't the Krag but another design of James Paris Lee, a six-millimeter (.236 caliber) five-shooter manufactured in New Haven, Connecticut, the Winchester-Lee Straight Pull rifle. It was a bolt-action weapon in which the bolt handle did not have to be turned upward prior to being pulled back, but was pulled straight back to open the breech, increasing the rate

18. Richard A. Hosmer, *The .58- and .50-Caliber Rifles and Carbines of the Springfield Armory, 1865-1872* (Tustin, Calif., 2006), 137.

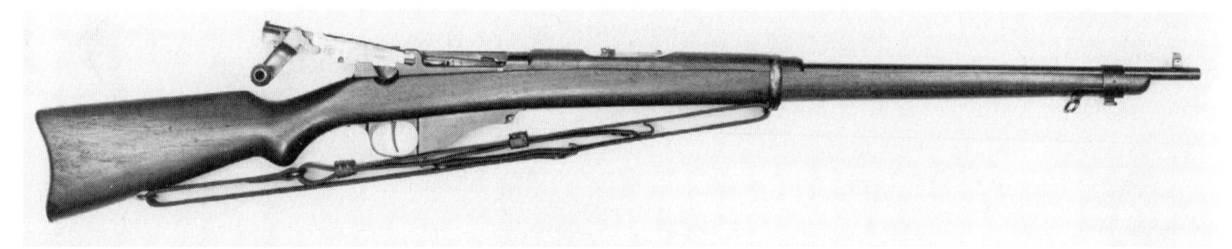

200. Winchester-Lee Straight Pull rifle, a standard issue weapon for the U.S. Navy and Marine Corps during the late 1890s.

of fire over an extended period. Manufacture began in October 1896, in time for the rifle's extensive use by navy and marines during the Spanish-American War. In 1900, however, both these branches of the military began a change over to the Krag (Figure 200).[19]

Although the navy never adopted the Springfield "trapdoor" as its regulation arm, a half dozen of these .45-70s were among the guns recovered from the battleship USS *Maine* which in 1898 was sunk in the harbor in Havana, Cuba. These Springfields probably were Marine Corps weapons and were sold by the navy in December 1899 to the famed New York surplus military goods dealer Francis Bannerman Sons, along with 54 Lee Straight Pull rifles also recovered from the sunken *Maine*.[20]

The marines didn't begin to replace their muzzleloaders with both Model 1868 and 1870 .50 Springfield breechloaders until 1870. Some of these were still in service as late as 1879 even though the corps had been receiving quantities of the .45 trapdoor Springfield rifle (plus a limited number of carbines) soon after the new model's adoption in 1873. Although the leathernecks tried a few other breechloaders, including the Ward-Burton, the Remington rolling block, and the Remington-Lee repeater, they relied primarily on the trapdoor until 1897 when they adopted the Lee Straight Pull rifle. Finally, as the twentieth century dawned, they moved to the .30-40 Krag.

Despite the navy's Civil War use of revolvers, it returned to the single-shot pistol afterward. In 1863 the commander of the ordnance bureau in his annual report to the secretary of war had recommended that the miscellany of small arms be reduced to the muzzle-loading Plymouth rifle and for boat crews the Sharps & Hankins breech-loading carbine. Revolvers, he reported, "are objectionable in the hands of seamen, and should be restricted to officers." Instead a large caliber single-shot breech-loading pistol using a metallic cartridge was suitable for boarders. He repeated his views on an appropriate navy handgun in his 1865 report.[21]

His preference was fulfilled with the navy's postwar adoption of the .50 caliber rimfire Model 1865 Remington rolling-block pistol It had no trigger guard and a cocked pistol could be discharged accidentally if thrust into a belt, so many of these were converted to the Model 1867 with the addition of a trigger guard. James Dahlgren in 1869, by then an admiral in charge of the navy's Bureau of Ordnance, in a letter emphasized that: "The new Navy breechloading pistol is intended to replace all other pistols in the hands of seamen. But the Bureau has no objection to supplying revolvers to Officers upon a requisition from the Commanders of ships." On November 14, 1873, in reply to a request for revolvers for ships' officers, Capt. William Jeffers, now the chief of the ord-

19. Herbert G. Houze, *Winchester Bolt Action Military and Sporting Rifles, 1877 to 1937* (Lincoln, R.I., 1998), 85. Franklin B. Mallory, *The Krag Rifle Story* (Dover, Del., 1979), 86.

20. Frances Bannerman Sons Military Goods Catalogue, January 1955, 32.

21. Ibid., 269-70.

nance bureau, peevishly reflected his own personal objection to revolvers as side arms: "Yes! If it will make anyone happy. There is no instance on record of a revolver having been fired. They are merely ornamental appendages." But after Captain Montgomery Sicard replaced Jeffers in 1881, revolvers no longer were viewed as ornaments and officers were required to undergo periodic target practice with them.[22]

The two decades after the Civil War was a time of severe fund shortages for the navy. Throughout this period, the navy's revolvers were .36 Remingtons and Colt 1851s and 1861s on hand after the war, some of which the Colt factory in 1873-76 and Remington in 1876 converted to use .38 centerfire metallic ammunition. The cost of alteration for the mixture of 2,097 Colt 1851s and 1861s was $3.50 each plus 75 cents to replace the stock (walnut grip) and either reblue or replate the trigger guard and backstrap. Money for the conversion work came in part from the navy's 1873 sale of surplus .36 Whitney, Colt, and Remington percussion revolvers for prices ranging from $1.25 to $2.50 each. The navy tested but couldn't afford to adopt the famed Colt Single Action, which served the army for a quarter century, and instead relied on its converted revolvers (Figure 201).[23]

In 1886 with a view to modernizing its arsenal, the navy requested of Winchester, Colt, Smith & Wesson, Remington, and Merwin & Hulbert samples of any .38 double-action revolver which these firms marketed and which would be suitable for naval use. Two years later it authorized the purchase of 5,000 Colts with a cylinder which swung out to the side and ejected all six empty cartridge cases with a single push on the ejector rod at the cylinder's front. Deliveries began in 1889 and these .38 double-action Colts and improved models thereof served throughout the remainder of the century, replacing the seamen's Remington pistols and officers' converted revolvers in service (Figure 202).

The Revenue Cutter Service, forerunner of the U.S. Coast Guard, had been created in 1790 to enforce customs laws. In the 1880s they too sought more up-to-date handguns. Their choice in 1881 was a .44 caliber improved version of the Smith & Wesson No. 3 large-frame revolver, known generally today as the New Model No. 3. Unlike the army which in 1878 had rejected samples, the cutter

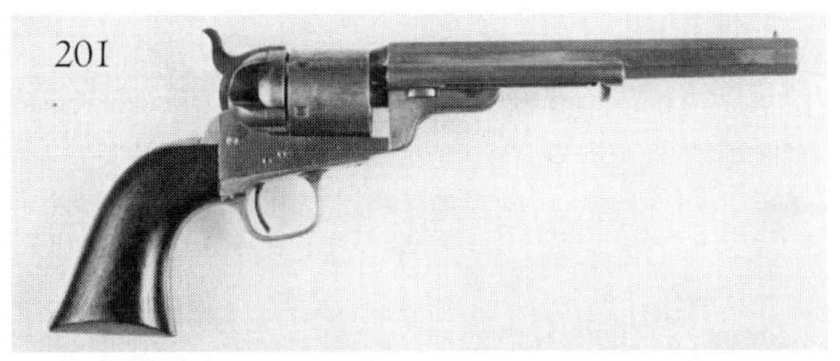

201. One of the navy's post-Civil War Colt Model 1851 Richards-Mason conversions to .38 centerfire. The tube on the side of the barrel houses the ejector rod. **202.** Colt .38 Model 1895 Navy revolver with black leather navy issue holster. (*Courtesy: Neil and Julia Gutterman*) **203.** Smith & Wesson .44 New Model No. 3 issued to the Revenue Cutter Service.

22. Ibid., 287-88, 430-31, 457.

23. C. Kenneth Moore, *Colt Revolvers and the U.S. Navy, 1865-1889* (Jenkintown, Pa., 1987), 128.

service didn't view the S&W's hinged frame as a weakness. Since there were only about two dozen vessels in the service, orders for these guns in the 1880s and 1890s were sporadic and in lots of two dozen or fewer at a time (Figure 203).[24]

24. Edward Scott Meadows, "Martial Smith & Wesson New Model 3 Revolvers, Part II," *The Gun Report,* July 2004, 14-15.

6 *Post-Civil War Civilian Arms*

(1866–1900)

The era from the close of the Civil War to the end of the century has been a fertile field for writers of fiction both in the form of novels and scripts for television and movies, often with the trans-Mississippi west as the locale. That is as it should be, for the era of "cowboys and Indians" has captured and held the public's attention for generations whether at Saturday movie matinees with Roy Rogers or Gene Autry or in tales of such real-life individuals as "Buffalo Bill" Cody, Wyatt Earp, "Wild Bill" Hickok, and others who gained heroic status, even if such status wasn't always deserved. In terms of firearms development, that period saw the establishment of the centerfire metallic cartridge and by the new century the introduction of smokeless powder, the first auto-loading pistols, and the army's adoption of a repeating rifle as standard (Figure 204).

Speaking for a moment concerning Hollywood's portrayal of the "old west," it's refreshing to occasionally see efforts made to properly portray firearms usage rather than arming all with Colt Single Actions and Winchesters of 1890s pattern regardless of the era portrayed. In the classic TV mini-series *Lonesome Dove,* which took place about 1876, man-burner Dan Suggs is armed with a reproduction Remington revolving rifle. Captain Augustus McCrae relies on a converted Colt Walker revolver and when he defends himself in a buffalo wallow from pursuing Indians, he and his Henry rifle outduel a renegade with a Sharps Model 1874. *The Last Hunt* was a rather accurate portrayal of professional bison hide hunters in which viewers could spot a large-frame double-action Smith & Wesson revolver as well as a Model 1876 Winchester. In the award-winning western *The Unforgiven,* Clint Eastwood portrayed a farmer with a dark past who relied on a double-action Starr revolver, Morgan Freeman demonstrated the use of a Spencer carbine, and their companion, the near-sighted "Schofield Kid," drew that nickname from his preference for that model Smith & Wesson .45.

Civilian Handguns

During the Civil War much of the handgun production by Colt, Remington, and some other makers had gone to fill wartime state and federal contracts, but some of their output was available for commercial sales. An extreme rarity today, for example, is one of the original cardboard boxes in which some civilian Remington .36s and .44s were packed along with a nipple wrench and bullet mold and with operating instructions printed in the lid. Beginning in the early 1870s, auction sales of surplus government arms began to provide dealers with large quantities of used guns to resell to civilians. Meanwhile Smith & Wesson still retained control of the Rollin White patent and although as early as late 1863 they'd been working on a .44 caliber version of their No. 2 revolver, they didn't introduce any handgun larger than .32 caliber until

204. Intriguing as nineteenth-century photos of armed individuals are, one has to remember that photographers often kept guns in their studio for use as props, as J. C. Burge of Holbrook, Arizona, probably did to equip such customers as this "dude" in fancy attire. All of these guns--the Colt Single Actions carelessly slung about his waist in mismatched belts and holsters and the Model 1873 or 1876 Winchester rifle--may fall in this category. (*Courtesy: Erich Baumann*)

1870. Earlier, after several years of attempted negotiations, Remington was able to secure an agreement with S&W by which Remington in 1868-69 was allowed to alter almost 5,000 of its .44 army revolvers for dealers Benjamin Kittredge of Cincinnati and J. W. Storrs of New York. These five-shot .46 rimfire conversions became the first large-caliber American cartridge revolvers legally produced in substantial numbers. Kittredge illustrated these revolvers on some of the firm's receipt forms, noting: "The[se] weapons are better than rifles for hunting. The Cartridges are inserted at rear end of [the] Cylinder without taking it out" (Figures 205, 206).[1]

The metallic cartridge was a major improvement over the fragile percussion cartridge and its separate percussion cap. Percussion arms converted to fire these improved cartridges bridged the gap before newly designed cartridge handguns became available following the 1869 expiration of the Rollin White patent. Conversions are found among various makes—Remington, Colt, Starr, and others. Sometimes it involved substituting a new cylinder with bored-through chambers and in other instances the percussion cylinder was retained but the rear portion which contained the nipples was removed and a spacer plate used to fill the gap. The percussion rammer assembly often was replaced with a form of ejector rod mounted in its place to punch out fired cases from the cylinder. Some arms, including some Colts, may appear to be conversions but actually weren't and left the factory in the early 1870s for the first time not as percussion weapons but in cartridge form during this transitional period. Other revolvers were altered not at the factory but individually by skilled gunsmiths. Regardless, conversions were far fewer in number for most model percussion revolvers than examples in original cap and ball form (Figures 207, 208).

Sam Colt throughout his career was an enthusiastic experimenter, seeking new production methods and improvements in his firearms. That tradition continued after the inventor's death. The Colt firm was one of the first to offer a conversion system applicable to several of its models. It didn't conflict with Smith & Wesson's patent rights and it was designed by Alexander Thuer and was available in late 1869. It fired a special tapered metallic cartridge which loaded from the front of the cylinder and the cartridge was seated using the existing per-

1. Ware, *Remington Army and Navy Revolvers, 1861-1888*, 360-61, 544.

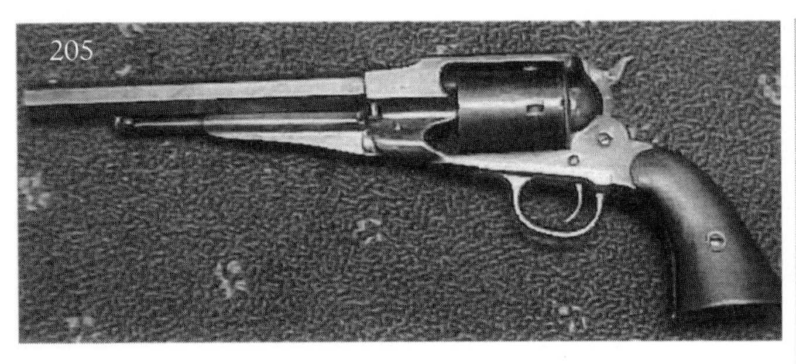

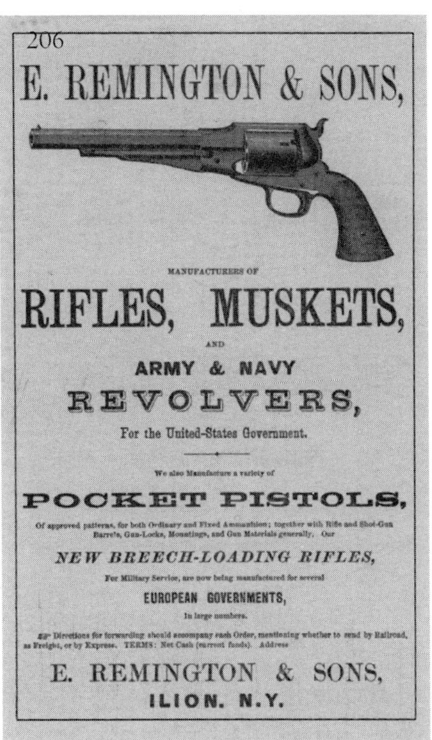

205. Remington .44 converted to a five-shot .46 rimfire revolver under special arrangement with Smith & Wesson. A plate fills the space created when the rear area of the percussion cylinder was removed. (*Courtesy: Neil and Julia Gutterman*) **206.** A Remington advertisement from the late 1860s. **207.** Remington advertisement for its .46 caliber five-shot altered revolver showing the ejector rod mounted beneath the barrel. (*Courtesy: Donald L. Ware*)

cussion rammer assembly. It was applied in very limited numbers to most of the then readily available models except .44 Dragoons and the 1855 side-hammer revolvers and revolving rifles. One selling point was the fact that a percussion cylinder could be substituted for the Thuer cylinder and the gun fired with loose powder and ball or percussion cartridge. However, sales were slow and modest in number at best.

Far more practical were the two Colt conversion systems designed by C. B. Richards and William Mason which loaded in the conventional fashion from the rear. The percussion rammer was replaced by a spring-loaded ejector rod mounted on the side of the barrel, although the ejector was left off some of the .36 pocket and police model Colts. One of these two systems was applied to all of the .36 and .44 models that were converted to rear loaders at the factory. For a brief period (1871-72) Colt manufactured a .44 rimfire six-shooter which closely resembled the Model 1860 with its open-top design but which was produced from the beginning as a cartridge revolver rather than a conversion. It soon would be replaced in the production line by the famed Single Action Colt (Figures 209, 210).

Meanwhile, Smith & Wesson in the summer of 1870 finally began marketing a large-frame .44 cartridge revolver, which they designated as the Model No. 3 (referring to its frame size), later in 1874 dubbed the American Model. This was a significant step forward, for it was the first large-caliber revolver made in the United States originally for metallic cartridges. The standard barrel length was eight inches, but when in 1871 the Nashville,

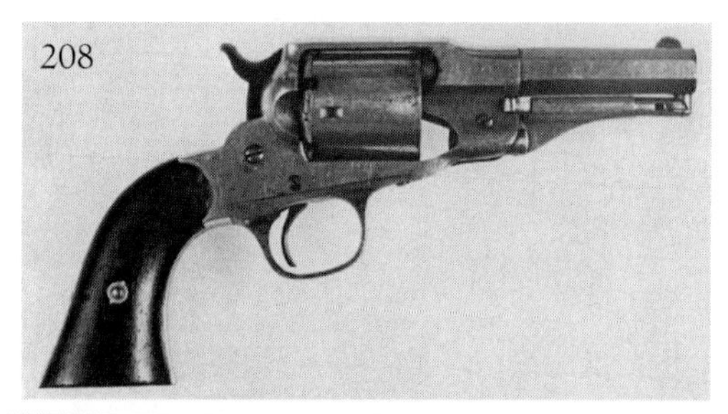

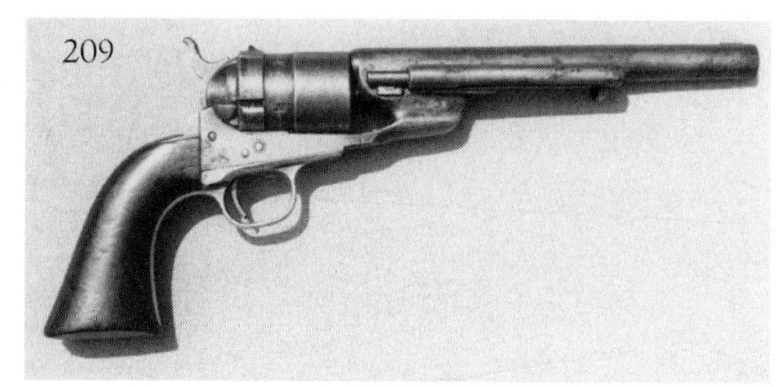

208. Factory alteration of a .36 caliber percussion Remington New Model Police revolver to .38 rimfire. The conversion cylinder has a removable spacer plate at the rear and a selling point was the ability to use a percussion cylinder if metallic cartridges weren't available. This specimen is marked as once being the property of the U.S. Express Company. **209.** A .44 Richards conversion of a Colt Model 1860. **210.** Colt .44 Open Top, at some point altered from rimfire to centerfire.

Tennessee, police department outfitted its force with thirty-two Americans, they specified a six-inch length. In 1873 Captain Robert H. Young, wrote to Smith & Wesson from Wyoming Territory complaining that the barrel of his .44 S&W often poked his horse in the neck, asking if he could order a .44 with a six-inch barrel.

Operation of the new S&Ws regardless of barrel length was simple. Releasing the barrel latch in front of the hammer and tipping the barrel downward operated the automatic ejection system, which emptied the cylinder of all the cartridge cases in one movement and readied the gun for reloading. This ejection system added to the gun's number of parts, and the difficulty in disassembly for cleaning was a factor in the army's rejection of this initial model after field trials were conducted. However, civilian sales of the new S&W .44 were strong, and the firm in May 1871 found an enthusiastic customer for the No. 3 in the Russian government, which by 1878 purchased almost 150,000 in several Russian model variations. Production of the American ended in 1874, but it inaugurated a long line of S&W No. 3 size revolvers sold domestically and overseas throughout the remainder of the century (Figures 211, 212).

Even though more advanced metallic cartridge handguns were available within a half dozen years after the Civil War ended, some people for reasons of economy, questionable quality of some early cartridges, personal preference, or their being in a remote area with an unreliable supply of ammunition continued to depend on older but now obsolete percussion revolvers. Colt, for example, was still producing percussion revolvers as late as 1873, although sales by then were modest. Arms dealer James Bown & Son's Enterprise Gun Works of Pittsburgh in their 1876 catalog offered the new Colt Single Action metallic cartridge revolver for $18 in .45 caliber, but one could still order a new percussion Model 1851 Navy for $14 or 1861 Navy at $15 from the same firm. In addition, sales of surplus Civil War military arms, both hand and long guns, made other obsolete but serviceable used weapons available at even lower figures, a tempting opportunity for price-conscious buyers.

A gunsmith's manual published in 1883 describing the handguns then in use had this to say about percussion Colt revolvers. "Though inconvenient, compared to the cartridge pistol of more modern make, the old [percussion] Colt's revolver is yet an excellent arm." The manual also noted that

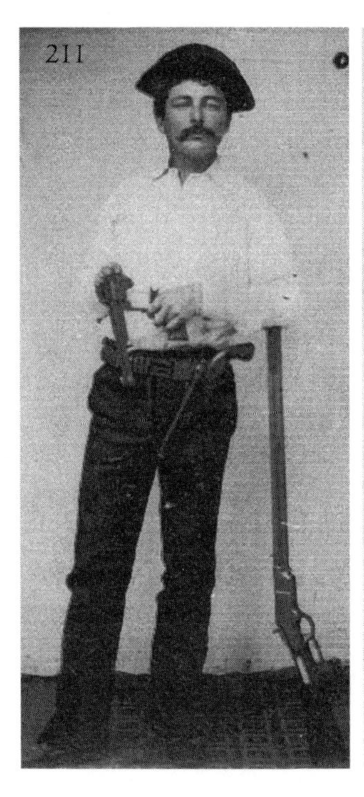

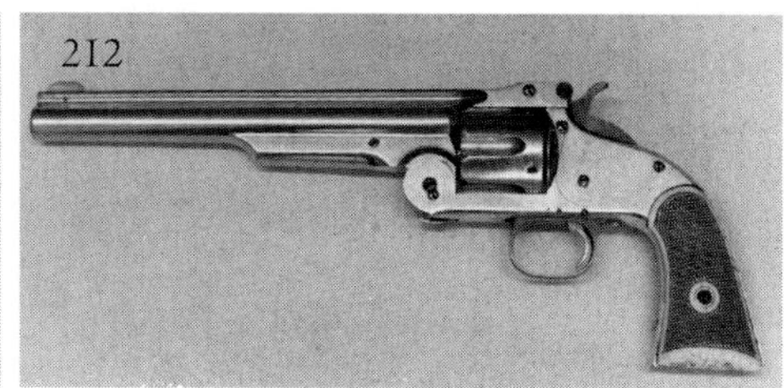

211. Well armed with a Henry .44 repeater and a pair of Smith & Wesson No. 3 revolvers. (Courtesy: Gary L. Delscamp) **212.** Smith & Wesson No. 3 .44 American owned by Gen. William J. Palmer, builder of the Denver and Rio Grande Railroad. The gun is nickel plated with checkered ivory grips. (Courtesy: Colorado Springs Pioneers' Museum) **213.** Popular as the combination of a Colt Single Action and a Winchester Model 1873 was in the west, hunters and other buyers in the east found the same marriage a convenience as this scene in a Chattanooga, Tennessee, photographer's studio indicates. (*Courtesy: Erich Baumann*)

"in rare instances a flint-lock 'horse pistol' . . . may put in an appearance."[2]

The year 1873 marked the introduction of what would become one of the world's most recognizable handguns, the Colt Single Action, or Model P as the factory designated it. Other nicknames included Peacemaker (in .45 caliber), Frontier Six-Shooter (in .44-40 caliber), as well as "hog leg," "thumb buster," and other colloquialisms. In the American trans-Mississippi west, it rather quickly became the most representative large-frame handgun of the last quarter of the nineteenth century. This popularity is reflected in photos and in writings of the period. The first production run reached almost 200,000 by 1900 and continued until 1940 when it halted at 357,000. Capitalizing on its appeal, in the late 1800s some Spanish- or Mexican-made copies were marketed in the southwest and elsewhere, sometimes so closely resembling the markings and appearance of the genuine article that from a careless glance they can pass for a Colt (Figure 213).[3]

Initially a customer could only buy a Single Action with a 7 1/2 inch barrel except on special order, but in 1875 5 1/2 and 4 3/4 inch lengths were added as standard. By special order, a barrel as short as 2 1/2 or up to 16 inches was available. The first Single Actions were marketed in .45 caliber, but in 1878 the manufacturer began chambering the Model P for .44-40 and in 1884 added .32-20 and .38-40. These offerings allowed an owner to pair his handgun with a Winchester Model 1873 rifle or carbine firing the same cartridge in .32, .38, or .44 caliber, a most convenient arrangement in view of the considerable popularity of the '73 Winchester. Eventually Single Actions were produced in a myriad of calibers from .22 to

2. Steele and Harrison, *The Gunsmith's Manual,* 37, 39.

3. J. P. Moore's Sons, one of the group of major Colt distributors known as the Allies, reportedly inaugurated use of the term "Frontier Six Shooter" for the Single Action in .44-40 caliber (Wilkerson, *Colt's Double-Action Revolver Model of 1878,* 28).

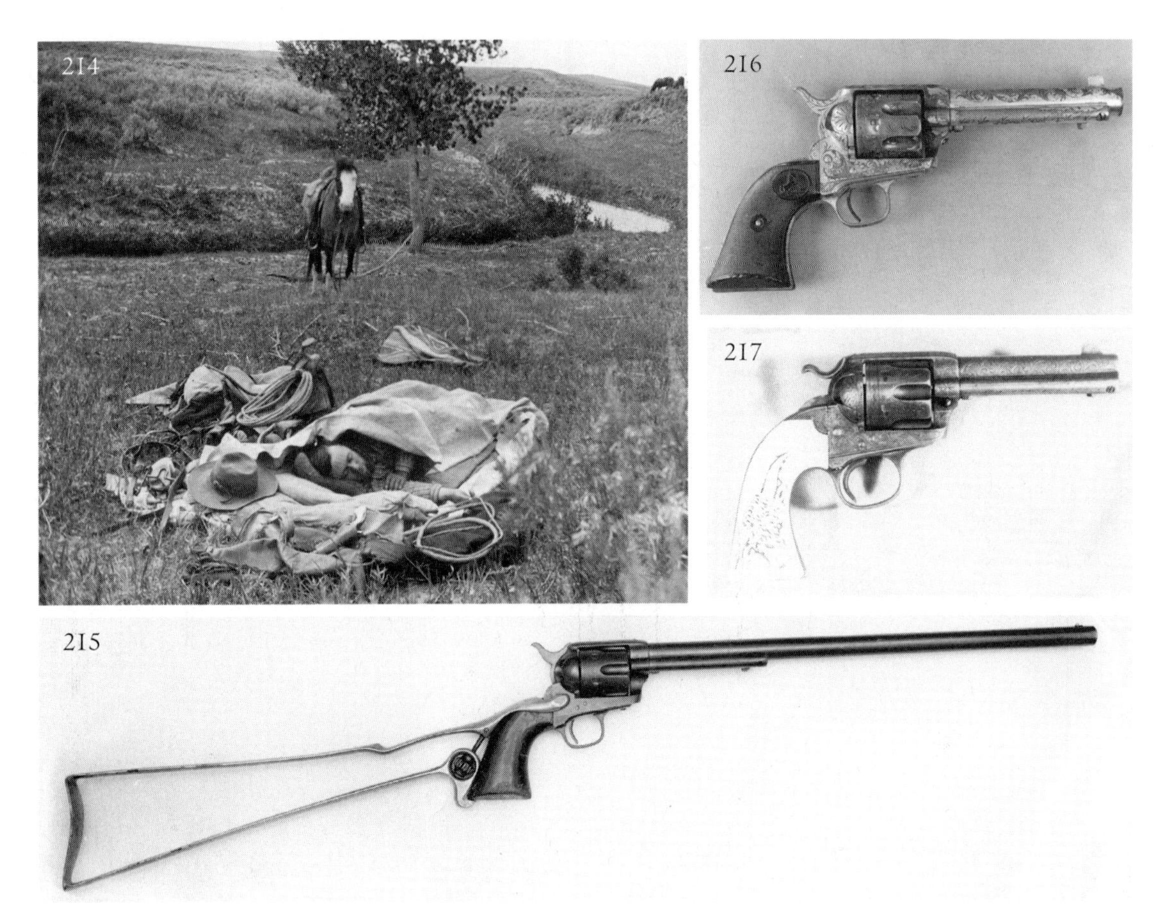

214. After a night guarding a cattle herd, a "night hawk" takes his rest. A holstered Colt Single Action is close by his head. "Drovers" was the common nineteenth-century term for those we call cowboys today. **215.** A Colt "Buntline Special" with 16-inch barrel and detachable skeleton shoulder stock. Factory records show only about a dozen were produced with this barrel length, most of them sold through B. Kittredge & Co. of Cincinnati. The concept obviously met with little enthusiasm. **216.** One of a number of Colt Single Actions purchased by Kansas bison hunter and later lawman William B. "Bat" Masterson, often as gifts for others. A factory letter confirms shipment of this .45 on October 23, 1879, with silver-plated finish and "W. B. Masterson" engraved on the backstrap. The original pearl grips were damaged and replaced with standard hard rubber checkered Colt grips. (*Courtesy: Kansas State Historical Society*) **217.** Colt Bisley, engraved with carved ivory grips and a 4 3/4-inch barrel, once the property of Montana cowboy and outlaw Dutch Henry. (*Courtesy: Montana Historical Society*)

British .476 Eley. In the frontier west of the nineteenth century, .44 and .45 calibers were most popular since they provided greater stopping power than did the smaller calibers offered later. Some who lived on the frontier had only contempt for small-caliber revolvers as practical sidearms, as shown by an incident reported in an 1886 Arizona newspaper. A citified tenderfoot drew a small .22 revolver on a drover (today we'd call him a cowboy) who had been harassing him in a saloon. The latter's response was to draw his own .45 revolver and demand: "Here, bring that damn thing over and let it suck" (Figure 214).[4]

4. Worman, *Gunsmoke and Saddle Leather*, 248.

A persistent tale involves western pulp fiction author E. C. "Ned Buntline" Judson's presentation of Single Actions with a special order 12-inch barrel to Wyatt Earp and several other Dodge City associates. Earp reportedly carried his gift as presented while the other recipients had the barrel shortened to a more convenient length. This story is based almost solely on Stuart Lake's highly controversial biography of Earp, and no factory records or nineteenth-century observations exist as confirmation (Figures 215, 216).

One major variation of the Single Action was introduced in 1894, a model designed with the target shooter in mind but available to all—the Bisley Model, named for the famed target range of that name in England. It had a lower and wider hammer spur, the trigger was widened and given increased curve, and the grip had a distinctive "humpback" configuration. It too could be had in numerous calibers, barrel lengths, and finishes—usually either blued with case-hardened frame and hammer or an optional nickel-plated finish which had been available on the Single Action after 1877 (Figure 217).

The Single Action was durable, easy to disassemble and repair, the grip was comfortably shaped, and even if the mainspring should break, it still was possible to fire the gun in a crude manner by striking the hammer with a rock or other hard object. Eventually it was offered in a caliber or barrel length to please almost anyone who wanted a large-frame handgun. To ready it for firing, one flipped the loading gate on the right side of the frame open with the thumb, placed the hammer on half-cock position to allow the cylinder to revolve freely, punched out the fired cases one at a time with the spring-loaded ejector rod mounted along the barrel, inserted fresh rounds (cartridges), closed the gate, and cocked the hammer. But if one was safety conscious, instead of loading six rounds, he loaded only five and kept the hammer resting on the empty chamber. Thus if the gun should fall and the hammer strike a hard surface, there wasn't a live round beneath the hammer to detonate. There was a safety notch on the hammer, but it wasn't foolproof so the colloquialism "loading five beans in the wheel" was the more appropriate safety measure.

Wyatt Earp on one occasion neglected to follow this safety precaution and had a potentially fatal reason to regret it. While seated at a table in a Wichita, Kansas, saloon in 1876 and as reported in the local newspaper, his revolver "slipped from its holster and in falling to the floor the hammer which was resting on the cap [primer] is supposed to have struck the chair, causing the discharge of one of the barrels [*sic*]. The ball passed through his coat, struck the north wall then glanced off and passed out through the ceiling."[5]

Frank Rollison wrote of his adventures as a Wyoming cowboy and recalled:

> We all preferred the Colt single-action six-shooter. Some liked the Bisley model [as Rollison did], others the Frontier model. Some were of different caliber, but all were built on a .45-caliber frame. I noticed that these men carried their gun with one empty shell in the cylinder, and five loaded cartridges. This was for safety's sake. The gun was carried with the hammer on the empty shell, and, when cocked, a loaded shell was ready to fire.[6]

In another incident, carelessness prevented the apprehension of several rustlers as drovers crept up upon their camp. Sam Tate's revolver slipped from its holster and the hammer apparently struck a rock, causing it to discharge. Despite the earlier warnings by a companion, he had continued to carry a full load of six cartridges in his revolver. He got a bullet through the brim of his Stetson hat and the rustlers leaped onto their horses and got safely away.[7]

5. Garavaglia and Worman, *Firearms of the American West, 1866-1894*, 295.

6. John K. Rollinson, *Pony Trails in Wyoming*, E. A. Brininstool, ed. (Caldwell, Idaho, 1941), 270, 287.

7. Worman, *Gunsmoke and Saddle Leather*, 255.

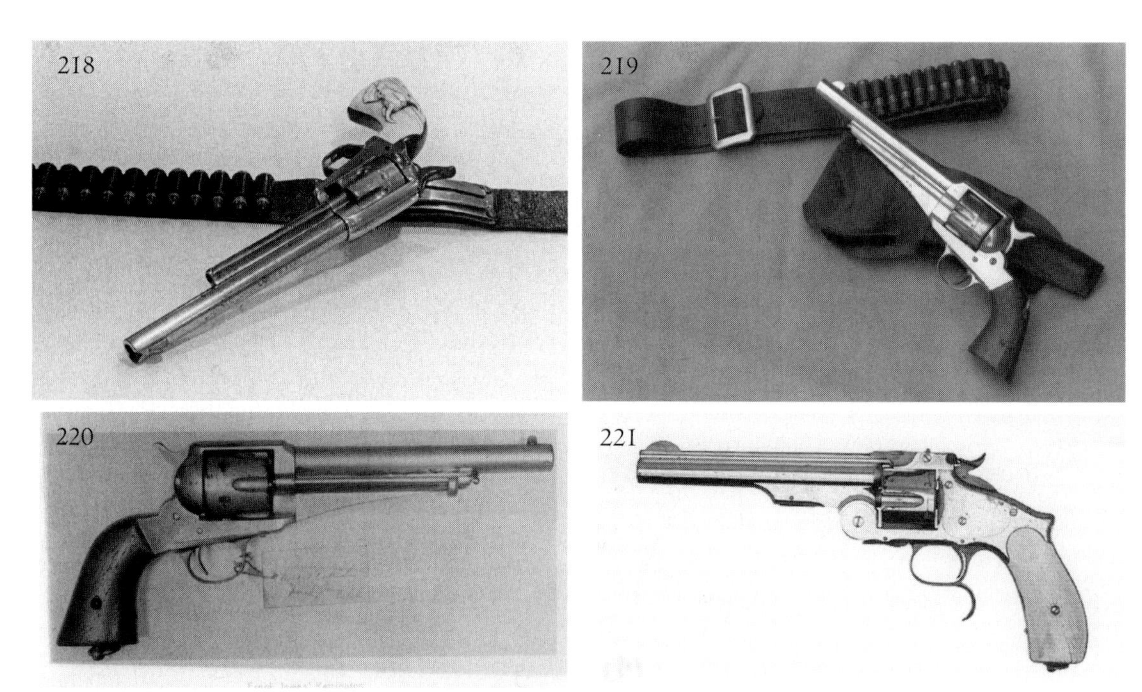

218. A Colt Single Action suspended from a Bridgeport rig, a substitute for a holster **219.** R. H. Bain joined the Alaskan gold rush and later settled in Dawson City, Yukon Territory, where he became a successful businessman. His son said his father had shot two claim jumpers with this Model 1875 Remington .44. The belt is stamped "Kennedy Hardware, Anchorage, Alas." (*Courtesy: Rod Smith*) **220.** Remington Model 1875 .44-40 surrendered by outlaw Frank James. **221.** Commercial version of the Smith & Wesson Third Model .44 Russian, manufactured between 1874 and 1878 for sale in the U.S. The factory had been operating at maximum capacity to produce more than 100,000 revolvers for the Russian government and dropped the earlier American No. 3 from the production line in 1874. The finger rest or "spur" on the trigger guard didn't meet with wide spread approval in the United States.

In December 1877, D. B. Wesson and James H. Bullard patented a rebounding hammer which came in contact with the cartridge primer only when the gun was fired. A blow to the hammer wouldn't discharge the revolver. Smith & Wesson first applied this to its spur trigger .32 Single Action pocket revolver and then to many of its New Model No. 3s, a safety feature not found on competing Colt and Remington single-action large-frame nineteenth-century handguns.

A novel means of carrying a Single Action was with the so-called Bridgeport rig, patented in 1882 by Louis S. Flatau as a "Pistol and Carbine Carrier." It consisted of a spring steel two-pronged belt clip from which a revolver was suspended by a special button-head screw which replaced the hammer screw, eliminating the need for a leather holster. One could slide the gun free from the clip or leave the gun in place but quickly swivel it into firing position and shoot from waist level. The army tested the invention but rejected it since it gave no protection from weather or sand and mud and didn't hold the gun securely, but it had modest appeal among some civilians. The rig also could be used with some revolvers other than the Single Action and appears in a photo of a Colorado deputy sheriff who carried a double action S&W in that manner (Figure 218).

Exclusive of those large-frame Remingtons converted from percussion, that firm's attempt at com-

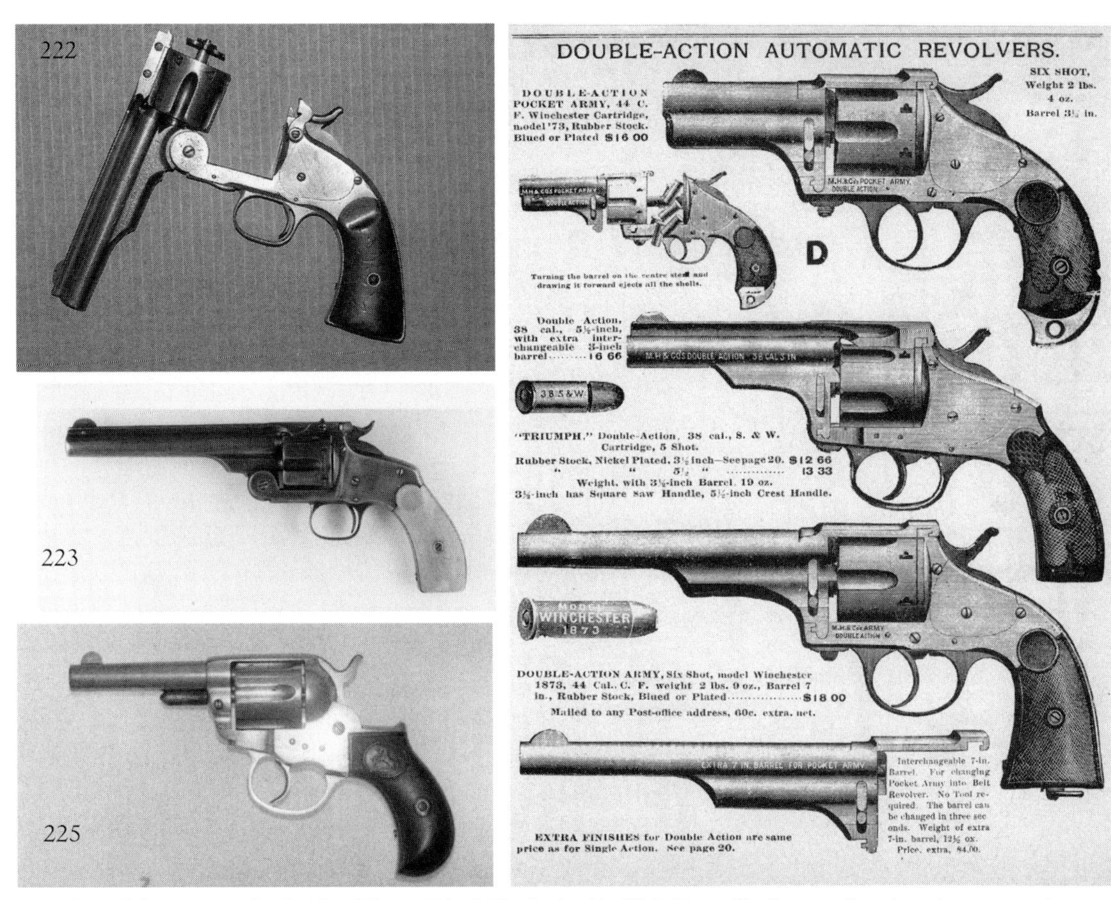

222. One of the army surplus Smith & Wesson Schofields obtained by Wells Fargo, "broken open" to show the extractor in operation **223.** Virgil Earp's Smith & Wesson .44 New Model No. 3 with ivory grips. Like his brother Wyatt, Virgil was a lawman in Dodge City and Tombstone before he was crippled by buckshot in an assassination attempt in Tombstone in December 1881. (*Courtesy: National Cowboy Hall of Fame*) **224.** An advertisement from the 1880s showing several models of Merwin & Hulbert revolvers. **225.** A peace officer's Model 1877 Colt .38 "Lightning," made about 1893 and carried by U.S. marshal W. G. Long. (*Courtesy: Charles L. Hill, Jr.*)

petition with the Single Action was introduced in 1875. It was quite similar in appearance to the Colt and operated in the same manner but was only offered in .44 and .45 calibers. Production between 1875 and 1889 totaled about 25,000, a far cry from the Model P, but much of Remington's production and sales effort was concentrated on foreign sales of hundreds of thousands of its rolling block single-shot shoulder arms. When Missouri outlaw Frank James surrendered to that state's governor Thomas Crittenden in 1882, James handed over a belt and holstered 1875 Remington in .44-40 caliber. When he did so, he announced: "My armament was two Remingtons and a Winchester rifle. The cartridges of one filled the chambers of the other" (Figures 219, 220).[8]

Smith & Wesson meanwhile in 1874 had ceased production of the .44 American and began marketing a succession of modified versions, some with such changes as a finger spur on the trigger guard

8. Worman, *Gunsmoke and Saddle Leather*, p. 181.

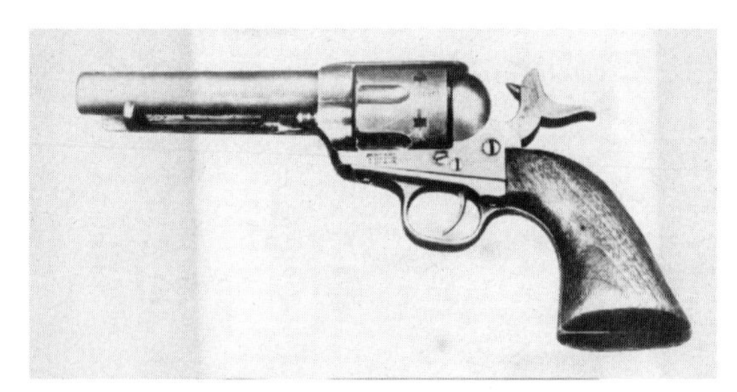

226. Colt .44 Single Action taken from Billy the Kid by Sheriff Pat Garrett and his posse at Stinking Springs, New Mexico Territory, in December 1880. The Kid and four companions withstood a siege of several days in an abandoned dwelling before surrendering. *(Courtesy: University of Oklahoma Library)*

or a modified grip shape as were applied to those guns being sold on the Russian contracts, changes that weren't readily accepted by some potential buyers in the United States. Although well made and quick to reload, the Russian models sold here commercially did not approach the Colt Single Action in popularity, although by the end of 1879 S&W had produced 215,000 No. 3s. Most were shipped overseas, to such countries as Japan and Turkey as well as Russia. About 700 .45 S&W Schofields, a model adopted by the army, reached the civilian market in the later 1870s. Some military Schofields were sold as surplus in the 1890s with a number of these going to the Wells Fargo express company with a barrel shortened to five inches. Outlaw Jesse James included a Schofield in his personal arsenal before his murder in 1882, despite the fact it required a special .45 cartridge slightly shorter than the more common .45 Colt round, but usable in the Colt (Figures 221, 222).

Smith & Wesson's New Model No. 3, introduced in 1878, was advertised for the next thirty-four years. Of the nearly 36,000 produced in almost a dozen calibers from .32 to Britain's .455, about one-third went to fill Japanese orders. Ira A. Paine, one of the premier target shooters of the day, used a target version of this weapon to establish national and international shooting records in the early 1880s. In doing so, he disproved the belief that the revolver couldn't match the accuracy possible with a single-shot target pistol (Figure 223).

Another major entrant into the arena of large-frame revolvers was the Merwin & Hulbert series of both single-action and double-action revolvers manufactured beginning about 1876 by Hopkins & Allen. Even though Hopkins & Allen produced lesser quality X-L and other model revolvers under the firm's own name, those bearing the Merwin & Hulbert label were well made and finely finished. The first Merwin & Hulbert .44s appeared with an open top but later models had a top strap over the cylinder. Most of the .44s were .44-40 and sometimes were marked "Calibre Winchester 1873," promoting the fact that these used the same cartridge as the popular 1873 Winchester rifle and carbine (Figure 224).

The ejection system, designed by Joseph Merwin, was unlike any other. By pressing a release button on the underside of the frame, the barrel and cylinder could be rotated a quarter turn to the side and then slid forward to expel the empty cartridge cases while leaving any unfired cartridges in place. This design also allowed easy removal of the barrel without tools so an owner could interchange barrel lengths of 7, 5 1/2, or 3 1/2 inches. A compact version was the Merwin & Hulbert Pocket Army, a .44 with a rounded "bird's head" grip and the shorter barrel length. Smaller frame .38 and .32 pocket models were available as well, utilizing the same ejection system and offering the same ease of barrel replacement.

As mentioned earlier, "self-cockers" or double-action revolvers were common in Great Britain in the 1850s, but such was not the case in the United States except in pepperbox form. Colt at the beginning of 1877 introduced its first such revolver. Major Colt distributor B. Kittredge & Co. of Cincinnati quickly nicknamed it the "Lightning" in .38 caliber "because it could shoot six Lightning bolts in two seconds." Later in that year Kittredge gave the name "Thunderer" to those in .41 caliber.

A few hundred .32s also were manufactured and in keeping with the meteorological theme, the Cincinnati firm dubbed those "Rainmaker." With some exaggeration, Kittredge advertised the .41 thusly: "It will serve good purpose for Buffalo shooting from horseback" (Figure 225).[9]

The Model 1877 looked like a smaller frame version of the Single Action but with a rounded "bird's head" grip shape. It came in various barrel lengths from 1 1/2 inches to ten with or without an ejector rod, although any length beyond six inches is a rarity. It was comfortable in the hand but it's surprising that it remained in production as late as 1912 with almost 167,000 made. Its double-action mechanism was delicate and many examples encountered today do not function properly. Attempts to repair them boost the sale of aspirin to gunsmiths! Nevertheless the American Express Company bought more than a thousand for its guards and messengers. In those days, when city policemen rarely carried a revolver openly, this model could easily be concealed in a hidden holster or in an overcoat pocket and equipped some metropolitan police departments including Kansas City, Washington, D.C., and Atlanta.[10]

The twenty-one-year-old killer "Billy the Kid" Bonney (Antrim) sometimes is said to have favored the 1877 Colt because it was a better fit in his small hand. Although he reportedly had a Thunderer in hand when he was shot to death in 1881, any preference he might have had for the Model 1877 was not exclusive. He surrendered a Single Action when he was captured at Stinking Springs, and in the only known identified photo of him he wears a holstered Single Action and holds an 1873 Winchester carbine. That same photo also led to the incorrect assumption that he was left-handed, for the photographic process produced a reverse image (Figure 226).

Colt in 1878 brought out its larger Double Action Model 1878, using the Single Action frame but retaining a rounded grip shape. Eventually it was available in many of the same calibers as the

227. Teenage outlaw Jennie Metcalf, also known as Little Britches, with a Colt Model 1878 "self cocker." Her career of crime in Indian Territory (Oklahoma) in the mid-1890s was short lived before she was arrested for selling whiskey to Indians and served a year in a Boston reformatory. (*Courtesy: University of Oklahoma Library*)

Single Action including the popular .45 and .44-40, but production only totaled about 51,000. Although its "self-cocking" mechanism was more durable that the Model 1877, the '78 had a relatively weak mainspring to counter a heavy trigger pull which sometimes resulted in a misfire. This was a major objection to it when the army tested and rejected it. Also its higher retail cost than the Single Action may have hindered sales, although it had its proponents including the Atlanta police department, army scout Jack Crawford (the "poet scout"), Medal of Honor recipient Lt. Marion Maus who led Apache scouts in pursuit of

9. Larry Hacker, "Thunderer & Lightning", *The Rampant Colt*, Spring 1999, 21-22

10. "Self-cocking" rather than "double-action" previously had been the common term. Smith & Wesson didn't adopt the latter terminology until about 1878.

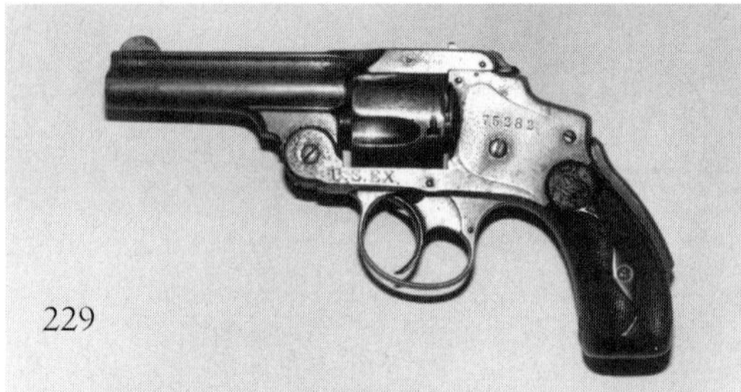

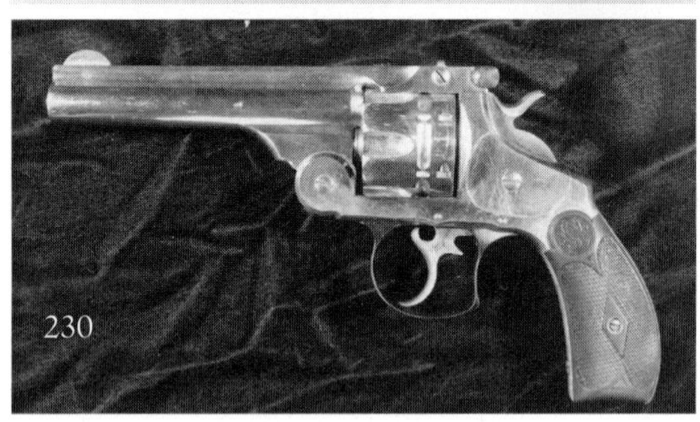

228. Cincinnati distributor B. Kittredge & Co. ad in the August 17, 1878, edition of the Army and Navy Journal illustrating the new Model 1878. Kittredge directed that the word "Omnipotent" be etched on the barrel of those .45 Model 1878 Colts he ordered with a 7 1/2-inch barrel. He also coined the nicknames for the Model 1877 Colt double action. **229.** Smith & Wesson .38 double-action Safety Hammerless or "lemon squeezer" with U.S. Express Company markings indicating its use by that firm's agents. (*Courtesy: Robert P. Palazzo*) **230.** Smith & Wesson .44 Double Action, introduced in 1881. Missouri governor Thomas Crittenden kept this revolver loaded in his desk as protection against retaliation by the James gang. (*Courtesy: Larry T. Shelton, Missouri State Capitol Museum, Jefferson City*)

Geronimo, the Canadian Department of Militia and Defence, and showmen Buffalo Bill Cody and G. W. "Pawnee Bill" Lilly. Internationally known pistol and rifle shot Walter Winans wrote that he won his first exhibition matches in England in the mid-1880s firing a Model 1878 chambered for the British .450 Eley cartridge (Figure 227).[11]

Colt relied on these two models, those of 1877 and 1878, as their only double-action offerings until 1889 when a new model appeared with improved mechanics. Now merely releasing a thumb latch on the left side of the frame allowed the cylinder to swing out to the left and with a single push on the ejector rod at the front of the cylinder, all six empty .38 or .41 cartridges were expelled at once. Eventually sales of these new double actions in several models to civilians and the government extended well into the new century and reached almost 300,000 (Figure 228).[12]

Remington manufactured no double-action cartridge revolver other than its four- or five-barrel Elliott ring-trigger derringer. Smith & Wesson had been receiving requests for a "self-cocker" as early as 1875, and in 1878 William C. Dodge wrote that his son had spent the winter in Hot Springs, Arkansas. "They won't buy any pistol there that is not self-cocking. . . . Colt's and F&W's [Forehand & Wadsworth] are exceedingly complicated and the former is very delicate." In early 1880, S&W did bring out its first double action, inaugurating what proved to be a very popular succession of pocket-size models of five-shot .38s and .32s. When production ended just after World War I, almost one

11. Wilkerson, *Colt's Double-Action Revolver Model of 1878*, 79.

12. Although the army rejected the Colt M1878, the government did purchase 5,000 in 1902 in .45 caliber with a six-inch barrel for issue to the Philippine Constabulary, a civilian police organization in the islands initially led by American army officers. The distinguishing characteristic of the arm is its long trigger and oversize trigger guard. These Colts apparently saw little use and until recent years were mistakenly called the Alaskan model by collectors.

million had been sold, each featuring an automatic cartridge ejection system similar to that first offered in the big American .44. After 1881, the buyer who wanted a large-frame double-action break-open S&W had access to one, a six-shot .38 or .44 (Figures 229, 230).

Much of the mystique and many of the myths of the larger frame six-guns relate to their use in the "wild west" which lay beyond the Mississippi. Hollywood in many instances has given a false impression. The character in a "B" western movie with a pair of Colts in a fancy double holster outfit slung almost at knee level probably would have brought coarse laughter or curious stares as he walked down a main street in Dodge City or Cheyenne or other frontier town in the 1870s or 1880s. The man who could fire two revolvers at the same time with any accuracy was a rarity, and any legitimate drover or other who spent hours in the saddle might have found his gun had slipped out of a low slung rig and fallen in the dust. Holsters were meant to protect a working tool like a revolver from dirt and to secure it comfortably in place while rendering it available while performing the hard work his job demanded of him. The typical drover made around $25 a month and a new Colt Single Action cost almost a month's wages so its loss wasn't insignificant. He often didn't have the extra money for the ammunition or the time to devote to hours of practice shooting so he rarely became as proficient as his counterparts seen on the movie screen. The true "gunfighter" was not a commonly encountered individual.[13]

Hollywood has vastly overdone the frequency with which gunfights occurred just as it has dramatically exaggerated the "fast draw." A rational man if he was expecting trouble was prepared for it with gun already in hand or perhaps even settled an argument with a shotgun blast fired from concealment in a dark alley. Movie makers also have exaggerated the number of violent deaths during the height of the cattle drives to the railroad shipping pens. One study of homicides between 1870 and 1885 in the five Kansas towns of Abilene, Ellsworth, Wichita, Dodge City, and Caldwell based on these cattle towns' newspapers showed only forty-five and that total included two lynchings. A year's high of five in any one town was reached only twice—Ellsworth in 1873 and Dodge City in 1878. Of these homicide victims, six were peace officers, nine were drovers (cowboys), and nine were gamblers.[14]

Wild Bill Hickok during his only term as a cattle town peace officer in 1871 in Abilene killed two men, one of whom was another policeman he shot by mistake. Wyatt Earp while serving as a lawman in Wichita and Dodge City may have mortally wounded one law breaker. "Bat" Masterson by one account killed twenty-six men during his lifetime but killed no one while in Dodge City, and his only actual slaying may have been Sergeant M. A. King of the 4th Cavalry, his rival for the attention of a saloon girl in Texas.[15]

One of the few documented face-to-face Hollywood-style western "shoot outs" occurred in July 1865 in Springfield, Missouri, between two gamblers and sometime drinking companions, James B. Hickok and Dave Tutt, before the former acquired notoriety as "Wild Bill." As the two men gambled, Tutt won all of Hickok's money as well as a gold pocket watch. As a favor, Hickok asked Tutt not to wear the watch in public so as not to embar-

13. Despite Hollywood's often romanticized image of Caucasian cowboys, drovers' ranks included African Americans, Mexicans, and some American Indians. For a detailed study of the various styles of holsters and cartridge belts used by them as well as other civilians and soldiers in the west, see Richard C. Rattenbury's comprehensive *Packing Iron: Gunleather of the Frontier West* (Millwood, N.Y., 1993).

14. Robert R. Dykstra, *The Cattle Towns* (New York, 1979), 145-47.

15. Ibid., 143. Joseph G. Rosa, *The Gunfighter: Man or Myth* (Norman, Okla., 1982), 67.

rass him. Tutt laughed at the request and Hickok threatened to shoot him if he did. The next day Tutt appeared on the street apparently wearing the watch or perhaps after boasting of his new timepiece. As the two men approached each other, Hickok drew and fired at a distance of about 75 yards, striking Tutt in the chest and killing him. Tutt reportedly was armed with a Colt Dragoon revolver and there is disagreement whether he fired or not before he was struck down. Hickok was tried but acquitted and although he later acquired a legitimate reputation for his skill with a revolver, his shot that day at that distance was a lucky one.[16]

Even with modern-day exaggeration, sufficient violence was present in the west to provide a writer of fiction with material. In Dodge City, a buffalo hunter witnessed a man standing at the bar place his revolver to another man's ear and literally blow his brains out. A dance-hall girl sitting on a billiard table jumped down and rubbed her hands in the pooling blood, clapped her hands together and shouted "cock-a-doodle-doo" splattering blood on her dress. Shaken by this callousness, the hunter left the scene and returned to camp.[17]

The respected western lawman Jefferson Davis "Jeff" Milton called bad-man-turned-attorney John Wesley Hardin the fastest man he'd ever seen with a gun. Some others whose life might depend on their dexterity with a revolver did become skilled although few achieved what's done before a television or movie camera. But in the 1930s a portly and unimposing exhibition shooter named Ed McGivern amazed spectators with his feats of fast and fancy revolver shooting, generally with twentieth-century double-action Smith & Wessons. He wasn't facing another opponent determined to kill him and he didn't perform within the confines of a smoky frontier saloon or with reflexes dulled by alcohol. However he consistently showed what could be done by "fanning" a Colt Single Action, holding the trigger back while repeatedly operating the hammer by striking it with the knife edge of the other hand. It's a fast way to empty a revolver although with little accuracy. But McGivern by fanning was able to fire five shots in 1.2 seconds and kept them in a group that could be covered by the palm of one's hand. He also drew and fired five shots into a man-sized target at ten feet in 1.6 seconds. In more recent years other demonstration shooters have performed similar amazing feats, but to achieve such success they have spent thousands of hours practicing.

While a revolver was handy on the range to dispatch a rattlesnake or crippled horse or cow, by the late 1870s some employers forbade the carrying of side arms to avoid accidental shootings or prevent arguments among employees from evolving into shooting frays. Even on lengthy cattle drives, it wasn't unusual for firearms to be kept in a wagon to be withdrawn only when needed. Employees of the giant Texas XIT ranch in the 1880s lived by a lengthy set of rules which included prohibitions on gambling and drinking intoxicants. Abusing or neglecting any mules, horses, or cattle could result in firing, and Rule 11 stated:

> No employee . . . or any contractor . . . is permitted to carry on or about his person or in his saddle bags, any pistol, dirk, dagger, sling shot, knuckles, bowie knife, or any other similar instruments for the purpose of offense or defense. Guests of the Company, and persons not employees of the ranch temporarily staying at any of its camps, are expected to comply with this rule, which is also a State law.[18]

16. In another Hollywood departure from reality, in any production involving Tombstone, Arizona, in the 1880s when the Earp brothers, Doc Holliday, and Curly Bill Brocius strode the dusty streets, scenes usually involve saloons, brothels, the Bird Cage Theater, or the O. K. Corral. In fact, the town really was rather sophisticated in that decade with a swimming pool, ice cream parlors, a bowling alley, and limited telephone service to nearby mines.

17. Ibid., 79-80.

18. Worman, *Gunsmoke and Saddle Leather*, 253.

By the 1870s, many organized settlements including the boisterous Kansas cattle towns, the railroad shipping point for many Texas herds, were enacting ordnances against carrying or discharging firearms, although not everyone abided by such restrictions. Abilene's city trustees in 1870 passed such an ordinance, but Texas drovers riddled the warning signs with bullets and tore down the first jail after "rescuing" its first prisoner, an African American cook. More of a sense of order followed with the appointment of Tom "Bear River" Smith as town marshal who kept the peace more often with his fists than a gun. After Smith was murdered, the controversial but skilled pistoleer "Wild Bill" Hickok enforced the law there for almost a year. In Caldwell, Kansas, during the 1880 cattle season of mid-April to mid-October, the police docket recorded 207 arrests. Prostitution or keeping a bawdy house totaled 62 arrests, drunkenness and causing a disturbance was second with 53, gambling 31, and fourth was 26 arrests for carrying or firing a weapon. Nevertheless to avoid antagonizing those whose herds of cattle brought welcome dollars to their towns, violations sometimes were overlooked or involved only a modest fine (Figure 231).[19]

A Caldwell newspaper article of September 25, 1879, undoubtedly caused one of the town's policemen embarrassment. Daniel W. Jones was accidentally locked in a bathroom in one of the town's better hotels.

> There is a seat in the room just opposite the door upon which Dan sat himself down, put his feet against the door, and with Heenan like strength pushed the door asunder, and at the same instant back went Dan's revolver, down to the bottomless—after which a light was brought into requisition—it was fished up, a tub of water, barrel of soft soap and scrubbing-brush were readily used up and the pistol looks as natural as ever, and if the street gossip don't mention this we will never say a word about it to Dan.[20]

231. Front Street in Dodge City, Kansas, in the 1880s. The sign above the one advertising prickly ash bitters warns: "THE CARRYING OF FIRE ARMS Strictly PROHIBITED." (*Courtesy: Kansas State Historical Society*)

Residents of Dodge City, Kansas, in late 1875 formed a city council which passed a law prohibiting the wearing of firearms within the town limits, similar to city ordnances in Abilene and Wichita. Guns were to be checked at stores, saloons, and elsewhere upon arriving in town and were not to be returned to drunks. By the 1880s, the practice of carrying a revolver on the range wasn't universal among drovers, although some persisted because of the sense of security it provided. Benjamin S. Miller recalled: "The Colt's '45' with its scabbard and belt of cartridges was a cumbersome affair, but so used did we become to it that we did not mind the extra weight and felt lost without it. Weapons of light caliber were an uncommon quality in cowland." Then he added: "One might carry a gun for ten years and never need it at all; then, again, he might need it like hell."[21]

James H. Cook wrote of a serious encounter of an unusual nature he had while herding cattle.

19. Ibid.

20. Ibid.

21. Ibid., 254.

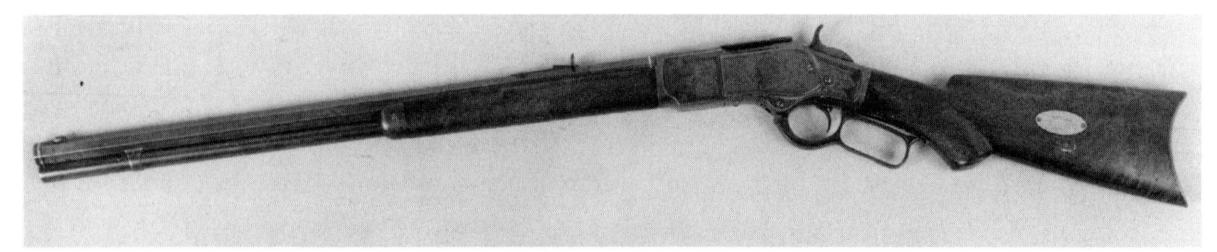

232. A deluxe engraved Winchester .44 Model 1873 rifle with a checkered past as well as a checkered pistol grip. On New Year's Day 1883 the citizens of Caldwell, Kansas, presented it to city marshal Henry Brown for his valuable services. But in the spring of 1884 he participated in the attempted robbery of a bank in Medicine Lodge, Kansas, in which the bandits shot and killed both the bank president and the cashier. Irate citizens captured the three robbers and later a mob broke into the jail. As he attempted to escape Brown was shot to death. His two companions died of "altitude sickness" as they swung beneath a convenient elm tree. (*Courtesy: Kansas State Historical Society*)

> [Jack Harris] had been having fun at my expense all along the trail. Every chance he could get he would ride up to me, suddenly draw his six-shooter, cock it, and aim it at me, saying, "Are you the sheriff that is looking for me?" Generally he would wind up this little act by taking his revolver by the barrel, his finger on the trigger guard, and reaching the butt toward me, and saying, "I'm tired of fighting; take my gun." Allowing some imaginary sheriff time to reach for it, he would reverse the weapon quick as a flash, cock it again, and aim straight for my face. All this, he said, was "just for practice." . . . One day before we reached Fort Griffin [Texas], he played his gun game on me once more. After he had put up his revolver, I said, "Jack, don't practice on me any more." "Why not?" "Because it would be dangerous for you," I replied. "If this sort of thing goes on, it will be only a matter of time before you will let your thumb slip, and I might go dead. Then you'd be sorry. I would rather be killed purposely than by accident. If you ever aim a gun at me again, you'd better shoot, for I surely will kill you if I can."[22]

Although the Pinkerton Detective Agency has received much publicity for its work in nineteenth-century America, another such firm of renown was that formed by Colorado lawman David J. Cook in the early 1870s, the Rocky Mountain Detective Association headquartered in Denver. Cook later put down a set of rules for survival by lawmen and which included various recommendations. Hollywood actors portraying western peace officers who "pistol whip" an antagonist violate Cook's first admonition: "Never hit a prisoner over the head with your pistol, because you may afterwards want to use your weapon and find it disabled." He went on to add such wisdom as to "have your pistol in your hand or be ready to draw" when attempting to arrest a desperado, "after your prisoner is arrested and disarmed, treat him as . . . kindly as his conduct will permit," and finally "never trust much to the honor of prisoners. Give them no liberties. . . . Nine out of ten of them have no honor."[23]

Whenever a shooting did take place, the injured's chances of survival in the nineteenth century as today depended upon a number of factors including proximity of medical assistance and the quality thereof, location of the wound, degree of damage done by the bullet, and health of the individual. A slow-moving .45 or .44 caliber slug could do substantial damage to one's vital organs and at close range the flame from the gunpowder could even burn clothing. But unless a bullet hit an artery and bleeding couldn't be stopped, wounds in an

22. Ibid.

23. Rosa, *The Gunfighter Man or Myth?* 58-59.

arm or leg generally didn't result in a fatality although infection was always a possibility and accounted for a number of deaths. Wounds to the torso or head were more serious. If no other antiseptic was present, whiskey sometimes was poured into a wound and occasionally a small amount of gunpowder was placed on the wound and lit to cauterize it. Although his injury didn't result from a gunshot, in the early 1870s an African American horse wrangler was bitten on the thumb by a rattlesnake. A companion cut the thumb around the wound, broke open a revolver cartridge, poured the gunpowder on the wound, and lit it with a match. The inflammation went no farther and the wrangler survived both the bite and treatment.[24]

For a comparison of prices of new handguns as the twentieth century dawned, Sears, Roebuck & Co. in 1902 offered Colt Single Actions beginning at $13.20, Smith & Wesson New Departure Safety Hammerless .32s for $11.75, Remington .41 Double Derringers were $5.00, and a .32 Forehand & Wadsworth "Double Action Automatic Police Revolver" for only $1.65. Falling in the category of an oddity was the nickel-plated .38 "New Harrington & Richardson Automatic [ejection] Bayonet Revolver" listed in the catalog for $4.65 or $1.10 more with pearl rather than hard rubber grips. The latter revolver is a rarity today and had a folding knife blade mounted beneath the barrel which when locked in place extended 2? inches in front of the muzzle. H&R made more than one million similar revolvers without the blade from about 1889 until just before World War II.

Civilian Rifles and Carbines

As the Colt Single Action became justly popular throughout the nation, so did Winchester's successor to its model of 1866, the Model 1873. It was one of the most popular civilian long guns of the late 1800s and remained in production for nearly a half century with a total production of a few thousand shy of 720,000. It operated in the same manner as the '66 but its steel frame was lighter and stronger than its brass-framed predecessor and it fired a reloadable centerfire .44-40 cartridge with 40 rather than only 28 grains of powder. Winchester added .38-40 caliber in 1880, .32-20 two years later, and finally .22 in 1884. However about 80 percent of the '73s sold were .44s. When Colt in 1878 began offering its Single Action revolver in .44-40 caliber (joined in 1884 by the .38-40 and .32-20), there was an instant marriage of convenience for anyone who wanted a handgun and a long gun firing the same ammunition—a Colt Single Action and a '73 Winchester (Figure 232).

One reason for the continuing popularity of the '73 Winchester may have been its ease of disassembly as illustrated by a Texas Ranger during a skirmish with Indians. As he was reloading his '73, he inadvertently slipped a .45 Colt cartridge into his .44-40 causing the gun to jam. While under fire he coolly used his knife blade to remove the screw holding the Winchester's side plates in place on the receiver, removed the offending cartridge, replaced the plates, and continued the fight. Nevertheless, all of the cartridges for which the '73 was chambered were revolver cartridges. If someone wanted long-range accuracy and shocking power, he continued to rely on a single-shot such as a Sharps, Remington, Ballard, or even a government .50 or .45 Springfield "trapdoor."

In the nation's centennial year of 1876, Winchester announced a new model sporting rifle which had been under development for some time. It closely resembled the '73 but with a larger frame designed to handle more powerful loadings. Production actually began in June 1877 in what was its most popular caliber, the .45-75. Additional offerings added in 1879 were the .45-60 with lighter recoil and the .50-95 Express Rifle tailored for the hunter seeking the most dangerous game in the United States or abroad. Competition

24. Edgar Beecher Bronson, *Cowboy Life on the Western Plains* (New York, 1910), 42-43.

233. Royal Canadian Northwest Mounted Police armed with Winchester Model 1876 carbines. **234.** Future president Teddy Roosevelt in hunting garb with a Winchester Model 1876 rifle across his lap. The rifle has a pistol grip and a half-length magazine.

from Marlin prompted the addition of Model 1876s chambered for the .40-60 cartridge in 1884. With a total production figure of about 65,000, the '76 never approached the '73 in sales. In 1878 the Canadian government bought the first of what would be almost 1,300 Model 1876 carbines in .45-75 caliber for issue to the Northwest Mounted Police. The use of those guns by that world-famous law enforcement body continued until just before World War I and brought added prestige to the Winchester name despite some problems the Mounties encountered with the '76 after long usage (Figure 233).[25]

Future president and ardent hunter Theodore Roosevelt was an enthusiastic Winchester promoter and owned three Model 1876s while ranching in Dakota Territory in the 1880s. But popular as TR became with many, his later hunting exploits sometimes drew criticism. When as president he began planning for retirement and a 1909 hunting trip to Africa, a British diplomat serving in South Africa, who was aware of changes taking place there, with commendable foresight suggested that instead of a rifle, Roosevelt should bring a camera and shoot with that. "Everywhere we witness the destruction of animals and birds indigenous to their native soils, and I am for preserving them rather than destroying them. Africa is no exception and big game there is slowly being exterminated." Nevertheless TR's party "collected" hundreds of mammals including 17 lions, 11 elephants, and 20 rhinos (Figure 234).[26]

25. Herbert G. Houze, *The Winchester Model 1876 "Centennial" Rifle* (Lincoln, R.I., 2001) 109, 126, 143, 156.

26. R. L. Wilson, *Theodore Roosevelt: Outdoorsman* (Agoura, Calif., 1994), 176.

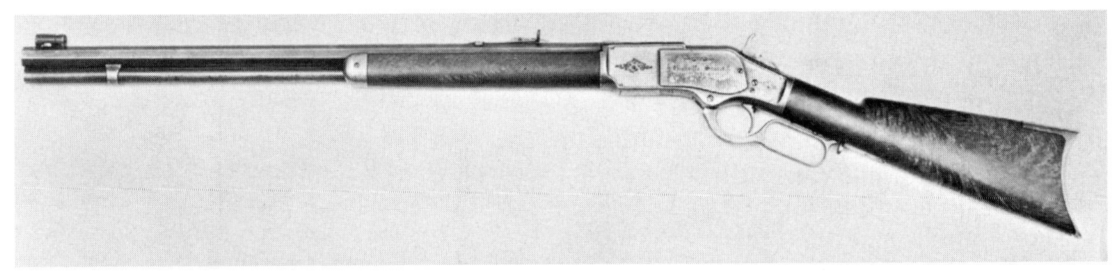

235. One of several Model 1873 Winchester "One of One Thousand" rifles ordered by rancher Granville Stuart for himself and friends. This .44-40 has his name and the date 1875 engraved on the left sideplate.

Winchester in 1875 embarked on what probably was an unwise marketing effort. In testing each batch of '73 sporting rifle barrels, those which were found to be particularly accurate were made up as higher grade guns and marked "One of One Thousand" and priced at $80 to $100. Those which didn't quite match that standard were marked "One of One Hundred" and sold at $60 to $75. Movie actor Jimmy Stewart decades ago starred in the film *Winchester '73* which featured one of the 136 "One of One Thousand" '73s produced. Sixty-one '76s were marketed under the same promotion, seven as "One of One Hundred" rifles, and the balance as "One of One Thousands." Eventually someone decided it wasn't wise to advertise that not every Winchester barrel shot as well as others and the campaign was scrapped. However today on the collectors' market genuine specimens of these Winchester rarities—and fakes are known—sometimes reach the six-figure range depending upon condition (Figure 235).[27]

To round out the story of Winchester repeaters during the nineteenth century, the firm broke with its tradition of lever-action guns and in 1877 acquired the right to manufacture the Hotchkiss .45 bolt-action repeater with a magazine in the hollow butt. Although the firm sold about 85,000 of these as sporting rifles and carbines beginning in the fall of 1879, their lever-action guns remained the best sellers. In 1886 Winchester added a new lever-action repeater which combined the design talents of the prolific John Moses Browning and William Mason. It was a more rugged action than the '76 and chambered several high power .40 and .45 caliber cartridges including the .45-70 Government and beginning in 1888 the .50-110 Express. One Wyoming cowboy bought a used .45-70 Model 1886 and took all winter to cut eight inches off the barrel using a meat saw. Theodore Roosevelt wrote:

> Now that the buffalo have gone and the Sharp's rifle by which they were destroyed is also gone, almost all ranchmen use some form of repeater. Personally I prefer the Winchester, using the new model [1886], with a forty-five caliber bullet of three hundred grains, backed by ninety grains of powder or else falling back on my faithful old standby, the [Model 1876] .45-75.[28] (Figure 236)

Two Winchester lever-action repeaters appeared in the 1890s which would each top the million mark in production. The first was a scaled down and lighter version of the '86 which appeared in 1892, firing the same three centerfire cartridges as the '73 and later the .25-20. Very similar in appearance to the '92 was the first Winchester civilian repeater designed to accept the higher pressure of smokeless powder ammunition, the Model 1894. This gun, most often sold in .30-30 caliber, prob-

27. Houze, *The Winchester Model 1876 "Centennial" Rifle*, 152.

28. Theodore Roosevelt, *The Wilderness Hunter* (New York, 1926), 370-72.

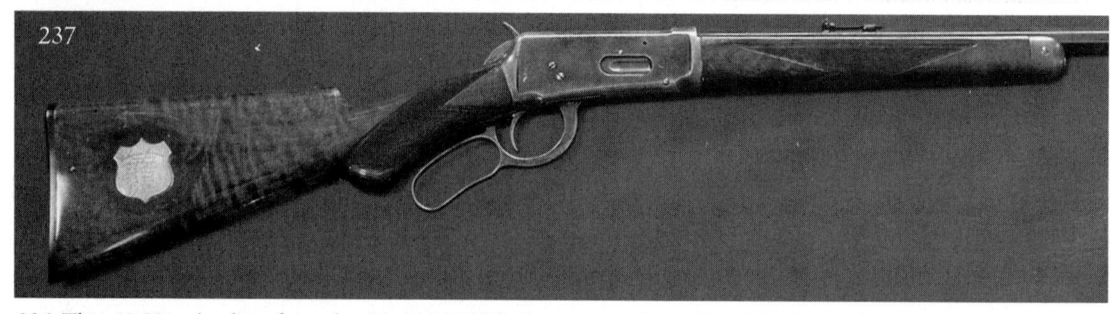

236. This .40-82 and at least four other Model 1886 Winchesters were shipped by Wells Fargo to H. J. Hagerman, foreman at John Chisum's South Spring Ranch & Cattle Company in Roswell, New Mexico Territory. Hagerman was a founder of the Denver & Western Railroad and obtained the rifles for possible use if force was necessary to settle a railroad workers' dispute. (*Courtesy: Ron Peterson, photo by Steven W. Walenta*) **237.** A deluxe Model 1894 .30-30 Winchester rifle with a political association in addition to its fancy burl walnut checkered stock, half-length magazine tube beneath the barrel, and silver presentation shield inscribed "Catherine Sherwood/Presented by/Wm. J. Bryan" followed by seven other names. Bryan was an unsuccessful four-time U.S. presidential candidate between 1896 and 1908 but perhaps is best known as the prosecutor in the famous 1926 "Scopes monkey trial" in which a Tennessee science teacher was accused of teaching Darwin's theory of evolution. Catherine Sherwood, a Philadelphia socialite and presumably a hunter as well, was active in the women's suffrage movement, an effort which Bryan supported strongly. In recognition of her actions, she was named Honorary Chairperson of the 1900 Democratic National Convention. The others named on the shield were also committee members. (*Courtesy: John F. Dussling, photo via Ronald Gabel*)

ably has killed more deer in the United States than any other firearm and by 1975 more than three million had been manufactured. Gone now was the telltale smoke and messy residue of black powder ammunition (Figure 237).

Close on the heels of the '94 came another design from John M. Browning, the lever-action Model 1895. Except for the Hotchkiss, previous Winchester repeaters had used a tubular magazine mounted beneath the barrel. However the lever action '95 was easily recognized with its distinctive box magazine located in front of the trigger guard. It was built to handle new smokeless powder loads, with either four or five cartridges in the magazine depending upon caliber. The Arizona Rangers were organized in 1901 and adopted the '95 in .30-40 caliber, the same ammunition as that used in the army's new Krag bolt-action repeater. Theodore Roosevelt following a hunting trip to Africa called his '95 in .405 caliber "the 'medicine gun' for lions." Although he could be a demanding customer, TR's unsolicited testimonials provided welcome advertising opportunities for the Winchester firm (Figure 238).[29]

29. When sharp-nosed bullets were used in a tubular magazine there was a risk that recoil or another sharp blow could cause a cartridge to discharge. A box magazine in which the cartridges were stacked on top of each other eliminated this risk.

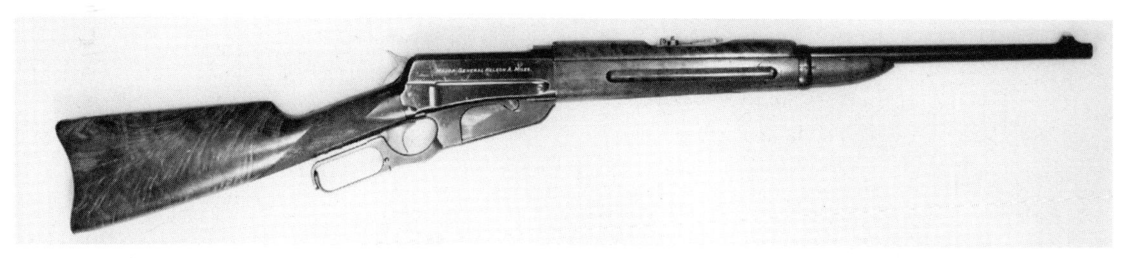

238. General Nelson A. Miles' deluxe Winchester 1895 carbine made in 1899. A silver plaque in the left side of the stock is inscribed "MAJOR GENERAL NELSON A. MILES/FROM HIS FRIEND/CAPT. J. R. HEGEMAN." The gun is in .30-40 caliber which would allow Miles to use standard army ammunition issued for the newly adopted Krag repeater. (*Courtesy: Norm Flayderman*)

In the west, "Winchester" had become largely synonymous for "repeating rifle." An unusual tribute to the popularity of the Winchester occurred in the late 1880s when a group of men sought to select a name for a new post office. Someone suggested naming it after the most representative firearm. Thus Winchester, Idaho, was named and still exists today, a small town near Lewiston. One former Wyoming cowboy recalled: "In 1891, at every small settlement and post office hitching rail in Wyoming, saddle ponies stood with a Winchester carbine in the scabbard. Every gun, whether rifle or carbine and regardless of make, was referred to as a 'Winchester.'"[30]

However another high-quality lever-action repeater that was to give Winchester its stiffest competition came from a cross-town neighbor in New Haven, Connecticut, the Marlin Firearms Company. From the first, the Model 1881 was chambered for the .45-70 Government cartridge followed by others in .32, .38, and .40 caliber loadings. Thus army officers with access to government ammunition and seeking a personal rifle probably gave a close look to the '81 Marlin with its magazine capacity of eight .45-70 rounds.

Subsequent models of lever-action Marlins included the lighter Model 1888 which fired the same .44, .38, and then .32 cartridges as the 1873 Winchester. The 1889 model was designed in part by Lewis Hepburn and was the forerunner of the modern Marlin lever-action repeaters in that the spent cartridge cases were ejected out of the right side of the receiver rather than the top, less of a distraction to the shooter and making it easier to mount a telescopic sight on top. Although Marlins were well-designed guns made to exacting standards, they didn't match Winchesters in sales. One feature that appeals to modern collectors is the brilliant colors of the case-hardened frame often found on the better condition antique Marlins encountered today (Figure 239).[31]

The buyer who wanted a lever action repeating rifle or carbine other than a Marlin or Winchester or the author today who wants to arm his character with a less common weapon has numerous other choices. The Whitney Arms Company, also located in New Haven, Connecticut, beginning about 1880 offered the Whitney-Kennedy in a small frame version which chambered those same three .32, .38, and .44 cartridges as the '73 Winchester and also a larger frame model in .40, .45, and .50 caliber. An historic example of a

30. Worman, *Gunsmoke and Saddle Leather*, 392.

31. While working as a Remington superintendent in 1879, Hepburn patented a rifle with a falling block action operated by a thumb lever on the right side of the frame, marketed in various styles as the Remington-Hepburn. In addition to being a talented designer, he also was a member of the American Creedmoor International Shooting Team and even made some of the barrels himself for the Creedmoor-style Hepburn target rifles.

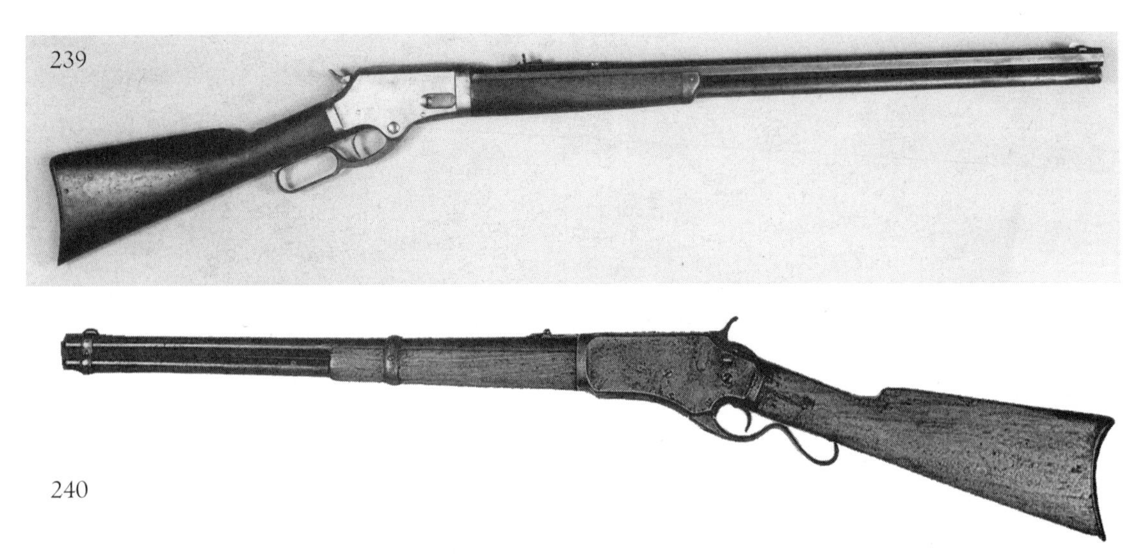

240

239. Marlin Model 1881 in .40-60 caliber brought to Dakota Territory by a sheep rancher in 1883. (*Courtesy: State Historical Society of North Dakota*) **240.** Whitney-Kennedy .44-40 carbine given to a lawman by Billy the Kid, authenticated by the former's daughter. Whitney eventually produced an estimated 15,000 of these repeaters in carbine, rifle, and musket lengths. (*Courtesy: Greg Martin*)

Whitney-Kennedy is a .44-40 carbine which "Billy the Kid" Bonney (Antrim) gave to a lawman. The unusual .44 Evans, introduced about 1873, not only was the only rifle mass-produced in Maine but also had the greatest magazine capacity of any nineteenth-century centerfire American repeater. The revolving magazine loaded through the butt plate and held up to thirty-four cartridges depending upon the model (Figure 240).

Colt in 1883 introduced its own lever-action .44-40 designed by Andrew Burgess, but production ended after only two years. It was followed in 1884 by Colt's "Lightning" slide-action (or pump-action) repeater, available in three sizes—first with a medium-size frame and chambered for those same three popular Winchester calibers, followed by a large frame in calibers up to .50-95 Express and a small frame .22 intended for small game hunting or casual target shooting or "plinking." Sliding the checkered hand guard mounted in front of the receiver backward and then forward extracted and ejected the fired casing, chambered a new round, and cocked the hammer (Figure 241).

It's often said that Colt ceased production of the Burgess after reaching a gentleman's agreement with Winchester whereby Colt wouldn't produce lever-action rifles and Winchester would shelve its own plans to manufacture revolvers. Winchester in 1876 had produced a dozen prototype double-action and single-action revolvers with a swing-out cylinder and had submitted samples to the U.S. Navy and also to Colonel Kasavery Ordinetz of the Russian army's ordnance department. Ordinetz was actively involved in his government's procurement of thousands of Smith & Wesson revolvers. In 1882, former Colt designer William Mason joined the Winchester firm and soon the model shop had turned out one .44-40 single action revolver which closely resembled the Colt Single Action, although neither it nor any of the earlier prototypes ever went into production. Thus circumstantial evidence supports the theory of an agreement between these two arms-making giants. It is also possible that Colt dropped the Burgess to concentrate its production and marketing resources on its slide-action Lightning model rather than compete with Winchester, Marlin, Whitney, and others already producing lever-action repeaters.

241. Cabinet card view of perhaps a hunter, armed with a Colt pump or slide-action "Lightning" rifle and what appears to be a Colt Single Action. The guns may have been the photographer's props in this St. Paul, Minnesota, studio. (*Courtesy: Erich Baumann*)
242. A nattily dressed and well-armed westerner with a Marlin-Ballard Pacific No. 5 sporting rifle. (*Courtesy: Idaho State Historical Society, neg. no. 79-127-1*)

By the early 1880s there were repeating rifles being produced in the U.S. which were capable of killing any game found in this country whether that be moose in Maine or grizzlies in Montana. However, there were still those shooters who preferred the single shot whether their intended use was target shooting or hunting. Production of Remington rifles at this time continued in a variety of styles. Although the manufacture of Sharps arms had ceased at the beginning of the decade, some arms dealers, such as E. C. Meacham of St. Louis, continued to offer their own Model 1874 Sharps style rifles, assembled by their gunsmiths. Dealers beginning in the late 1870s also sometimes advertised "sporting Springfields," rifles assembled using a mixture of surplus obsolescent government parts and those which were newly manufactured by the dealers. The quality of these "trapdoor" rifles varied, but some were of superior workmanship. Winchester in 1883 acquired the rights to another John M. Browning design, a single-shot rifle which Winchester manufactured until just after World War I in calibers from .22 to .50. Known today as the High Wall or Low Wall depending upon frame type, one could secure an example of this first Winchester single shot in any of numerous styles from a plain sporting carbine to a schuetzen-style target rifle.

Certainly not to be ignored among the many single shots of the late 1800s were the Ballard and the Stevens, favorites of many shooters. Initial manufacture of Charles H. Ballard's military and sporting long guns began during the Civil War, and they were among the earliest and most successful of metallic cartridge rifles. They were made by five different manufacturers before 1875 when John Marlin (after 1881 the Marlin Firearms Co.) took over their production. An unusual feature found on

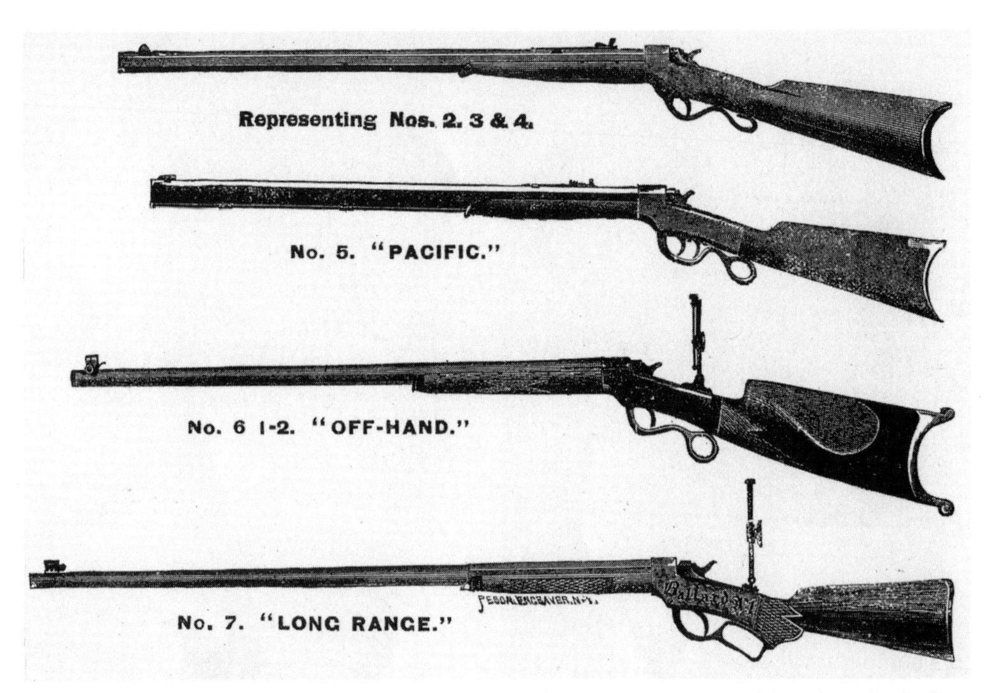

243. Homer Fisher of New York City in 1880 illustrated a variety of sporting and target model Marlin-Ballard rifles.

the early Marlin-Ballards was a reversible firing pin which allowed one to use rimfire or centerfire ammunition. Some pre-Marlin-Ballard sporting rifles had featured a similar dual ignition system with a swivel striker on the hammer allowing use of rimfire ammunition or the percussion nipple mounted on the breechblock. Both sporting models such as the Hunter's Rifle or the Pacific Rifle and such target models as the Union Hill and the Schuetzen Off-Hand bore the popular Marlin and Ballard names (Figures 242, 243).

Although his efforts were short lived, James J. Bullard marketed another competitor to Winchester and Marlin lever-action repeating rifles. Bullard left the employ of Smith & Wesson and in 1880 formed his own company which first began manufacturing single-shot breech-loading target and hunting rifles. Six years later he added a lever-action repeater which quickly gained a reputation among sportsmen for superior quality workmanship and unusual smoothness of operation. It too used a tubular magazine beneath the barrel, but it was rather unusual in that it loaded from the underside of the frame rather than through a loading gate on the side. The gun also could be loaded as a single shot even with a full magazine. One could purchase a Bullard repeater in .32 or .38 caliber in a small frame version or a larger frame model in calibers from .40 to .50 including the popular .45-70 government size. Unfortunately the Bullard Repeating Arms Company couldn't withstand competition from Winchester and Marlin and ceased production in 1890 with only about 11,500 repeating rifles completed.[32]

Joshua Stevens had worked for various gun makers including Colt and Eli Whitney, Jr., before he began making single-shot metallic cartridge pistols in the mid-1860s. These soon evolved into a lengthy series of popular single-shot pistols with a release button on the left side of the frame which allowed the barrel to tip down to load. The Stevens line extended into the twentieth century in various

32. McDowell, *Evolution of the Winchester,* 167.

244. A trio of young hunters. The lad in the center holds a Stevens single shot tip-up rifle, manufactured from the 1870s until the mid-1890s in calibers from .22 to .44. (*Courtesy: Wade Lucas*) **245.** An example of a J. Stevens .22 offhand target pistol. Pressing the release button on the left side of the frame allows the barrel to tip down.

sizes and calibers of pistols and rifles, styles for serious target shooting or casual sporting use (often for hunting small game), and even "pocket rifles" with various size frames and calibers with a detachable metal shoulder stock. Stevens never made a nineteenth-century military arm but the "tip-up" design carried over into full-size rifles for the civilian market, some of which were designated in an 1888 factory catalog as ladies' rifles. In 1894 the first Stevens Ideal lever-action falling-block single-shot rifles appeared, and as often was the case were offered in various styles from rather plain sporting models to those aimed at the serious target shooter including the "Walnut Hill" and schuetzen models (Figures 244, 245).

The J. Stevens Arms and Tool Company by the early 1900s also was becoming one of the most prolific makers of inexpensive small caliber "boys' [and girls'] rifles", most in .22 caliber, marketed as the Maynard Junior, Crack Shot, Favorite, and other models. Many a youngster learned to shoot with one of these moderately priced Stevens. As a boy, my own first shooting gun was a used .22 Stevens Favorite. It hadn't been particularly well cared for and the bore was worn, and although I had no desire to shoot birds or animals, it did perforate many a tin can despite its age.

Shotguns in Many Forms

Thousands of breech-loading shotguns were imported from England and continental Europe into the United States during the later years of the nineteenth century, competing with those produced domestically. Bore sizes of 10 or 12 gauge were quite popular and among the better known imports of the 1870s and 1880s were those double-barrel guns by Greener and C. G. Bonehill of England at around $35 and the less expensive European LeFaucheux. Among the many lower or middle grade double guns are those imported or manufactured by such firms as H. & D. Folsom Arms Company of New York. This particular company is known to have marked guns it handled with any of more than 100 different trade names requested by hardware companies, distributors, and other quantity buyers including Sears, Roebuck & Company. Among American makers of better quality breech-loading doubles in the last quarter of the nineteenth century in addition to Colt and Remington were the Baker Gun Company, Charles Daly, Ithaca Gun Company, Parker Brothers, and L. C. Smith. Lesser quality guns bore such names as Crescent, Great Western Gun Works, Hood, and many others (Figure 246).

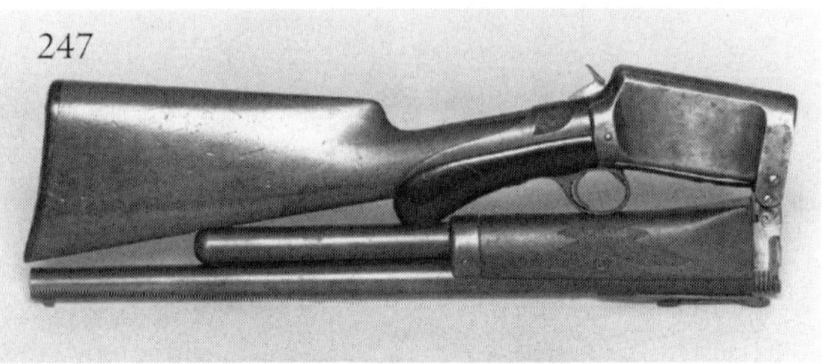

246. A pair of shotgun-armed lady hunters from Massachusetts pose during a hunting excursion on the Northern Pacific railroad in the late 1870s. (*Courtesy: Montana Historical Society*) **247.** Burgess folding shotgun. (*Courtesy: Mark Wright*)

Repeating single-barrel shotguns gained acceptance in the last two decades of the nineteenth century. After 1884 the Spencer slide-action repeater, holding five shells in a tubular magazine beneath the barrel, was marketed rather widely. This sporting 12-gauge shotgun was the product of Christopher Spencer, designer of the famed Civil War seven-shot rifles and carbines, in conjunction with Sylvester Roper. Winchester's Model 1887 lever-action repeating 10- and 12-gauge shotguns, all with a five-shot capacity, were designed by the famed John M. Browning, and in 1893 Winchester added a slide-action repeater to its shotgun line. Just as Colt's "Lightning" rifle operated, sliding a wooden handguard beneath the barrel backward and then forward ejected the spent casing and chambered a new shell. A rather unusual weapon warranting mention is one version of the Burgess slide-action 12-gauge of the 1890s. The barrel and magazine group was hinged at the forward end of the receiver (frame) and could be folded back to lie against the underside of the butt stock making a compact weapon measuring only about 20 inches in length when folded. The company promoted it as particularly well adapted for use by lawmen, express company messengers, and prison and bank guards and offered a "quick draw" belt holster for it (Figure 247).

A feature applicable to shotguns was the idea of adding a "choke" or inward taper to the inside of the barrel. The barrel could be left as an unrestricted "cylinder" or "straight" bore but a choke restricted the spread of the shot as it left the muzzle and kept more pellets within a given circle adding distance to the killing range. In a British patent of 1866, W. R. Pape mentioned tapering a shotgun barrel inward, but gave no more explanation of the procedure. Sylvester H. Roper and Dexter Smith received U.S. patents on forms of detachable choke devices in 1868 and 1871, respectively. Although W. W. Greener's first ad for a choke-bored gun appeared in an English sporting paper in late 1874, it apparently wasn't until the late 1870s that the idea of shotgun choke began to gain favor in the United States (Figure 248).[33]

33. There also is evidence that some form of choke boring was attempted in Britain by the 1790s.

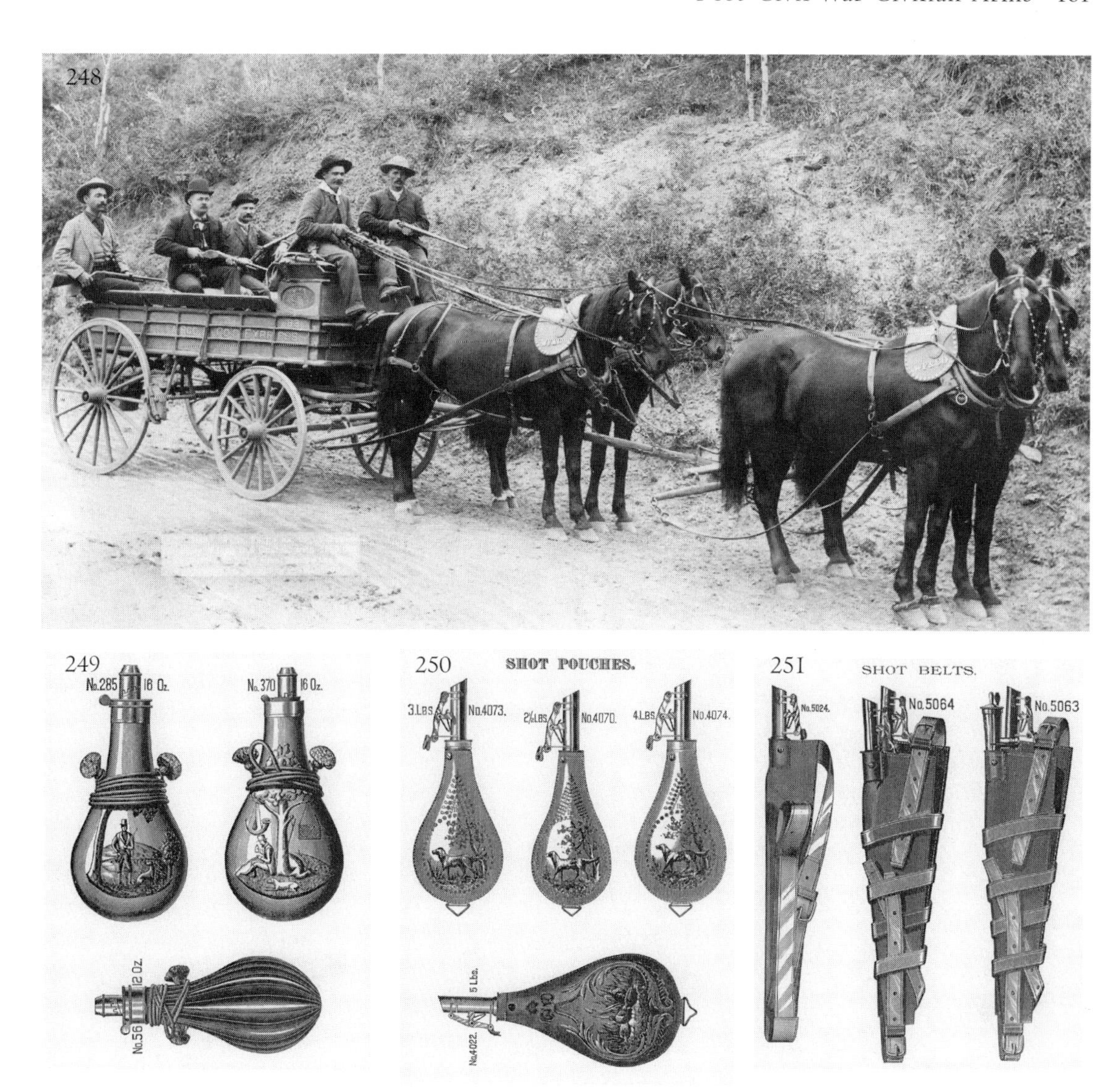

248. Wells Fargo & Co's. Express wagon guarded by four shotgun-armed messengers. The two guards at left are carrying five-shot Winchester 1887 or 1901 lever-action shotguns, the other two are equipped with double-barrel guns with external hammers. (*Courtesy: Library of Congress*) **249, 250, 251.** The ongoing use of muzzle-loading percussion shotguns in the 1880s and 1890s supported continued sales of such accessories as leather shot pouches and shot belts. Powder flasks advertised in these same years were necessary for both percussion rifles and shotguns still in use.

There always was a market for inexpensive but serviceable single-barrel shotguns. In the post-Civil War years, many of the rifle muskets sold by the government as military surplus were bought by dealers who converted them into low-price muzzle-loading shotguns by removing the rifling and often cutting back the forestock to lighten the arm. In its 1893 catalog, M. F. Kennedy & Bros. of St. Paul, Minnesota, was still offered such shotguns for $4 to $5 in comparison to $6 to $19 for muzzle-loading doubles, $30 for a plain breech-loading Greener double, $40 up for a similar Remington,

253

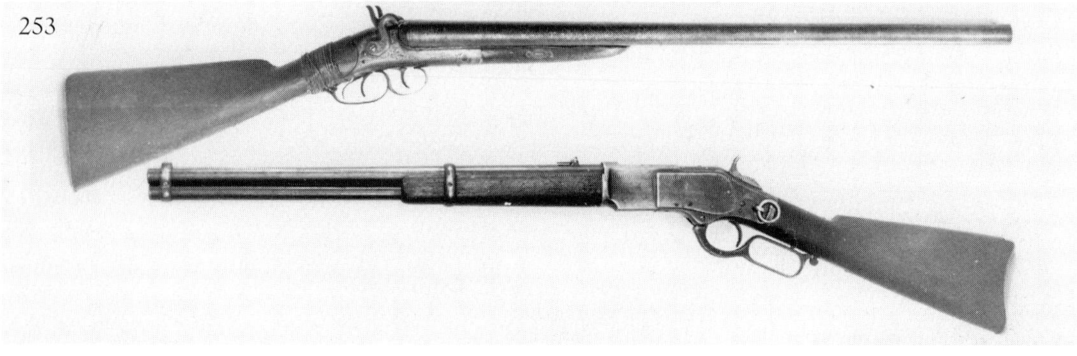

254

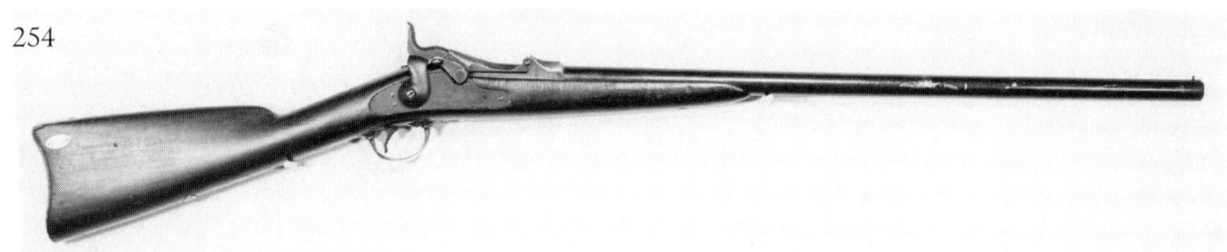

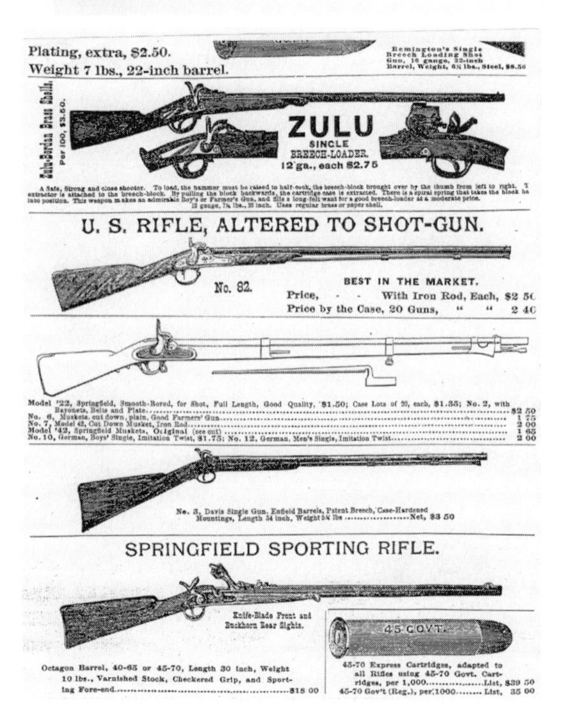

Plating, extra, $2.50.
Weight 7 lbs., 22-inch barrel.

Remington's Single Breech Loading Shot Gun, 16 gauge, 32-inch Barrel, Weight, 6¾ lbs., Steel, $8.50

Zulu-Berdan Brass Shells. Per 100, $3.50.

ZULU
SINGLE
BREECH-LOADER.
12 ga., each $2.75

A Safe, Strong and close shooter. To load, the hammer must be raised to half-cock, the breech-block brought over by the thumb from left to right. T extractor is attached to the breech-block. By pulling the block backwards, the cartridge case is extracted. There is a spiral spring that takes the block ba into position. This weapon makes an admirable Boy's or Farmer's Gun, and fills a long-felt want for a good breech-loader at a moderate price.
12 gauge, 7¾ lbs., 32 inch. Uses regular brass or paper shell.

U. S. RIFLE, ALTERED TO SHOT-GUN.

No. 82.

BEST IN THE MARKET.
Price, - - With Iron Rod, Each, $2 50
Price by the Case, 20 Guns, " " 2 40

Model '22, Springfield, Smooth-Bored, for Shot, Full Length, Good Quality, $1.50; Case Lots of 20, each, $1.35; No. 2, with Bayonets, Belts and Plate $2 50
No. 6, Muskets, cut down, plain, Good Farmers' Gun 1 75
No. 7, Model 42, Cut Down Musket, Iron Rod 2 00
Model '42, Springfield Muskets, Original (see cut) 1 65
No. 10, German, Boys' Single, Imitation Twist, $1.75; No. 12, German, Men's Single, Imitation Twist 2 00

No. 3, Davis Single Gun, Enfield Barrels, Patent Breech, Case-Hardened Mountings, Length 34 inch, Weight 5¾ lbs .... Net, $3 50

SPRINGFIELD SPORTING RIFLE.

Knife-Blade Front and Buckhorn Rear Sights.

45 GOVT.

Octagon Barrel, 40-65 or 45-70, Length 30 inch, Weight 10 lbs., Varnished Stock, Checkered Grip, and Sporting Fore-end $18 00

45-70 Express Cartridges, adapted to all Rifles using 45-70 Govt. Cartridges, per 1,000 .... List, $39 50
45-70 Gov't (Reg.), per 1000 .... List, 35 00

252, left. Page from an 1884 catalog by E. C. Meacham of St. Louis advertising the inexpensive Zulu breech-loading shotgun as well as muzzle-loading shotguns made from former U.S. military muskets and rifle muskets and also Springfield pattern "trapdoor" sporting rifles sometimes assembled by dealers. **253.** Whitney 10-gauge double barrel shotgun used by "Billy the Kid" Bonney (Antrim) to kill deputy sheriff Bob Ollinger during his escape from the jail in Lincoln, New Mexico, in 1881. The Winchester Model 1873 carbine (below) was the Kid's and was authenticated by Sheriff Pat Garrett in 1883, two years after he killed the twenty-year-old outlaw. (*Courtesy: Robert McNellis*) **254.** Springfield 20-gauge Model 1881 shotgun. (*Courtesy: National Park Service, Fuller Gun Collection, Chickamauga and Chattanooga National Military Park*)

$25 for a basic Winchester lever-action repeating shotgun, or for the truly affluent, a $450 "Premier Quality" hammerless English double by W. and C. Scott & Sons. The Sears, Roebuck & Co. catalog in 1902 advertised cut-down Springfield Model 1863 rifle muskets altered to shotguns for $2.75—as well as Spencer carbines with 25 rounds of ammunition for $3.65 (Figures 249, 250, 251).

Another example of an inexpensive conversion of what originally was a military piece was the "Zulu" single-barrel breech-loading shotgun which began life as an 1857 French muzzle-loading rifle musket. Thousands of these were marketed in the U.S. and elsewhere after conversion to shotguns beginning in the mid-1870s. Homely though the Zulu was, it met a poor farmer's or homesteader's need for a serviceable shotgun at a low price (Figure 252).

While the shotgun has been used most often for hunting, it is a formidable weapon when loaded with buckshot. As mentioned earlier, trappers in the west in the early 1800s described its usefulness against night marauders, stagecoach and express companies in the west often equipped their guards with them, and lawmen for generations and even today have relied on shotguns as potent pacifiers when a spread of shot is appropriate. Of course a shotgun could be just as intimidating when held in a bandit's grip. A Wells Fargo employee in the late 1870s later recalled that "we stage drivers were furnished . . . sawed-off shotguns especially made in the East for the company. The shot-gun barrels were charged with 7 1/2 grams [115 grains] of powder and loaded with 16 buckshot in four layers, with four shot to the layer." Another Wells Fargo messenger of the early 1880s in Arizona Territory instead of riding beside the driver crouched with a sawed-off shotgun in the coach's rear boot where luggage and other baggage was stowed. "The bandits shot some good men off the high seat without warning but they quit [harassing] the Dripping Springs line after they had been given a salute from the great leather boot at the rear" (Figure 253).[34]

One of the most respected lawmen in the west was U.S. marshal Bill Tilghman. In January 1896 he traveled to Eureka Springs, Arkansas, following a report that outlaw Bill Doolin had been seen there. The marshal prepared for the anticipated confrontation in a most unusual way. He went to a carpenter and had a wooden box constructed which he could carry beneath his arm and in which he could conceal a shotgun. With a slight movement of his hand, the box would fall away leaving him with the scattergun in his grasp. But before he could put this ruse to use, he found Doolin reading a newspaper at a mineral bath and made the arrest without incident.[35]

Another western lawman, Chauncey B. Whitney, had a shorter reign as a peace officer before he fell victim to a shotgun blast. He'd survived the battle of Beecher's Island in the fall of 1868 and after several years as a constable in the cow town of Ellsworth, Kansas, was elected county sheriff in November 1871. In the summer heat of August 1873 tempers flared between brothers Ben and Billy Thompson from Texas and two others over a gambling debt. A drunken Billy, armed with Ben's English Gibbs breech-loading shotgun, accidentally discharged one barrel almost injuring several bystanders. Whitney arrived to pacify the group and even though there apparently was no animosity between the sheriff and the Thompsons, he received a load of buckshot from Billy's shotgun and would die an agonizing death several days later. In reply to Ben's admonishment for what he'd done, Billy reportedly said he didn't give a damn and would have shot "if it had been Jesus Christ." Billy was arrested in Texas in 1877 and stood trial in Ellsworth but justly or unjustly was acquitted on the grounds that the shooting was accidental. As an example of how legends become reported fact, it's said that Wyatt Earp at the time of the shooting stepped forward and offered to arrest Thompson, but there's no evidence placing Earp in town at the time.[36]

John H. "Doc" Holliday was a tubercular dentist, gambler, saloon keeper, and associate of Wyatt Earp. He reportedly owned a Belgian double-barrel breechloader with barrels cut to only a foot in length and the stock shortened to give the gun an overall measurement of a mere 21 inches. A brass ring affixed to the rib between the barrels allowed it to hang from a leather shoulder strap, concealed

34. Worman, *Gunsmoke and Saddle Leather*, 305.

35. Ibid., 312-14.

36. Rosa, *The Gunfighter Man or Myth?* 101-3.

beneath a duster or overcoat. Such a weapon despite its powerful recoil would have been wicked at close range, but there are suspicions that this tale, like many other facets of western lore, is fiction.[37]

Holliday's associate Wyatt Earp used a shotgun to dispatch Curly Bill Brocius, a shooting that has been questioned by some but which appears to have happened. In retaliation for the murder of his younger brother Morgan, Earp (accompanied by Holliday and several others) in March 1882 also shotgunned Frank Stilwell, a suspected train robber, in Tucson's railyard as the train carrying Morgan's coffin pulled away.[38]

The army too found shotguns useful and in the mid-1870s obtained a number of foreign as well as some American-made Parker 10-gauge doubles for use by guards at the federal prison at Fort Leavenworth, Kansas, and by army payroll escorts. Another Ordnance Department purchase involved 354 Spencer repeating shotguns between 1886 and 1893, apparently secured for use in guarding prisoners. In 1881-85, the Springfield Armory produced almost 1,400 20-gauge single-shot breechloading "trapdoor" shotguns, made using barrels, stocks, and some other parts from obsolete Civil War rifle muskets. These were distributed at various military posts in the west and Alaska to hunt small game (Figure 254).

Later Derringers and Other Pocket Persuaders

By the late 1860s, a wide variety of pocket handguns was available, not just to gamblers but anyone who wanted the security one offered. An example of a true "poor man's derringer" was the Manhattan Fire Arms Company's percussion "Hero" which retailed for $1 or so. It had no trigger guard, a forged brass frame, simple wood grips held by a wood screw, and a smoothbore screw-off barrel of about .34 caliber. But new and improved breechloading derringers accepting metallic ammunition were challenging the percussion guns. Some were single shot, others multi-shot. One of the most practical of the former design was that patented by David Williamson in 1866. The barrel slid forward to load and although it fired a .41 rimfire cartridge, insertion of an auxiliary steel chamber with a nipple at the rear allowed it to be loaded with loose powder and ball and fired with a percussion cap if cartridges weren't available. Wild Bill Hickok reportedly owned a pair (Figures 255, 256).

Christian Sharps had ceased his involvement with rifle manufacture but in 1859 patented and soon had in production a compact four-barrel pepperbox-type derringer. Rather than the barrels revolving, each time the hammer was cocked the firing pin located in the hammer or frame rotated one quarter turn to strike another cartridge in the stationary barrel cluster. Pressing a button beneath or on the left side of the frame allowed the cluster to slide forward to load. First made in .22 caliber, later models in .30 and .32 caliber were offered and eventually more than 100,000 were produced. Eben Starr in 1864 patented a similar .32 rimfire four-barrel pepperbox derringer but the barrel cluster tipped downward to load (Figures 257, 258).

Colt didn't offer a derringer until 1870 when it bought out Daniel Moore's National Arms Company which since about 1864 had marketed a .41 rimfire single-shot derringer with a barrel which tilted sideways to load. Earliest specimens were all metal but a second model had wooden grips. Colt in 1875 added a third model derringer, one in which the barrel pivoted horizontally. It would be advertised in company literature as late as 1912. Like Moore's it fired a .41 short rimfire cartridge, a load which a number of firms adopted for their derringers. Despite the popularity of this cartridge, it didn't have enough punch to satisfy one western lawman when he shot a murder suspect resisting arrest. The bullet struck the accused in the forehead,

37. Lee A. Silva, *Wyatt Earp, A Biography of the Legend: Volume I: The Cowtown Years* (Santa Ana, Calif., 2002), 468.

38. Ibid.

255. James Bown & Son of Pittsburgh in 1876 was selling the percussion "Uncle Sam" derringer for as little as 80 cents. It was sent by mail, "nicely boxed so that no one knows what is inside of the box." **256.** The practical Williamson .41 derringer with the barrel slid forward to load. Use of the auxiliary steel chamber shown above allows the gun to be loaded with loose powder and ball and fired with a percussion cap. **257.** C. Sharps four-barrel .30 rimfire derringer. The release button to allow the barrels to slide forward is seen on the underside of the frame. **258.** Starr .32 four-shot derringer.

but didn't penetrate the skull. After a dozen hours, the suspect regained consciousness and the disgusted lawman threw the Colt derringer away and never carried another one (Figures 259, 260).

Remington by about 1866 was offering a rather extensive line of pistols which by their size fall into the category of derringers or pocket guns. Smallest was the aptly named .22 Vest Pocket which measured only four inches overall, diminutive enough for a lady's purse or gentleman's watch pocket. Lieutenant Colonel George A. Custer's wife, Elizabeth, probably was referring to one of these when discussing another army wife's armament.

> She produced a tiny Remington pistol . . . that she had carried in her pocket when traveling in the States. It was not much larger than a lead pencil, and we could not help doubting its power to damage. She did not insist that it would kill, but . . . we had to laugh at the vehement manner in which she declared that she could disable the leg of an enemy.[39]

Similar to the Sharps and Starr "pepperbox" derringers was the compact Remington Elliott ring-trigger pocket pistol, a five-barrel .22 or four-barrel .32, each with a revolving firing pin. There were other Remington derringers, however the longest lasting of these was their .41 rimfire Double Derringer, of which more than 150,000 were manufactured at Ilion, New York, between

39. Garavaglia and Worman, *Firearms of the American West 1866-1894*, 269.

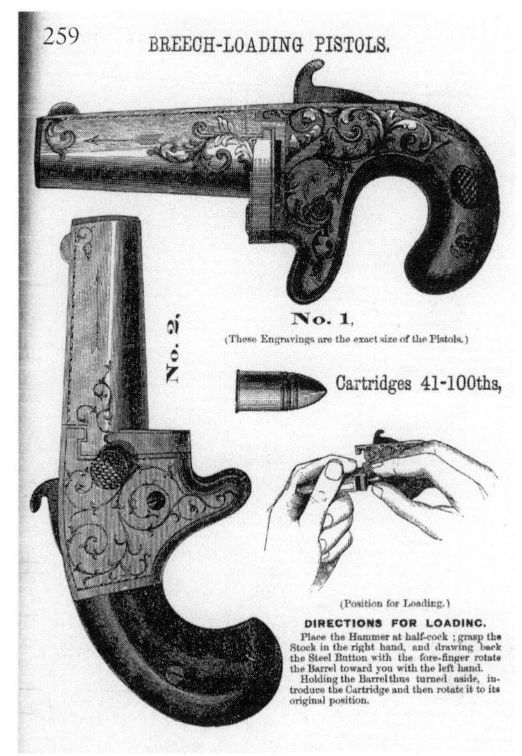

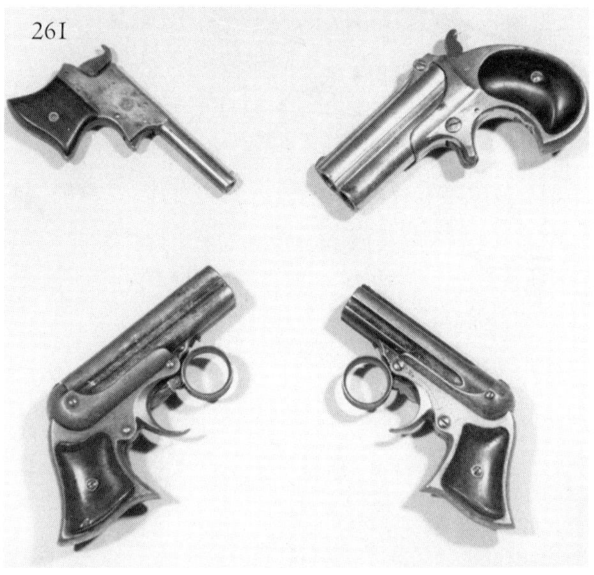

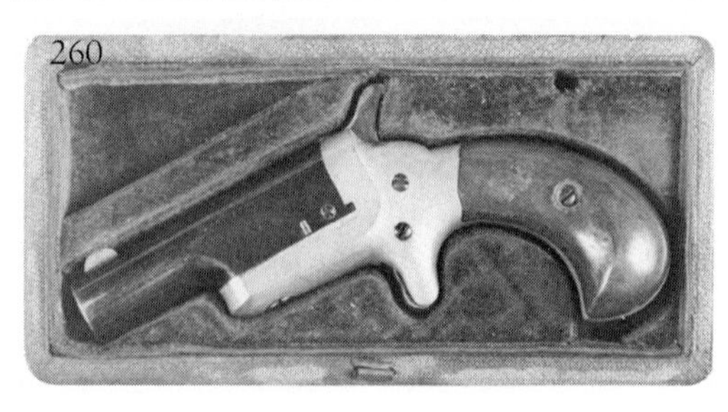

259. Page from the 1865 National Arms Co. catalog illustrating their 1st and 2nd Model derringers. The Colt firm in 1870 bought out the National Arms Company and sold those remaining in stock as well as ones of their own manufacture. The release button on the right side allows the barrel to be pivoted to the side to load. **260.** Mounted in a leather-covered case, this Colt .41 Third Model derringer bears English proof marks indicating its probable sale through Colt's London sales agency. Another known example of a Third Model was owned by President Theodore Roosevelt's wife Edith. The barrel swings to the side to load. **261.** A quartet of Remington cartridge derringers. Clockwise from one o'clock: .41 Double Derringer, five-barrel .22 Elliott, four-barrel .32 Elliott, and .22 Vest Pocket.

1866 and 1935. Releasing a latch on the right side of the frame allowed the over-and-under barrels to swing upward to load. The firing pin alternated between the two barrels each time the hammer was cocked. General Douglas MacArthur is known to have owned one of these. A number of modern manufacturers have produced improved copies of this design in various calibers from .22 to a model with one barrel a .410 and the other in .45 caliber (Figures 261, 262).[40]

As cartridge revolvers gained popularity during the 1860s and after, numerous makers offered handguns which aren't small enough to be classed as derringers per se, but still were compact enough to carry in a pocket or purse. The "big three" revolver makers—Colt, Remington, and Smith & Wesson—each had its entries in this field. Colt in 1871 introduced a seven-shot .22 "open top" without a top strap over the cylinder and beginning in 1874 its New Line series of solid-frame spur-trigger pocket guns in .22, .30, .32, .38, and .41 caliber. Remington soon countered with its Smoot

40. One of the author's friends, a very knowledgeable gun designer, owns a .357 Magnum double derringer. He only fired it once and then described the recoil as "murder."

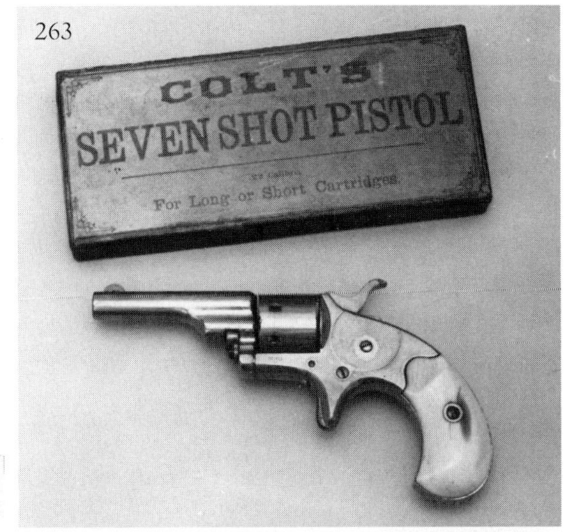

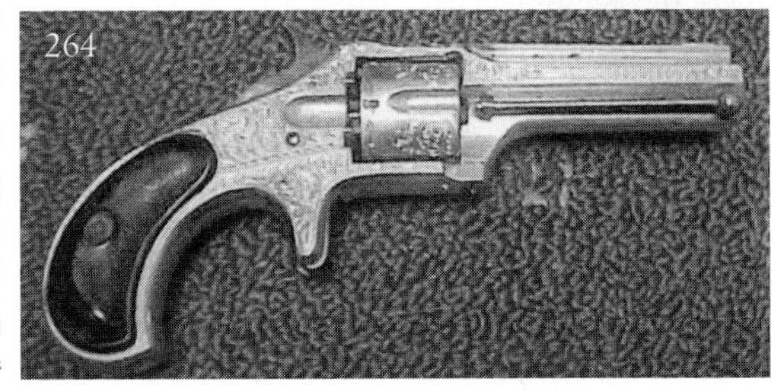

262. Remington-Rider magazine pistol in its original cardboard box. Made in the 1870s and 1880s, the tubular magazine beneath the barrel held five extra short .32 rimfire cartridges. **263.** Colt .22 Open Top with ivory grips and its original cardboard box. **264.** Factory engraved .30 caliber five-shot Remington Smoot. Made from the mid-1870s until the late 1880s, these competed with Colt and other pocket revolvers. (*Courtesy: Neil and Julia Gutterman*) **265.** Page advertising Webley's British Bull Dog from the 1880 catalog of Homer Fisher of New York City.

Patent pocket revolvers, unusual in that some were produced with the barrel and frame forged in one piece (Figures 263, 264).

It wasn't available until 1887, first in .38 caliber then .32, but one of the most popular such guns proved to be Smith & Wesson's double-action Safety Hammerless with its hammer concealed within the frame. A less formal nickname was "lemon squeezer" and a variation with only a two-inch barrel was called the Bicycle Model to promote popularity with cyclists at the turn of the century. A major selling point was its grip safety, a bar set in the back strap which prevented the gun from being fired if it wasn't gripped firmly. Another advantageous innovation was the trigger design with a discernable hesitation just before the trigger released the hammer. Even though these revolvers could only be fired double action, this feature allowed them to be shot almost as accurately as a single action. Production continued into the 1930s, indicative of the model's sales appeal.

These three makers were faced with competing models from such firms as Forehand & Wadsworth, Iver Johnson, and Hopkins and Allen. The latter firm in the 1880s also was manufacturing high-quality single- and double-action .32 and .38 pocket revolvers bearing the Merwin & Hulbert name. Often overlooked today are those imported English

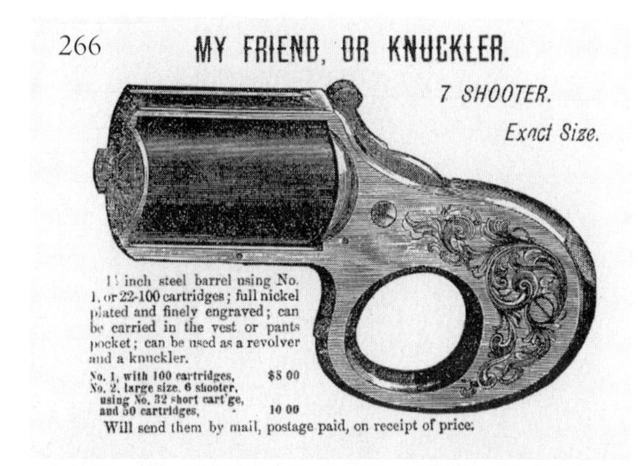

266. An 1876 ad by James Bown & Son of Pittsburgh for the Reid "MY FRIEND OR KNUCKLER," the seven-shot .22 for $8 and the five-shot .32 for $2 more, although Bown incorrectly lists the .32 as six-shot. **267.** Chicago Firearms Co. Protector palm pistol. (*Courtesy: Neil and Julia Gutterman*)

"bulldog" double-action pocket guns by Webley—sometimes marked "BRITISH BULLDOG"—and American copies, all of which found a rather substantial market in this country. Some were available in calibers as large as .45, making them a rather potent yet fairly compact pocket protector for buyers from the late 1870s on. In mid-1881 with borrowed money, Charles Guiteau purchased a .44 British Bulldog from which he fired two shots, wounding President James Garfield at the Washington, D.C., train depot. Garfield lingered in agony for more than two months before succumbing. The assassin stated one factor in his selection of the nickel-plated revolver was how it would look on exhibition in a museum, where he was sure it eventually would be placed (Figure 265).

Another British Bulldog found its way to a museum in Coffeyville, Kansas. An audacious daylight attempt in 1892 to rob two of the town's banks at the same time resulted in the destruction of the Dalton gang of outlaws in a bloody shootout with townspeople armed mostly with rifles and shotguns. Four citizens and four of the robbers finally lay dead, although Emmett Dalton survived despite having been wounded more than a dozen times in the melee. From the body of Bob Dalton, one of the townsfolk retrieved his "hide out" gun, a .38 British Bulldog which is exhibited today in Coffeyville's Dalton Defenders Museum.

Among the more distinctive pocket guns of the later 1800s were those by James Reid and the palm pistols made by the Chicago Firearms Company and the Minneapolis Firearms Company. Reid's My Friend "knuckle dusters" were produced from the late 1860s until the early 1880s, most as seven-shot .22s, a lesser number as five-shot .32s, and a very few as larger size .41s. These had an all-metal frame shaped conveniently to strike a blow as a form of brass knuckles in a hand-to-hand scuffle. Most recognizable of the Reid revolvers are those without a barrel, the bullets exiting directly from the cylinder's chambers as with the percussion pepperboxes of a generation earlier. The Chicago and Minneapolis Protector palm pistols were based on French and 1883 American patents and were held in the palm of the hand with the short barrel protruding from between the fingers. They were fired double action by squeezing the firing lever at the rear. To load with the .32 extra short cartridges, a circular side plate was rotated and removed, followed by the withdrawal of the flat turret cylinder (Figures 266, 267).

Not as unusual in design or appearance was the four-shot version of Colt's House Pistol manufactured between 1871 and 1876, the Colt Cloverleaf. When viewed head-on, the four chambered cylinder indeed did resemble that traditional symbol of good luck and when the hammer was positioned between two chambers it presented a rather flat

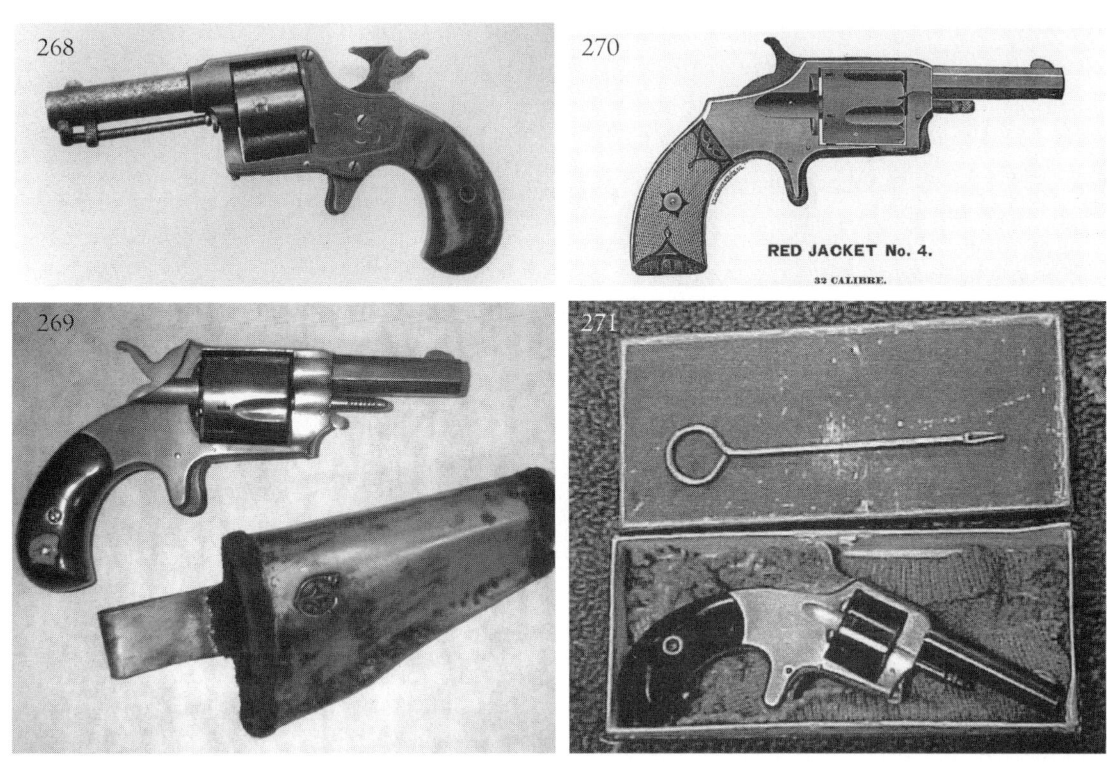

268. Four-shot .41 Colt "Cloverleaf" House Model revolver made between 1871 and 1876. **269.** Forehand & Wadsworth .38 rimfire "Bull Dog" owned by a Denver madam. A piece of turquoise stone is inlaid in the right grip and the tin holster is lined with red velvet. (*Courtesy: Jerry Pitstick*) **270.** An 1884 advertisement by C. J. Chapin Arms Company of St. Louis for various inexpensive "suicide specials." **271.** A .22 caliber "Alert" in its original pasteboard box with cleaning rod. (*Courtesy: Neil and Julia Gutterman*)

profile for convenient stowage in a pocket. It fired the same rather anemic .41 rimfire cartridge as did many derringers. Probably it's best known today for its use by the flamboyant James Fisk to murder his former associate Edward Stokes on the main stairway of New York City's Broadway Central Hotel, his rival for the attention of actress Josie Mansfield. Fisk was one of the most notorious of the "robber barons" and in 1868-69 had attempted to corner the market on gold (Figure 268).

At the lower end of the scale of pocket or home protection handguns in terms of overall quality of finish and workmanship were those solid-frame, single-action spur trigger revolvers sometimes referred to by today's collectors as "suicide specials." They were lower price competition to similar pocket handguns like the Colt New Line or Remington Smoot. Made from the 1870s forward, these pocket guns sometimes bore the maker's name, but often were unmarked except for such fanciful names as Robin Hood, Blue Jacket, Tramps Terror, Defender, Penetrator, Wide Awake, Smoker, Bang Up, Cowboy Ranger, Earthquake, or any of numerous other imaginative terms. In 1879 a young man attempted to assassinate famed American actor Edwin Booth, brother of John Wilkes, during a performance of *Richard the Third.* He fired two shots from what the newspaper described as a .32 True Blue but missed with each.[41] The great variety of names found among these handguns has attracted collector interest in recent

41. Charles Becker, "Assassinating Booth," *Surratt Courier* (Clinton Md.), vol. 32 no. 3, March 2007, p. 4.

years, particularly for those examples in nearly new condition. Hardware dealer Homer Fisher of New York City in his 1880 catalog offered the seven-shot .22 Pathfinder for only $1.25, a five-shot .32 Blue Whistler for $2.50, or a .38 Pioneer for $1 more. In comparison he priced the .22 Colt New Line at $4 and a Smith & Wesson .38 at $11. He didn't ignore the Webley British Bull Dog, including a drawing of one and listing them at $9.50 to $15 (Figures 269, 270, 271).[42]

Despite the uncomplimentary term "suicide special," one who fired one of these inexpensive handguns was not risking his life. If one didn't expect the quality to equal that found in a revolver by one of the major manufacturers, a .32 Defender purchased at retail for $1.75 or a similar handgun was a functional and generally reliable weapon. If used with ammunition for which they were intended they were safe. If fired at a reasonable distance of perhaps ten feet or so the shots could be kept in a 12-inch or smaller group, certainly sufficient for defense against a mugger on a dark New York City street or a home intruder.

Homer Fisher did offer some straightforward advice to a potential buyer of a "suicide special," however his intent, like that of other dealers who maligned these guns, probably was to steer the customer toward a more expensive purchase just as a car salesman today would:

> All Revolvers using the No. 1 or 22-Caliber Cartridges are Seven Shot. Those of the cheapest grade, such as Robin Hood, Defender, Protector, Rover, Buffalo Bill, Little Giant, &c, are about one and the same pistol with different names, nearly every dealer using a different name. If my advice is asked about the purchase of this grade of pistol, I should say "don't." They will do well enough to fire blank cartridges [on] Fourth of July, but for accuracy and penetration for ball cartridges they are good for nothing. Better to invest a dollar more, and get a Colt or a Guardian, and you will then have a well made and accurate arm.

Cane Guns and Other Oddities

One advantage the fiction writer has is the opportunity to arm a character with an unusual yet historically correct "oddball" weapon. Bizarre firearms have been the product of those with a fertile mind for centuries. The following examples reflect inventor ingenuity and while some like the palm pistols described earlier achieved modest popularity, many designs were proven impractical (Figure 272).

Guns concealed as canes employed the same concept as a sword cane and were available throughout the mid- and late 1800s. Thus a stately-looking gentleman strolling in a park may have been capable of defending himself against a snarling dog or threatening thug not merely with what appeared to be a walking stick but with a potentially lethal single-shot cane gun. These were available in percussion form or later firing a metallic cartridge. One rarely encountered percussion example was that patented by Roger Lambert in 1832. The hollow wooden shaft of the cane concealed the gun's barrel; the brass tip had a swiveling muzzle cover. The knob at the top of the cane pulled up and swiveled to reveal a flat bar hammer and trigger. Some of these were sold as far west as Missouri, where an 1834 newspaper advertisement listed six of them for sale as "believed to be the first of their kind offered in St. Louis."

More common although still rare today were the percussion .31 and .44 caliber cane guns produced by Remington beginning about 1858. In a December 1865 newspaper ad, Remington listed the arms the firm was marketing to the public and

42. The catalog issued about 1882 by gunsmith and dealer Joseph L. Rawbone of Toronto, Ontario, offered a .32 Toronto Belle for $2.75, which in the illustration closely resembles a Colt New Line. "The hammer has also a long sweep, which ensures the perfect explosion of the cartridge." Gooding, *The Canadian Gunsmiths, 1608-1900,* 154.

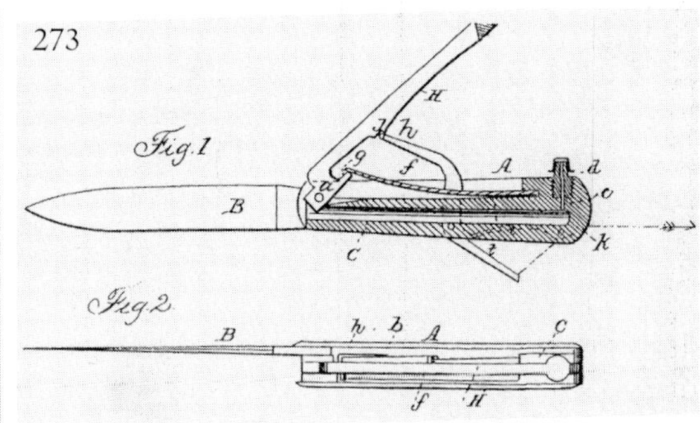

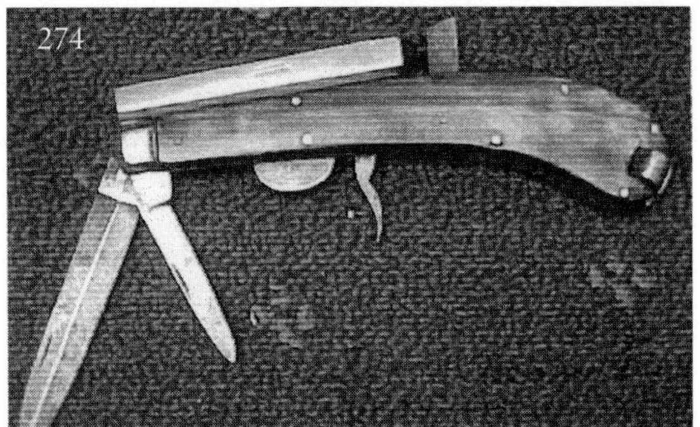

272. Flintlock trap gun. A network of trip wires was connected to the trigger of such a device and when an animal or poacher tripped one of the wires, the gun swiveled in that direction and fired. (*Exhibited at Springfield Armory National Historic Site*) **273.** An 1865 patent drawing of the Peavey percussion knife-pistol. **274.** English percussion Unwin & Rogers knife pistol with horn handles. The original tweezers and bullet mold are contained in the grip. (*Courtesy: Neil and Julia Gutterman*)

included an updated cane gun firing a .32 metallic cartridge. The ad pointed out that the barrel was covered with rubber and combined the advantages of a walking stick and rifle. "It is light and portable, but at the same time is nearly as efficient in point of Range, Accuracy, and Penetration, as a Rifle of the same length." The model was produced until the late 1880s in .22 and .32 caliber. James Bown & Son in 1876 was advertising its own "rifle cane guns" in either .22 or .32 caliber for $11. "Every person who is out at night should have one for protection against dogs and rowdies," their catalog urged. As late as 1903, New York dealer Schoverling, Daly & Gales' catalog illustrated a smoothbore gun cane at $10, "arranged to shoot a .44 W.C.F. [Winchester centerfire] shot cartridge with excellent results."[43]

There were other attempts to conceal a firearm in an umbrella or bicycle handlebar or combine one with a sword or dagger. Among those combined with a knife, the Elgin cutlass pistol has been mentioned earlier. Probably the most widely sold pocket-size knife pistols were those by Unwin & Rodgers and imported from England, sometimes containing several blades. These were first produced as a muzzleloader and one unusual variation had a bullet mold and tweezers held by springs in the handle, reminding one of the popular modern Swiss army knife with its many accessories. Later versions fired a .22 cartridge. A. J. Peavey of Maine marketed a combination single-shot pistol and clasp knife, in 1865 patenting one in percussion form and a year later a .22 rimfire cartridge version. The single blade was set off to the side and the barrel ran the length of the handle beneath the long hammer bar (Figures 273, 274).

An oddity among the multitude of American military arms was the Ellis-Jennings four-shot repeating flintlock rifle, patented by Isiah Jennings in 1821. It employed superposed loads, one on top of the other, with a swivel cover over each separate

43. Ware, *Remington Army and Navy Revolvers, 1861-1888*, 340. Enterprise Gun Works 1876 catalog, 57.

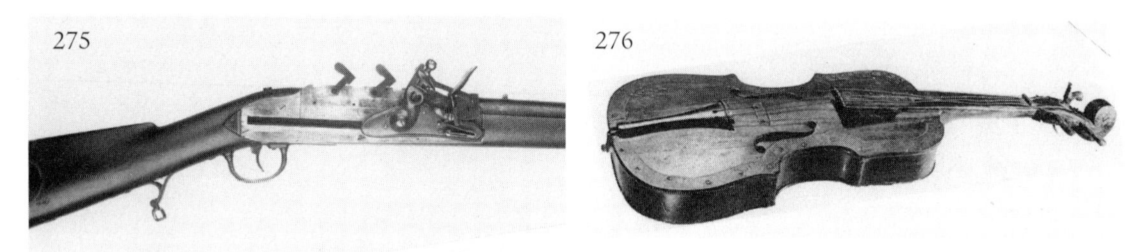

275. Ellis-Jennings four-shot repeating rifle. **276.** Southern mountain fiddle with a single-shot percussion pistol built into the neck.

touchhole and a sliding lock. Reuben Ellis obtained a government contract for 520 of these .54 caliber rifles for delivery to the state of New York; production was carried out in 1829 by R. & J. D. Johnson of Middletown, Connecticut. A rare ten-shot version is known but probably was experimental (Figure 275).

Almost certainly a one-of-a-kind curiosity was advertised for sale a half century ago, a .44 caliber fiddle! As the late Colonel Leon "Red" Jackson described it, you could classify it as an oddity or a prime example of mountain ingenuity. The .44 single-shot percussion pistol was built into the neck of the instrument with the hammer on top and trigger on the underside. It appeared to have been fired many times but still was a practical musical instrument. It came from Pike County, Kentucky, and reportedly was owned by one of the "Blackberry McCoys" of the Hatfield-McCoy feud (Figure 276).[44]

An unusual two-shot percussion pistol with a single barrel was designed by John P. Lindsay and sold in the early 1860s in both pocket and larger size as the "Young America." It fired superimposed loads, one charge of powder and ball on top of the other, with two nipples, two hammers and either two triggers or a single trigger arranged to release the hammers in the proper sequence. Probably no more than 500 or so Lindsay pistols were made in all. In the 1870s Jacob Rupertus tried a rather simple approach with a double-barrel .22 rimfire pocket pistol. There was a single hammer with a firing pin in the nose which could be adjusted to strike either barrel.

Just as rare was the single action 14-shot rimfire cartridge revolver with two cylinders mounted on a single base pin, patented by Charles E. Sneider of Baltimore in 1862. A disc separated the cylinders and the long hammer extended over the rear cylinder to discharge the cartridges in the front one. When the forward cylinder was emptied, one tipped the hinged barrel down and replaced the empty cylinder with the loaded one. An improved version, an eleven-shot .32, continued to use the dual-cylinder design but after the front cylinder was emptied, bullets from the rear one passed through an empty chamber in the forward cylinder. Very few such guns of either design were made (Figure 277).

Somewhat more successful were the ten- and twelve-shot (.31 and .36 caliber, respectively) percussion revolvers patented in 1859 by John Walch. Each of the chambers was double loaded, one charge of powder and ball in front of the other. There were two hammers and two sets of nipples arranged in two concentric circles. The design was such that the hammer which fell first struck the forward row of nipples. During the Civil War, some members of Company "I" of the 9th Michigan Infantry privately purchased Walch revolvers and reportedly were pleased with them (Figure 278).

The Osgood Gun Works of Norwich, Connecticut, in the 1880s adopted one element of the Civil War LeMat "grapeshot" revolver design for its own nine-shot Duplex revolver. Its eight-shot .22 cylinder revolved about the .32 rimfire

44. Jackson Arms catalog No. 11, Dallas, Texas.

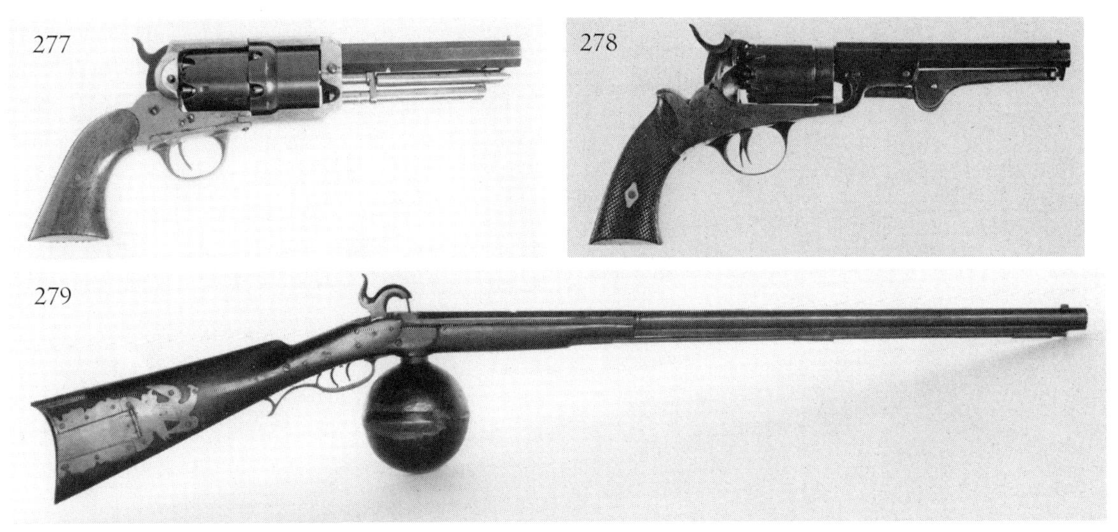

277. Another innovative but impractical revolver was that patented in 1865 by G. H. Gardner. The rear cylinder had six chambers, the forward cylinder had five plus a sixth bored completely through to serve as a rearward extension of the barrel. (*Courtesy: Smithsonian Institution, neg. no. 72-4547*) **278.** Twelve-shot Walch .36 revolver. **279.** An airgun of .35 caliber, probably made in the 1830s and perhaps in Philadelphia. The ball-like reservoir holds air compressed by a separate pump. (*Courtesy: Smithsonian Institution, neg. no. 49289A*)

center barrel and a hammer striker could be pivoted to select to fire the caliber of one's choice. Novel though the design was among cartridge revolvers, it didn't attract many buyers.

Guns which fired a projectile such as a dart or a ball by means of compressed air were known as early as the 1600s. They were not common and at first could be considered as oddities. The Lewis and Clark expedition to the Pacific (1804-6) carried an air rifle which impressed those American Indians who witnessed its demonstration. In January 1806 Captain Meriwether Lewis fired it for a group of astonished Clatsop Indians who could not "comprehend it shooting so often and without powder." By then air guns were available which employed an air reservoir such as a copper ball mounted beneath the breech or a reservoir contained in a hollow butt stock. Using a separate pump, air pressure could be built up to allow multiple shots. Such guns were virtually noiseless and were useful for short-range target practice or shooting small game, but by the late 1800s they lost any semblance of novelty (Figure 279).

In 1870 Henry Quackenbush patented an air pistol, the Eureka, in which air was compressed by pushing the barrel backward. It was an inexpensive substitute for more expensive shoulder-fired spring-power air guns of the 1850s and 1860s firing darts and used in shooting galleries. These were cocked either by means of turning a crank or moving the trigger guard. Quackenbush went on to mass-produce well into the 1900s thousands of pistols and rifles in calibers from .17 to .22 firing darts or slugs. Then in 1886 the Markham Air Rifle Company of Plymouth, Michigan, was established, promoting the Challenger. Crude though this mostly wooden single-shot rifle was, it became the first commercially successful toy BB gun.

Soon after, in 1888 the struggling Plymouth Iron Windmill Company of Michigan undertook the manufacture of toy air rifles (smoothbores although called rifles) designed by Clarence J. Hamilton with the intent of using them as promotional gifts. The single-shot prototype was an all metal nickel-plated gun with a wire skeleton stock and a long cocking lever which lay along the top of

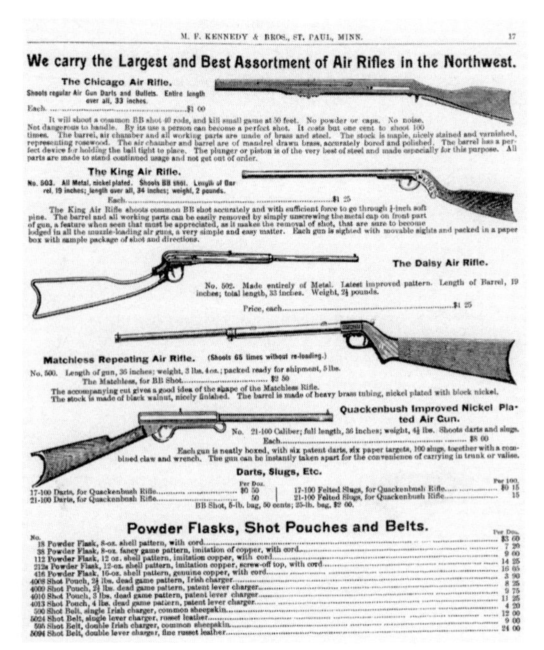

M. F. KENNEDY & BROS., ST. PAUL, MINN. 17

We carry the Largest and Best Assortment of Air Rifles in the Northwest.

The Chicago Air Rifle.

Shoots regular Air Gun Darts and Bullets. Entire length over all, 33 inches.

Each. $1 00

It will shoot a common BB shot 40 rods, and kill small game at 50 feet. No powder or caps. No noise. Not dangerous to handle. By its use a person can become a perfect shot. It costs but one cent to shoot 100 times. The barrel, air chamber and all working parts are made of brass and steel. The stock is maple, nicely stained and varnished, representing rosewood. The air chamber and barrel are of mandrel drawn brass, accurately bored and polished. The barrel has a perfect device for holding the ball tight to place. The plunger or piston is of the very best of steel and made especially for this purpose. All parts are made to stand continued usage and not get out of order.

The King Air Rifle.

No. 503. All Metal, nickel plated. Shoots BB shot. Length of Barrel, 19 inches; length over all, 34 inches; weight, 2 pounds.

Each...$1 25

The King Air Rifle shoots common BB shot accurately and with sufficient force to go through ½-inch soft pine. The barrel and all working parts can be easily removed by simply unscrewing the metal cap on front part of gun, a feature when seen that must be appreciated, as it makes the removal of shot, that are sure to become lodged in all the muzzle-loading air guns, a very simple and easy matter. Each gun is sighted with movable sights and packed in a paper box with sample package of shot and directions.

The Daisy Air Rifle.

No. 502. Made entirely of Metal. Latest improved pattern. Length of Barrel, 19 inches; total length, 33 inches. Weight, 2½ pounds.

Price, each...$1 25

Matchless Repeating Air Rifle. (Shoots 65 times without re-loading.)

No. 500. Length of gun, 36 inches; weight, 3 lbs. 4 oz.; packed ready for shipment, 5 lbs.

The Matchless, for BB Shot.............................. $2 50

The accompanying cut gives a good idea of the shape of the Matchless Rifle. The stock is made of black walnut, nicely finished. The barrel is made of heavy brass tubing, nickel plated with block nickel.

Quackenbush Improved Nickel Plated Air Gun.

No. 21-100 Caliber; full length, 36 inches; weight, 4½ lbs. Shoots darts and slugs.

Each... $8 00

Each gun is neatly boxed, with six patent darts, six paper targets, 100 slugs, together with a combined claw and wrench. The gun can be instantly taken apart for the convenience of carrying in trunk or valise.

Darts, Slugs, Etc.

	Per Doz.		Per 100.
17-100 Darts, for Quackenbush Rifle	$0 50	17-100 Felted Slugs, for Quackenbush Rifle	$0 15
21-100 Darts, for Quackenbush Rifle	50	21-100 Felted Slugs, for Quackenbush Rifle	15

BB Shot, 5-lb. bag, 50 cents; 25-lb. bag, $2 00.

Powder Flasks, Shot Pouches and Belts.

No.		Per Doz.
18	Powder Flask, 8-oz. shell pattern, with cord	$3 60
38	Powder Flask, 8-oz. fancy game pattern, imitation of copper, with cord	7 20
112	Powder Flask, 12 oz. shell pattern, imitation copper, with cord	9 00
212a	Powder Flask, 12-oz. shell pattern, imitation copper, screw-off top, with cord	14 25
416	Powder Flask, 16-oz. shell pattern, genuine copper, with cord	16 65
4008	Shot Pouch, 2½ lbs. dead game pattern, Irish charger	3 90
4009	Shot Pouch, 2½ lbs. dead game pattern, patent lever charger	8 25
4010	Shot Pouch, 3 lbs. dead game pattern, patent lever charger	9 75
4013	Shot Pouch, 4 lbs. dead game pattern, patent lever charger	11 25
590	Shot Belt, single Irish charger, common sheepskin	4 20
5024	Shot Belt, single lever charger, russet leather	12 00
595	Shot Belt, double Irish charger, common sheepskin	9 00
5094	Shot Belt, double lever charger, fine russet leather	24 00

280. Miscellaneous air rifles illustrated in a St. Paul, Minnesota, sporting goods dealer's 1893 catalog.

the barrel. Within a year, the firm changed its name to the Daisy Manufacturing Company and devoted all its resources to the manufacture of these guns. Daisy in 1890 introduced its first break-open design and about 1900 its first repeater. Soon Daisy was selling about 85,000 air guns a year, the beginning of more than a century of not only BB air rifles and handguns but water pistols, cork-firing guns, sonic air-blasting toy guns, and of course in 1940 the lever-action Red Ryder carbine made even more famous by Ralphie Parker's feverish desire for one in the classic film *A Christmas Story* (Figure 280).

Buffalo Guns and the Slaughter of the Bison

Before the 1870s, the herds of bison which roamed the great plains of the west seemed endless. They had been estimated to number between 60 and 75 million in the early 1800s. For generations, hunting these shaggy monarchs with lance and bow had been a life-sustaining element of the plains Indians' economy, providing food, clothing, shelter, tools, boats, and other necessities of life. Some tribes in the 1830s were establishing a lucrative trade with whites for tanned bison robes, skins tanned with the hair intact, a trade which eventually surpassed commerce in beaver pelts. In 1843 Pierre Chouteau Jr. & Company purchased some 17,000 hides from Indians paying $2 to $4 for ones painted by American Indian women, $2.50 to $3.50 for winter robes, $1.50 to $3 for summer hides, and $3 to $6 per dozen for bison tongues.[45]

Nineteenth-century whites entering the plains in steadily increasing numbers found the bison was a ready source of meat and often a source of sport of a most cruel nature, such as seeing how much lead a bison could withstand before dying. As early as 1845, one of John C. Frémont's companions noted that American Indians couldn't be blamed for hating the "whitefaces" who slaughtered their "cattle" on which they depended for their subsistence. As railroads began to bisect the plains after the Civil War, some passengers took pleasure in their misguided idea of "sport" as they fired at bison from moving trains with no expectation of recovering the meat or of putting wounded animals out of their misery. Railroads such as the Union Pacific and Kansas Pacific sometimes took advantage of the enthusiasm for the hunt and offered hunting excursions into bison country.

Hunting for food was a legitimate use of the bison, wasteful though it was, for in those days before refrigeration, hunters often merely took the hump, tongue, or other choice pieces of meat, leav-

45. Trapper Osborne Russell mentioned an 1835 bison hunt by Bannock Indians during which "upwards of a Thousand Cows were killed without burning one single grain of gun powder." Russell, *Journal of a Trapper,* 36. As late as the mid-1870s some Mexican *ciboleros* continued to hunt bison with the lance.

ing most of the carcass to rot in the sun. Young William F. Cody in late 1867 signed a contract to provide meat to construction workers on the Kansas Pacific Railroad. He received $500 a month to kill a dozen bison a day—twenty-four hams and a dozen humps, the only portions of the animals taken. He reportedly killed 4,280 for his employer, relying on "Lucretia Borgia," as he had nicknamed his .50 caliber Springfield single-shot breechloader.[46]

While hunting for the Kansas Pacific, Cody defended his title of "Buffalo Bill" when he engaged in a day-long contest with scout Billy Comstock. Although the latter used a Henry .44 repeater, Cody astride his buffalo horse Brigham and relying on his more powerful Springfield "trapdoor" killed 69 to his opponent's 46 (Figure 281).[47]

Before the days of the professional hide hunters, whites might have crept up on a herd, to kill with a rifle, musket, or even a shotgun. More exhilarating was the hunt from the back of a horse or a mule, riding close to a chosen animal and perhaps dispatching it with a rifle, shotgun, or large-bore pistol or revolver. Exciting though this undoubtedly was, it was dangerous for a maddened bison might hook horse or rider with a horn or one's mount might shy at a rabbit or step into a hole and throw its rider to the ground. Captain John C. Frémont during his 1843 western exploring expedition almost lost his life in such a mounted hunt when his pistol accidentally discharged, the ball passing close to his head. Some years later Lieutenant Colonel George A. Custer had the embarrassment of accidentally killing his own horse with a revolver during such a chase. Upon first spying a herd of bison, emigrants crossing the plains by covered wagon sometimes threw caution to the wind in their excitement for a hunt, carelessly firing wildly without regard to the risk to others.

The market hunting of bison for meat persisted, particularly in winter when spoilage wasn't such a problem. However, it wasn't until a practical means of converting untanned hides into high-

281. Buffalo Bill Cody's .44 percussion Remington which he sent in 1906 to his friend and business associate Charles Trego and his wife, Carrie. Accompanying it was a business card on which Cody wrote: "This old Remington revolver I carried and used for many years in Indian wars and buffalo killing and it never failed me. W. F. Cody." (*Courtesy: Charles Trego family via Norm Flayderman*)

grade industrial leather such as machinery drive belts was developed about 1870 that white hunters came in sufficiently large numbers to decimate the herds not just for meat or robes but for their hides. Unbelievable though it seems, between about 1872 and the late 1880s those gigantic herds disappeared and by 1900 there were estimated to be fewer than a thousand or so free-roaming bison in the United States.[48]

Hide hunting by whites began in the winter of 1871-72 in Kansas. Colonel Richard Dodge estimated that in the three years of 1872-74, railroads hauled more than one million dry untanned "flint"

46. Cody's "Lucretia Borgia" is preserved in the firearms museum which constitutes part of the Buffalo Bill Historical Center in Cody, Wyoming.

47. Although Cody didn't inaugurate his wild west shows until 1882, by the late 1870s his exploits were elevating him to celebrity status. In May 1877 while visiting the Evans rifle factory in Maine he was presented with a handsomely finished example of their .44 caliber lever action sporting rifle, an unusual arm holding up to 38 cartridges in a rotary magazine in the butt stock fed through the butt.

48. Some theorize that hunters alone may not have been responsible for the destruction of the free-roaming bison. Diseases introduced into the herds by domestic cattle may have been a contributing factor as well.

282. Hunters butchering a bison. A Sharps carbine leans against the carcass. One hopes the gun was unloaded since the hammer was cocked. (*Courtesy: University of Oklahoma Library*)

hides, almost 7 million pounds of bison meat, and more than 16,000 tons of bones. The number of hides shipped represented only a portion of the total number of animals killed, since some hides were ruined by careless skinning or by insects. Billy Dixon estimated that in the winter of 1872-73 as many as 75,000 bison were killed within 60 to 75 miles of Dodge City alone. "The noise of the guns of the hunters could be heard on all sides, rumbling and booming hour after hour, as if a heavy battle were being fought," he recalled. F. C. Zimmerman, first located in Kit Carson, Colorado, was a major supplier of guns and other needed goods to the Kansas hunters. After moving to Dodge City, he advertised himself as a dealer in firearms, ammunition, hardware, tinware, groceries, grain, and lumber and as an agent for Sharps sporting rifles and Oriental gunpowder (Figure 282).[49]

In 1874, hunters were moving south to Texas as Kansas herds were thinned out. Fort Griffin became the supply and marketing center for the Texas hide trade and an area nearby known as the Flat soon became a notorious recreation center for hunters, skinners, cowboys, and others. The outfitting firm formed by Frank Conrad and Charles Rath at Fort Griffin reportedly had on hand thirty tons of lead and five tons of gunpowder. General Phil Sheridan, vehemently arguing against a proposed Texas law to protect the bison, credited the hunters with doing more to "settle the vexing Indian question" than the army had done in thirty years. By 1880, the hunters had done their job on the Texas herd, leaving bones and orphaned calves in their wake, and they turned northward to Montana and the northern plains where they aimed their rifles at those bison remaining there.

Those men who joined in the hunt were as varied as the guns they used initially. Some individuals saw it as an easy way to turn a profit and thought a surplus Civil War carbine or rifle would suffice, but they soon learned it was a demanding, dirty, and dangerous business. American Indians, aroused by the decimation of the herds and the intrusion into hunting grounds sometimes thought to be guaranteed them by treaty, posed a serious threat, as did winter blizzards or the risk of an accident miles from any assistance. Nevertheless a listing of known hunters includes many hundreds of names—Wyatt Earp; brothers Ed, Jim, and "Bat" Masterson; future famed lawman Bill Tilghman; J. Wright Mooar; Napoleon Bonaparte "Arkansas Jack" Greathouse; John R. Cook; Billy Dixon; James and Robert Cator from England; and the notorious horse thieving "Dutch Henry" Born among them.[50]

After giving up the bloody business, "Bat" Masterson later became a gambler and a lawman. A persistent myth is that he acquired his nickname from his habit of battering law breakers with his walking stick, but it actually came from his given

49. T. Lindsay Baker and Billy R. Harrison, *Adobe Walls: The History and Archeology of the 1874 Trading Post* (College Station, Tex., 1986), 8.

50. Bones were gathered and shipped east where some were made into handles for knives and combs and the rest were ground up for fertilizer or used in refining sugar. The bone business was an economic blessing for many a poor homesteader family.

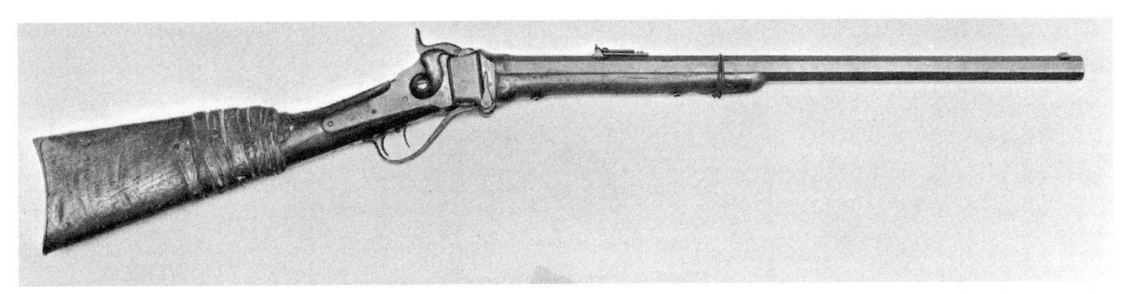

283. Sharps Model 1874 sporting rifle used by Bill Tilghman, a professional bison hide hunter. Factory records indicate it originally was shipped to dealer F. C. Zimmerman of Dodge City, Kansas, in June 1874 in .40 caliber. When the rifling became worn, Tilghman sent it back to the factory and had it rebored to .50 caliber. The break in the stock occurred when his horse fell. The application of wet rawhide was a common means of repairing a broken stock for as the hide dried, it shrank creating a strong repair. (*Courtesy: National Cowboy Hall of Fame, Oklahoma City*)

name of Bartholomew. After a varied career in the west, in his final years he became a sports writer and died at his desk in a New York City newspaper office.

Repeating rifles such as the Henry, Spencer, or early model Winchesters lacked the range and killing power for efficient hunting. "The Winchester is a laughing stock among the hunters," remembered one professional in 1876. Single-shot rifles chambered for more powerful cartridges were the answer. The .50-70 Springfield army "needle gun" was an early choice of many. Remington by late 1871 had begun offering their large frame No. 1 size sporting rifle chambered for .40, .44, .45, and .50 centerfire cartridges, and George A. Custer, although not a professional bison hunter, in 1872 obtained a .50-70 Remington and gave it high praise. A substantial number of hide hunters selected a Remington rolling block and some a Maynard, Peabody, Ballard, or other sporting rifle. But based on the written accounts by hunters, the Sharps became the leading bison killer in the hands of the professionals, available in various weights, barrel lengths, and calibers and sometimes fitted with a $40 Malcolm or Potter telescopic sight.

The price for a military style Sharps Model 1874 in 1871 began at $20, but a hunter could easily pay $40 and up for a sporting model and the necessary reloading equipment. Wright Mooar claimed he killed 6,500 with a 14-pound Sharps and 14,000 more with his 11-pounder of the same make. Many different loadings of .40, .44, .45, and .50 centerfire cartridges became available for the Sharps and other sporting rifles, some using a cartridge almost three inches long. Included among these was what sometimes was known as the Sharps "Big Fifty," a .50 caliber cartridge introduced in the summer of 1872 containing 100 grains of powder (Figure 283).[51]

A typical hide hunting party consisted of four to six men—a hunter or two, skinners, and a man to cook, stretch out the hides to dry, and take care of the camp. Chasing bison on horseback scattered the carcasses over a large area; the most efficient hunting method involved what was known as "getting a stand." This meant locating a grazing herd and creeping to within a hundred yards or so if possible, all the while dragging one or two heavy rifles, several belts containing perhaps forty or so

51. Don't be confused by the Model 1874 designation for this Sharps model since it actually appeared several years earlier. A cartridge designation such as .50-100-425 indicated .50 caliber with 100 grains of powder and a bullet weighing 425 grains. Winchester, Union Metallic Cartridge Company, and other ammunition makers offered a wide variety of centerfire metallic cartridges. Some cartridges popular with bison hunters were the .44-77, .50-70, and .45-70.

cartridges each, and water. The hunter first shot what appeared to the leader of the band, often a cow. If others nearby were made restless by the smell of blood, the hunter attempted to concentrate his fire on those which appeared ready to move away.

If a hunter was skilled and lucky, a successful stand might allow him to kill scores of bison without moving from his original location if there was enough wind to dispel the powder smoke and allow him to see his quarry. One hunter killed 79 bison with 91 shots in a stand and Tom Nixon reportedly shot 204 at another site. Since it often took fifteen to twenty minutes for a skinner to complete his bloody task, such numbers kept a hunter's skinners busy for long hours. A pair of crossed "shooting sticks" or a rest made from the crotch of a tree branch could help a shooter steady his aim and as one rifle became hot and its accuracy diminished, he might cool it with water or switch to the spare as the first cooled. Factory-loaded ammunition was an expense, about five or six cents each, so he retained his empty cartridge cases and reloaded them in camp. One hunter recalled paying 50 cents each to have his hides hauled by mule train, then receiving $2 per cow hide and $3 for those from bulls.

Almost any discussion of the bison slaughter involves mention of what often is known as Billy Dixon's "long shot." On June 27, 1874, a group of 29 whites including one woman were besieged by 400 to 600 Kiowas, Comanches, Cheyennes, and Arapahoes at a trading post known as Adobe Walls in northern Texas, erected at the site of an earlier trading outpost built about 1830 by Charles and William Bent and Ceran St. Vrain. Fortunately the group included about nine buffalo hunters and it was their skill with their rifles which enabled the whites to hold out during the day-long siege. After the battle, about a dozen Indians appeared on a bluff and Billy Dixon was challenged to fire at them with his Sharps. He did so and one of the warriors fell from his horse, struck at a distance reported to have been about 1,500 yards, more than three-quarters of a mile. Even though Dixon was a skilled hunter, well versed in estimating distance and was firing at a group of men rather than a single individual, it was a lucky shot and he admitted it. Some have disputed the shot, yet modern tests have shown that such a Sharps is capable of killing at that distance.[52]

But Dixon's encounters with hostile Indians weren't over for that year of 1874. In September while delivering dispatches for Colonel Nelson Miles, he along with army scout Amos Chapman and four 6th Cavalry troopers were caught on the open plain by about 100 Kiowa and Comanche warriors. Eventually they were able to take refuge in a shallow buffalo wallow where they held out the remainder of that day and the next before the five desperate men still alive were rescued. Dixon was the only one of them who hadn't received a serious wound, and his skill with his Sharps rifle presumably was a factor in their survival in what became known as the Buffalo Wallow Fight. All the men were awarded the Medal of Honor, but Congress later revoked Dixon's and Chapman's since they were civilian scouts. Dixon refused to surrender his medal and it's on display today at the Panhandle Plains Historical Museum in Canyon, Texas.

Target, Exhibition, and Trap Shooting

The later decades of the nineteenth century were marked by increasing interest in both short- and long-range target shooting, in part spurred by the Civil War and its revelation of the lack of marksmanship training within the army. There may have been somewhat greater familiarity with firearms among southerners from the largely rural south, but many northern and some Confederate recruits

52. In October 1874, Billy Dixon returned to the abandoned trading post, by then largely destroyed by Indians. He was astonished and delighted when his setter Fannie, missing since the day of the battle, greeted him proudly with her litter of four pups.

were thrown into combat with virtually no training in the use of the firearms recently thrust into their inexperienced hands. This postwar period also saw the birth and somewhat hesitant growth of the National Rifle Association of America (NRA).

Colonel William Church as editor of the widely circulated *Army and Navy Journal* loudly decried in editorials that lack of marksmanship ability within the volunteer militia. Captain (later general) George W. Wingate of the New York militia seconded this call and in 1870 began the preparation of what became the first detailed formal manual on military rifle training in this country, published in book form in 1872. It was directed primarily at state militia units (which after 1877 generally were known as national guard) and was innovative in that it encouraged aiming practice and "dry" firing with an unloaded gun before advancing to actual live ammunition. These efforts by Wingate, Church, and other associates drew largely on Great Britain's successful methods in developing rifle skills among its regular soldiers and volunteers throughout the 1860s, a program which included regional and national matches at ranges of up to 1,000 yards, an unthinkable distance among many of America's shooters.

In the United States, the initial aim was directed at improving the training of New York's militia regiments, then armed with .50 caliber Remington single-shot rolling-block rifles. But interest in the effort spread among several other states and in November 1871, the National Rifle Association was incorporated as a unifying body with Major General Ambrose Burnside as its first president. Although the general's Civil War accomplishments as commander of the Army of the Potomac had been less than remarkable, his bravery and honesty were never questioned. By mid-1872 the organization had acquired land known as Creed's farm on Long Island, property which would become famous as Creedmoor, the site of numerous national and international rifle matches. Eventually the Creedmoor name would be applied to some of

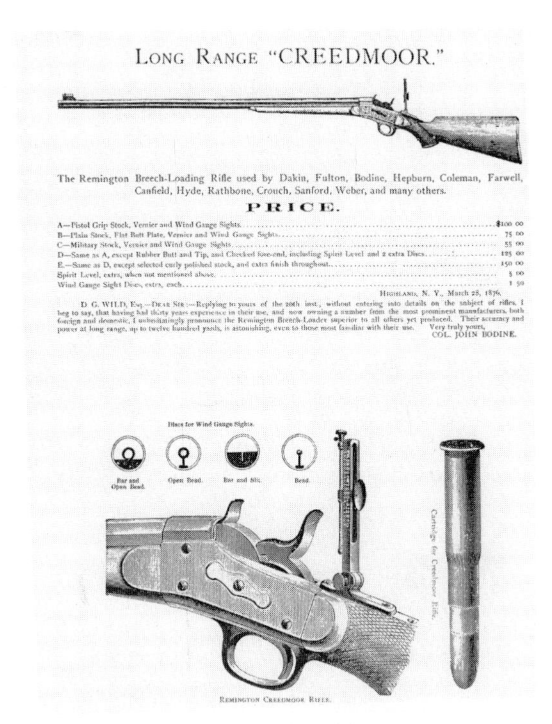

284. Ad from a Remington catalog of about 1880 promoting their Long Range Creedmoor target rifle, priced from $75 to $150.

the highest grade target rifles made by Sharps, Maynard, Remington, Marlin-Ballard, and others (Figure 284).

Targets at Creedmoor at first didn't use the concentric circle design in common use today but instead were patterned after those of the National Rifle Association of Great Britain. They were heavy slabs of iron which could be bolted together to increase target width at longer distances. The bull's-eye at ranges up to 300 yards was an eight-inch black square on a white two-foot center ("inner"). Shots striking outside the center but in a rectangle six feet high and four feet wide scored as an "outer." The bull's-eye increased to two feet square at ranges between 400 and 700 yards with a four-foot center and six-foot square outer. At distances of 800 yards or greater, the target was twelve feet wide by six feet high with a three-foot square in the center. Scoring beyond 700 yards counted hits on

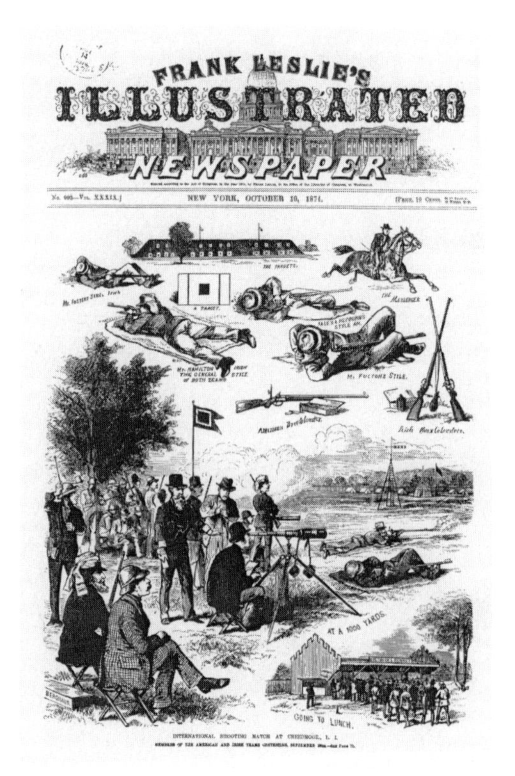

285. The first international match at Creedmoor received front-page coverage in the October 10, 1874, issue of the widely circulated *Frank Leslie's Illustrated Newspaper.*

either side of the center but not above or below. The theory was that in a military situation, one probably would be firing at a massed group of soldiers or an artillery battery where one could still strike an enemy on either side of the intended target if the shooter was correct in elevation. Hits were signaled not by flags but by colored discs on long poles. Bullet marks on the target were erased by a paint brush secured to the back of the disc.

The initial match at Creedmoor was held on June 21, 1873, with representative teams from the New York and New Jersey state militia plus two regular army teams. One of the prizes was a $100 Winchester 1866 rifle embellished with gold. In October the first annual match took place with such prizes as rifles by Remington and Frank Wesson plus a Gatling gun complete with carriage. Meanwhile the Irish rifle team, Great Britain's champions led by famed gunsmith John Rigby, in late 1873 offered a challenge to Americans for a long-range match. The Amateur Rifle Club accepted and was given full use of the Creedmoor facility. One of the rules was that American shooters would use only American-made rifles, thus setting the stage for a test between American breechloaders and the Irish team's premier muzzleloaders including those by Rigby himself.

Americans who competed successfully for a spot on the team were not experienced in long-range competition and had much to learn and little time to do so. Both E. Remington & Sons and the Sharps Rifle Manufacturing Company agreed to build special .44-77 caliber single-shot breechloading rifles for the meet without charge. On September 26, 1874, before a crowd of almost 8,000 spectators, the American team clad in business suits and firing from positions on their back or side, won a narrow and unexpected victory 934 to 931 with their final shot at the 1,000 yard range to become champions of the English-speaking world. Colonel John Bodine made the winning bull's-eye despite having cut his hand badly when a bottle of sun-warmed soda exploded shortly before he took his place on the firing line. The contest was too close to determine the supremacy of muzzle loader or breechloader, however (Figure 285).[53]

The return match the following summer at Dollymount in Ireland drew an unexpected crowd of some 30,000 spectators who saw an American team again come out on top, this time firing on newly adopted circular targets but still backed by the iron slabs with their encouraging clang when a bullet hit. The victory helped spark a rising interest in target shooting in the U.S. and the formation of numerous non-military clubs promoting rifle practice. New York City sporting goods dealer Homer

53. Russell S. Gilmore, "The Golden Age of American Target Shooting 1840 to 1900," *Remington Society of America Journal*, Fourth Quarter 2003, 26.

Fisher soon was advertising his 100-yard indoor practice range, "Creedmoor Jr.," located in a tunnel beneath Broadway. By 1890, there were close to a dozen public short-range shooting galleries in New York City alone. Women too were becoming increasingly drawn to the shooting sport and by 1880 several cities had all-women clubs.

The international match held in the United States' centennial year of 1876 again saw the American team victorious. Although the English-speaking foreign teams again relied on their muzzleloaders, many members of these teams from abroad ordered Sharps or Remington Creedmoor rifles before departing. Competition between Sharps and Remington owners was fierce and there was no clear-cut supremacy of one make over the other on the target range until the Model 1878 Borchardt Sharps came into frequent use. The Sharps Rifle Company was always quick to take advertising advantage of any victory by their rifles over those by Remington (Figure 286).

286. On the firing line at the centennial rifle match at Creedmoor on September 6, 1876.

Unfortunately the continued dominance of American shooters in the 1877 international match discouraged any foreign teams from competing the following year, a year in which the new hammerless Sharps Borchardt rifles were selected by most shooters, a choice which persisted for the next several years. An informal return to international competition came in 1880 with American shooters, using one Ballard rifle and the remainder Sharps Borchardts, again topping their Irish competitors, some of whom used the new Rigby breech-loading target rifles. A follow-on match in England brought defeat to an American team shooting against the Brit's Martini-Henry breechloaders.

Partly influenced by the popularity of competitive shooting and the favorable showing made by national guard teams, a newfound interest in individual marksmanship developed within the regular army. After some controversy, new training manuals were adopted beginning in 1879, and in 1884 the army produced a pocket-size "Soldier's Handbook" giving instructions on how to care for one's weapons and included a place to record the soldier's annual qualifying target scores. Army teams firing the .45 Springfield service rifle soon were competing in interdepartmental matches. Despite the addition of several more colorful military skirmish-style shooting contests by the NRA in its matches at Creedmoor, the regular army's participation was sporadic. But by 1890 the army was conducting rifle matches within its organization at all levels and was giving substantially more attention to the soldier's individual shooting skills. Freeman R. Bull, a machinist and gauge maker at the Springfield Armory, began shooting Springfield trapdoor rifles in competition as early as 1875. He developed precision adjustable sights for .45 Springfields modified for match competition, and his successes included earning the diamond-stud-

287. An infantryman demonstrating the "Texas position" for long-range target shooting. (*Courtesy: Colorado Historical Society*)
288. Members of a San Antonio, Texas, German shooting group or *schutzen verein*. The target shooter at left favors a Remington rolling block, his companion a Sharps. Such shooting clubs were particularly popular in German communities in the United States. One organized in New Braunfels, Texas, a town settled in 1840 by German immigrants, may be the oldest schutzen verein in the United States, dating back to 1848. (*Courtesy: San Antonio Conservation Society*)

ded World Champion Shooter Medal in 1887, adding to that weapon's reputation. His brother Milan W., also a Springfield Armory employee, shot with Freeman on the Massachusetts Volunteer Militia team in the 1880s, at a time when most competitors were using Sharps or Remington target rifles (Figure 287).[54]

In 1892, national matches were held at Sea Girt, New Jersey, a new range developed by the New Jersey State Rifle Association and which would succeed Creedmoor due to urban sprawl and a lack of support by New York state officials. Sea Girt in 1896 witnessed the first competition in which repeating rifles firing smokeless powder .30-40 caliber ammunition were used, Winchester's lever-action Model 1895 and the army's newly adopted Krag-Jorgensen bolt-action rifle. Additional matches offered at Sea Girt included those for handguns as well as popular German and Swiss-inspired *schuetzen* matches as interest in the long-range contests waned. In Europe's Germanic countries, the *schuetzenfest* had been an elaborate combination of festival and shooting contest. Targets sometimes were innovative, such as a full-size metal deer being chased by an iron dog, the pair being pulled along rails from concealment in bushes. Striking the deer might win a $5 cash prize, but hitting the dog forfeited the match.[55]

Similar shooting contests became increasingly popular in the United States in the mid-1800s in those communities where German influence was strong including Buffalo, Cincinnati, and Chicago. In the late 1800s, such matches in America often involved the use of customized small-caliber rifles, often .22, fired from a standing position at targets up to 200 yards away; the "Creedmoor" or long-range rifles were shot at distances of up to 1,000

54. McAulay, *U.S. Military Carbines*, 191.

55. Ibid., 26.

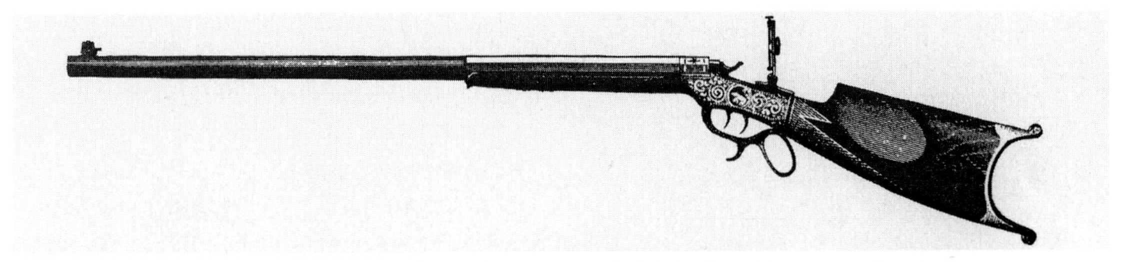

289. Marlin-Ballard No. 6 Schuetzen rifle as illustrated in an 1888 Marlin Fire Arms Company catalog.

yards. The later *schuetzen* rifles often were fitted with a pronged butt plate which partially enclosed the shoulder, elaborate sights, and sometimes an adjustable rest for the palm of the supporting hand in front of the trigger guard. Such makers as the Massachusetts Arms Company (producer of Maynard rifles), J. Stevens Arms and Tool Company, and Remington offered such rifles.

To appeal to target handgun shooters, Colt beginning about 1888 began production of a version of its Single Action with a flat-top frame, adjustable sights, and often oversize target grips. Fewer than a thousand were made but in 1894 Colt introduced a special target variation of its Bisley model Single Action. The standard Bisley already offering some features that appealed to target shooters. Most of these target Colts were sold with the 7 1/2-inch barrel length. Smith & Wesson, in addition to selling its New Model No. 3 with target sights, in 1893 began offering single-shot barrels in .22, .32, and .38 caliber which could be substituted for the barrel and cylinder of some of its .38 Single Action revolvers, converting them into single-shot target pistols. These were followed by a limited number of single-shot pistols, most with a 10-inch barrel, but made on revolver frames. Remington too produced a small number of its rolling-block single-shot pistols in target form in the 1890s, but more of its .50 caliber military pistols over the years have been altered by individual gunsmiths to the whim of a target-shooting customer (Figures 288, 289).

The new century dawned as a rejuvenated NRA, given access to the facilities at Sea Girt by the New Jersey association, in 1901 organized a renewal of international competition. The organization's increased interest in formal handgun competition was another element of the NRA's forward movement as it expanded into a new role in promoting the many national guard, army, and civilian shooting programs.

Akin to the interest in formal competitive shooting at stationary targets throughout the later 1800s was the development of the sport of trapshooting. This shotgun sport began in England in the late 1700s initially using live birds as targets at first thrown from pits or from behind a wall and later released from cages called traps. Cincinnati in 1831 may have been the site of the initial organized trapshooting contest in the United States. The sport grew in popularity and in 1866 George Portlock of Boston introduced the glass target ball from England, not as challenging and as erratic in its flight as a live bird but certainly more humane. A mechanical throwing device using a wagon spring for propulsion was invented in 1877 and allowed the thin glass spheres to be launched into the air. There even was an inexpensive launcher patented in Covington, Kentucky, which used a large rubber band for propulsion and sold for less than $1.

Amber became a popular choice of color for these targets for increased visibility with various shades of blue common too. Famed tenor-turned-exhibition-shooter Ira Paine in that same year patented a ball filled with powder or feathers to give a more realistic appearance when shattered in flight. Factories turned out glass target balls in large numbers, many with an embossed surface pattern so the

290

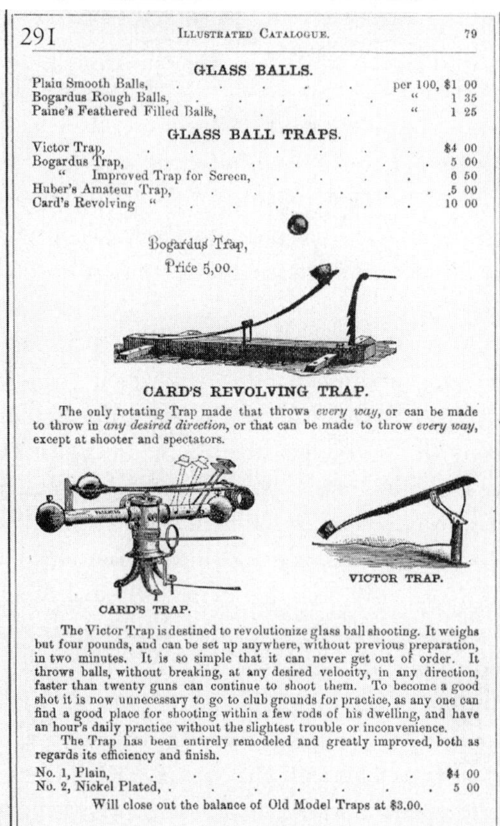

291 ILLUSTRATED CATALOGUE. 79

GLASS BALLS.

Plain Smooth Balls,	per 100, $1 00
Bogardus Rough Balls,	" 1 35
Paine's Feathered Filled Balls,	" 1 25

GLASS BALL TRAPS.

Victor Trap,	$4 00
Bogardus Trap,	5 00
" Improved Trap for Screen,	6 50
Huber's Amateur Trap,	.5 00
Card's Revolving "	10 00

CARD'S REVOLVING TRAP.

The only rotating Trap made that throws *every way*, or can be made to throw in *any desired direction*, or that can be made to throw *every way*, except at shooter and spectators.

The Victor Trap is destined to revolutionize glass ball shooting. It weighs but four pounds, and can be set up anywhere, without previous preparation, in two minutes. It is so simple that it can never get out of order. It throws balls, without breaking, at any desired velocity, in any direction, faster than twenty guns can continue to shoot them. To become a good shot it is now unnecessary to go to club grounds for practice, as any one can find a good place for shooting within a few rods of his dwelling, and have an hour's daily practice without the slightest trouble or inconvenience.

The Trap has been entirely remodeled and greatly improved, both as regards its efficiency and finish.

No. 1, Plain,	$4 00
No. 2, Nickel Plated,	5 00

Will close out the balance of Old Model Traps at $3.00.

290. A pair of glass target balls, one with a plain surface (right) and the other with an embossed pattern to facilitate breakage. **291.** Homer Fisher in 1880, located at 299 Broadway in New York City, offered glass target balls as well as traps to throw them including the trap designed by exhibition shooter Adam Bogardus.

shot didn't glance off as readily thus increasing the chances of breakage. Sometimes the manufacturer's name or the name of a sporting goods house was embossed on the surface. Some wound up as ornaments on Christmas trees. Although originally they sold for about a penny each, today's collectors eagerly pay far more, on occasion a price in four figures for a rare specimen (Figures 290, 291).

In the 1880s the popularity of the glass balls was threatened in 1880 with the invention by George Ligowsky of Cincinnati of a baked clay target, followed about 1884 by Fred Kimble's and Charles Stock's "Peoria Black Bird," a target similar to the clay "birds" used today, which included coal tar and pitch in its composition and shattered easily when struck by pellets. The partners added to their success when they invented a hand "trap" by which their targets could be thrown. The year 1889 saw the first trapshooting association formed, followed in 1893 by the first Grand American tournament. Despite the growing pressure from humane associations, live birds were used as targets at the Grand American until the beginning of the twentieth century.

A discussion of target shooting can't ignore the popularity of exhibition shooters who entertained thousands of spectators with their success in blasting aerial targets as well as such unorthodox ones as playing cards, coins, or the ash on the end of a cigarette. Two men and a diminutive dark-haired young lady stood out among these entertainers. Adam H. Bogardus was a market hunter, slaughtering wild birds in Illinois when he began to enter live bird shooting competitions, matches in which he quickly rose to national prominence for his skill. He progressed to exhibition shooting before paying audiences and in 1869 using a muzzle-loading shotgun which he loaded himself reportedly killed 500 pigeons in 528 minutes. Wild passenger pigeons, now extinct, were the preferred targets, but common pigeons, meadowlarks, quail, and sparrows were used in his exhibitions until humane laws in many cities forced him to turn to glass ball targets.

In 1877 with retirement in mind, Bogardus announced he would attempt to break 5,000 glass balls in 500 minutes. The event took place in New York City in January 1878, and this time he was using a breech-loading double-barrel shotgun on which he changed the pairs of barrels regularly during the event. Although exhaustion finally forced him to shoot from a chair, before a cheering crowd he completed his 5,156th shot in just under 481 minutes. He soon was off to Europe where he performed at the Paris Exposition.[56]

Meanwhile farther west, W. F. "Doc" Carver was another meat hunter and had hunted bison with Buffalo Bill Cody. He began to put on demonstrations of his skill with a rifle, including one before amazed Winchester representatives in New Haven, hitting glass balls and silver dollars in the air. In New York on July 13, 1878, using six Winchester rifles, he smashed 5,500 glass balls in seven hours with 6,212 shots. In 1882, Carver and Cody became partners in a touring "wild west" outdoor show, but egos and professional jealousy caused a dissolution in the fall of 1883.[57]

In the spring of 1883, Carver and Bogardus finally met for a series of matches in various eastern and mid-west cities, champion against champion, shooting at live birds and clay targets. Both men used an English-made 12-gauge double-barrel shotgun, Carver a W. W. Greener and Bogardus a W. C. Scott & Son. "Doc" Carver proved his superiority by winning nineteen of their matches, losing three, and tying three.[58]

The popular twentieth-century musical *Annie Get Your Gun* was based loosely on the life of the champion lady markswoman of the previous century. Barely five feet tall as an adult, Phoebe Ann Mosey was born near Greenville, Ohio. After her mother was twice widowed, the youngster was sent to the county poor farm and spent some time as a servant to a local family who abused her and whom Annie later characterized as "the wolves." She later reunited with her own family and hunted to help support them. She sold the surplus game to hotels and restaurants and by age sixteen had earned enough to pay off the mortgage on the family farm. She gained a local reputation as a sharpshooter and in 1881 at age twenty-one she accepted a challenge by exhibition shooter Francis "Frank" Butler, who offered $100 to anyone who could beat him. Frank missed one of his twenty-five shots and lost to her, but a year later the pair married and began a lifelong partnership both on and off the stage (Figure 292).

292. Annie Oakley with either a Model 1892 or 1894 Winchester.

Annie took the stage name of Oakley and initially acted as Butler's assistant in his shooting act. When it became clear she was the better shot, he

56. James B. Trefethen, "They Were All Sure Shots," *American Heritage Magazine*, April 1962 (no page numbers shown on Internet archived article).

57. Ibid.

58. Richard Hamilton, "The Sporting Event That Captured America in the Spring of 1883," www.traphof.org, Trapshooting Hall of Fame and Museum Web site.

relinquished the limelight to her and became her manager and assistant. In 1885 they joined Buffalo Bill Cody's wild west show, where fellow performer Sitting Bull gave her a nickname which was translated and billed as "Little Sure Shot." Always the demure lady, she became a female superstar, enchanting spectators here and abroad and once shot the ashes from a cigarette held by Germany's future Kaiser Wilhelm II. On one occasion with a .22 rifle she hit 4,472 of 5,000 glass balls in midair. The Butlers left the show, apparently in part over publicity given a youthful and raucous lady shooter, Lillian Smith, but later returned. Annie was a performer with the Cody show for almost twenty years during which time she won scores of shooting awards. After William Randolph Hearst published a false story that she had stolen to support a cocaine habit, she won fifty-four of fifty-five libel lawsuits against newspapers which published the story. She continued to shoot even in her sixties after a crippling auto accident, and at age sixty-two broke 100 straight clay birds from the sixteen-yard line. In her later years she was active and generous in financial support of various charitable causes. She was a strong advocate for the idea that women should learn to shoot and be capable of defending themselves with firearms if necessary.

7 Indian Guns

The gradual acquisition of Spanish horses by the plains Indians in the 1500s had a profound effect on their culture, giving them mobility and other advantages over those American Indians who lacked them. The same was true as some American Indians began to acquire and use firearms secured through trade (either legal or illegal) or capture. The Dutch may have been the first to exchange guns for furs. Mohawks in 1614 used Dutch muskets against their enemies the Delawares as the former ranged far from their villages in New York's Mohawk Valley. The Dutch established a trading post at today's Albany, New York, in 1618 and before long they were charging twenty beaver pelts per gun and in 1633 alone may have distributed 1,500 long guns including matchlocks, primarily through the Iroquois League.[1]

The Spanish had what became permanent settlements in Florida and New Mexico long before the English settled Jamestown in 1607, but they were reluctant to provide Indians with firearms. The influx of other European guns among the various tribes with which they came in contact caused Spanish officials substantial concern. As early as 1613 the French too had prohibited the distribution of guns to American Indians and as of 1620 such illicit trade carried the death penalty. But about 1640 this policy was revoked and a legitimate French gun trade began, a trade which expanded quickly. England's Hudson's Bay Company in 1670 received its charter to trade in North America and during nineteen of the next thirty years records indicate the firm sold approximately 10,100 long guns and 100 pairs of pistols. It's not unreasonable to assume that the French had similar successes. Often gunsmiths offered repair services at both nations' trading posts and depots and some even lived in Indian villages.[2]

American Indians in the southwest who did acquire Spanish guns by capture had difficulty replenishing their supplies of gunpowder and other needed components before they gained access to French traders in the later 1600s. As an example of French trade influence in the west, Comanche Indians with French guns were a persistent threat to Spanish outposts in Texas. In Florida, Spanish officials in 1731 captured Robert Jenkins and his ship and accused him of crimes including smuggling guns and other contraband to Seminole Indians, runaway slaves, and others. He claimed the Spanish coast guard cut off his ear as a warning to others, and six years later when the ear was displayed to Parliament, it aroused such resentment that the British government in 1739 reluctantly declared war, a conflict caused by economic rivalry and known as the War of Jenkins' Ear. But by the early 1700s, both English and French traders were bartering guns west of the Mississippi for horses and pelts.[3]

1. Ahearn, *Flintlock Muskets in the American Revolution*, 132. Brown, *Firearms in Colonial America*, 152.

2. Gooding, *The Canadian Gunsmiths 1608 to 1900*, 8-9, 25.

3. Brown, *Firearms in Colonial America*, 90, 119, 173.

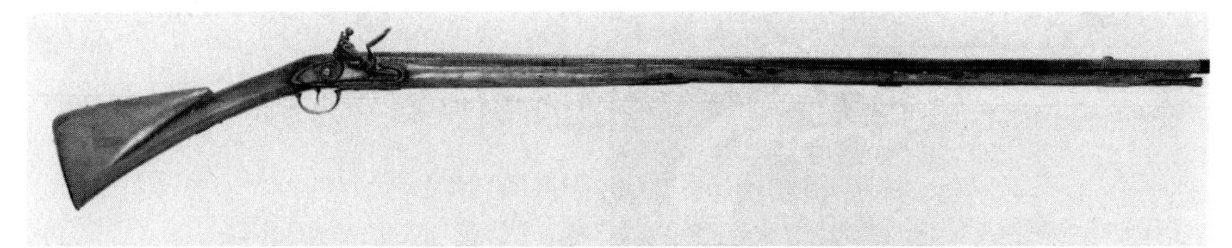

293. French smoothbore trade gun of the mid-1700s, about .55 caliber with a 48-inch barrel. (*Courtesy: Museum of the Fur Trade, Chadron, Nebraska*)

Based in part on archaeological evidence, French flintlock trade guns of the late 1600s and early 1700s appear to have been lightweight smoothbores, about .50 to .70 caliber, and with a barrel varying in length from three feet to about eighteen inches longer and secured to the full-length stock by transverse pins. Some presented as gifts to American Indians of prominence were a step above in terms of quality and ornamentation. Some Indians had access to repair facilities, gunsmiths employed by the French and by England's Hudson's Bay Company in the late 1600s. But many didn't, and often American Indian firearms were used hard so examples of these early trade guns today are not common (Figure 293).

While a smoothbore trade gun offered the versatility of firing shot or a single ball, by the 1750s in Pennsylvania and surrounding area Indians there were making some use of rifles. Robert Kirkwood was a young Scotsman who began a decade of service in North America with the British army when he "accepted the king's shilling" in 1757. In September 1758 he was captured by Shawnee Indians after being wounded by buckshot in a skirmish near Fort Duquesne, where Pittsburgh now stands. His life was spared when he was adopted as a brother by the warrior who had taken him prisoner. He became a proficient hunter but the following June he slipped away and later reached a British outpost. In describing a two-month hunt with the Shawnees, he wrote: "Some French traders coming up the Ohio [River], exchanged powder and shot with us for furrs and skins. I had succeeded so well in this party that I bought a riffle gun, some powder, and two new blankets."[4]

Naturalist John Bartram in 1756 wrote of his observations of Indians in Pennsylvania: "they commonly now shoot with rifles with which they will at a great distance from behind a tree . . . take such sure aim as seldom miseth their mark." A year earlier a witness to General Braddock's march against the French in western Pennsylvania described some of the Indians who accompanied the British as being "tall, well made, and active, but not strong, but very dexterous with a rifle barreled gun, and their tomahawk."[5]

Such use caused growing concern among British officials and in 1764 a draft of regulations governing Indian trade was proposed:

> Rifled Barreled Guns should certainly be prohibited; the Shawanese and Delawares, with many of their neighbours are become very fond of them [rifles], and use them with such dexterity, that they are capable of doing infinite damage, and as they are made in some of the frontier Towns, where the Indians will procure them at any Price . . . all white persons should be restricted on a very severe penalty from selling them to any Indians.

The proposal was adopted but enforcement was difficult, particularly since rifles were quite often dispensed as official gifts with the intent of

4. McCulloch and Todish, *Through So Many Dangers,* 48.

5. Bailey, *British Military Flintlock Rifles,* 75.

cementing British-Indian relations. Some such Indian rifles before the Revolutionary War would have been American made. However, during the later years of that war, some British rifles were made on contract in England by William Grice, William Wilson & Company, and Robert Barnett and styled after the Pennsylvania rifle. Surviving specimens of these rifles are rare, but most are thought to have been about .55 caliber, brass mounted, and often with a sliding wood rather than a hinged brass patch box cover. Each with a bullet mold and a gun case, these were exported to North America for distribution to Indian allies as gifts or trade goods along with the ever popular and less expensive smoothbore trade guns.[6]

Americans weren't slow in entering the lucrative trade with Indians for furs, but between 1796 and 1822, they had competition from the U.S. government itself, trading with American Indians through a chain of "factories" or trading posts where guns, blankets, kettles, and many other goods were sold at prices only slightly above cost. American ships appeared off the northwest coast of North America in the 1790s and early 1800s and their trade in guns brought complaints from officials in Russian America. Regardless of the source, the gun with its longer range than the bow and arrow and its psychological advantages of smoke, noise, and seemingly mystic characteristics initially provided the "haves" with awesome power over the "have nots."

On the western plains, the lance and bow had been primary Indian weapons for centuries, used for warfare and hunting. In northern Mexico in the mid-1500s, settlers described the Chichimeca Indians using oranges tossed into the air as targets for their arrows. Even as firearms became more readily available, the bow and arrow remained a potent short-range weapon well into the 1870s for it could be fired rapidly and did not give off sound or smoke to reveal the shooter's position. Anthony Glass, a trader, kept a journal of his travels in 1808-9 and reported that he'd seen an Indian on the Red River in Texas using a bow made of Osage

294. An 1845 painting by J. M. Stanley entitled A Buffalo Hunt on the Southwestern Prairies. (*Courtesy: Smithsonian Office of Anthropology, neg. no. 2860-gg-1*)

orange drive an arrow entirely through a bison. More than a few whites were impressed by the accuracy with which Indians could use a bow and arrow at moderate distances, knocking coins from sticks stuck in the ground, for example. In warfare, the lower velocity arrow didn't have the shocking power of a bullet, but an arrow head left in the body could disable an opponent and perhaps kill by infection later (Figure 294).

Western movies sometimes show an arrow being pulled free from a wound but this is more fiction than fact. Unless the wound was shallow, such an action could easily separate shaft from arrowhead making the latter's removal even more difficult. Furthermore, arrowheads often were made of soft hoop iron, and if they struck bone or otherwise became bent, removal was even more of a challenge.

Those American Indians who did acquire firearms were reliant upon white traders for gunpowder, later metallic cartridges, and other needed accessories. They frequently lacked access to a means of repairing a broken gun so their firearms often were kept in use far longer if worn or damaged than if they had been in white hands. If a gun barrel burst, it might be cut off and the gun remain in use, perhaps even converted into a crude but

6. Ibid., 77, 81, 203.

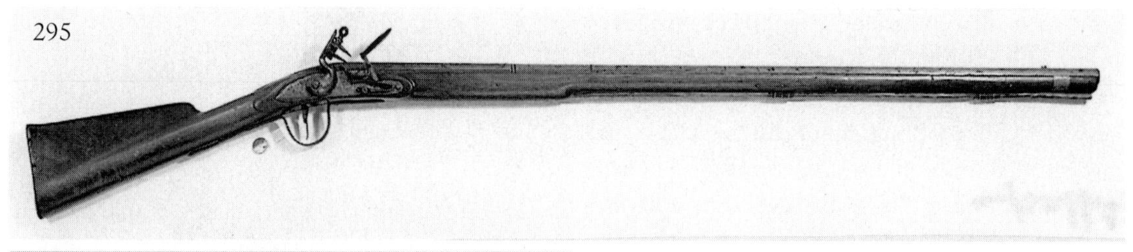

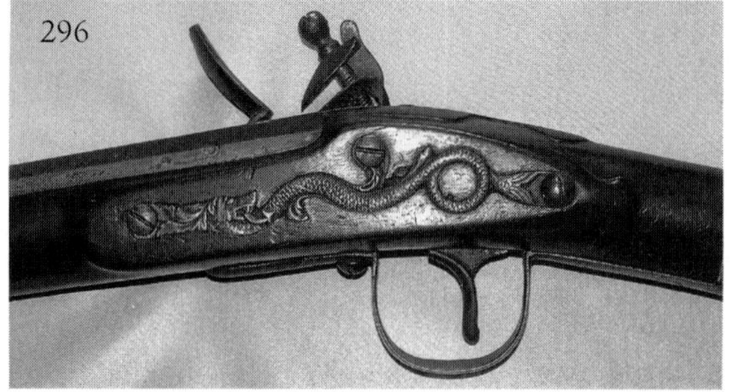

295. Northwest trade gun made by Henry Leman of Lancaster, Pennsylvania, and discovered in Laguna Pueblo in New Mexico. (*Courtesy: James D. Gordon, photo by Steven W. Walenta*)
296. Two of the characteristics common to typical Northwest trade guns were the dragon or serpent side plate and large iron trigger guard.

functional pistol. If a gunstock was shattered in a fall, wet rawhide wrapped around the break shrank as it dried to provide a strong repair. Sights when lost might be restored with a crude replacement or since powder often was in short supply, they might be left off entirely rather than waste powder on long-range firing. Few Indians had the resources necessary to practice often, so few became proficient shots at longer distances.

The gun's brass or iron butt plate sometimes was removed and used as a hide scraper, or a section of damaged barrel could be beaten flat and sharpened to use for the same purpose. One chief's pipe bowl was seen to have been fashioned from a section of rifle barrel. For convenient use one-handed on horseback the barrel might intentionally be cut back to a length of a foot or so and the butt stock shortened. Such drastically shortened guns also could easily be concealed beneath clothing or a blanket and today sometimes are referred to as "blanket guns."

While American Indians might acquire any type of hand or shoulder gun available to them, one gun is unique to the Indian trade, often referred to as the Northwest gun or sometimes in writings of the day as "fuke," "fusee," "Mackinaw gun," or "fusil." It appeared by the mid-1700s and in the west remained in frequent use as late as the 1870s, even later in Alaska and remote Canadian areas. It was a light smoothbore gun, usually of about .58 caliber, capable of firing either shot or a round ball. Its large iron trigger guard could admit a gloved finger, but according to a Hudson's Bay Company request as early as 1740 it was to allow two fingers to contact the trigger. Barrel lengths of 42 or 48 inches were common in the 1700s but later lengths of 30 and 36 inches were frequent choices. One of its distinguishing characteristics was a brass sideplate set into the stock opposite the lock, bearing the form of a dragon or serpent. A common misconception is that the price of such a gun intended for trade to an Indian was determined by stacking furs to a height equal to its length. Such was not the case. As an example, the official Hudson's Bay Company rate of exchange for a five-foot trade gun in 1684 was twelve stretched and dried beaver pelts (Figures 295, 296).

Most Northwest guns were of English or Belgian manufacture, although Henry Deringer of Philadelphia produced some as early as 1815; others by Americans J. J. Henry, Edward K. Tryon, and Henry Leman followed in the 1840s and 1850s. Barnett, Wheeler, and W. Chance were common English makers of these smoothbores as were

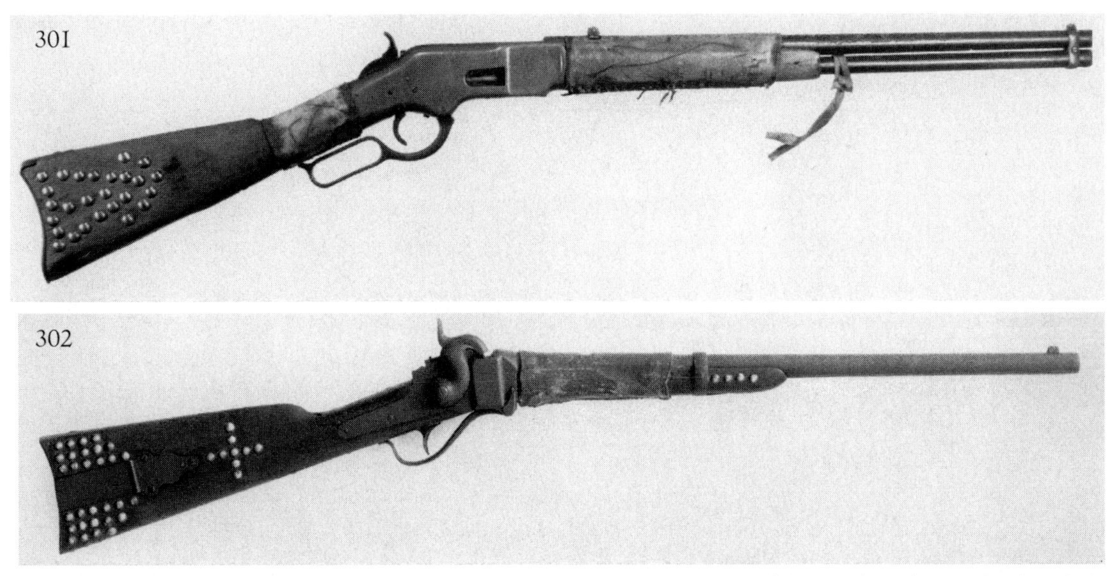

301. An 1866 Winchester carbine from the Crow Reservation in Montana decorated with brass tacks and repaired with rawhide wrappings. (*Courtesy: Ralph Heinz*) **302.** Decorative brass tacks and a sinew sewn rawhide repair to the forearm indicate this Model 1863 percussion Sharps carbine probably was an Indian gun. It was found in the 1930s beneath the wooden floor of an adobe house that was being restored near the waterfront in Vallejo, California. (*Courtesy: Fred Fellows and Ralph Heinz*)

In this 1879 listing, breechloaders included an assortment of army Springfield rifles and carbines, Spencers, Sharps, Winchesters, and various single-shot Civil War–era carbines. Many of these guns would have been considered unserviceable, with broken or worn stocks, faulty locks, and missing or crude replacement sights. But even though worn, many "could be used by so enterprising an enemy as the American Indian" the report stated. Handguns tallied were just as varied in make and model, although army Colt Single Actions and Smith & Wesson revolvers were notably absent. Of course it was easier to hide the best revolvers such as these beneath an Indian woman's clothing, for example, than it was a long arm (Figures 300, 301).

Brass tacks, used to hold a cover on a trunk or fabric on furniture, were popular trade items for Indians who sometimes added them to the butt stock or forearm of a gun for decorative purposes or religious symbolism. One Sioux warrior with a strong physique—he was described as six feet four inches tall and weighing 240 pounds—acquired the name "Pawnee Killer" and is said to have had 140 such tacks in his Winchester rifle, reportedly one for each enemy he had dispatched (Figures 302, 303).

American Indians on average were not viewed as skilled marksmen, but there were exceptions. During the army's pursuit of Chief Joseph and the Nez Perce toward safe refuge in Canada in 1877, in one engagement a warrior with a telescopic sighted rifle picked off a number of soldiers until he finally was dispatched. Granville Stuart, an early Montana pioneer who was proud of his success with his new breech-loading Maynard rifle, in 1861 lost an embarrassing shooting match against an Indian named Pushigan who out shot him using a muzzle-loading rifle with a crude hoop iron sight. The bets at first were modest, $1 or so, but as the match continued they increased as did the range. The elderly Indian continually bested Stuart, who later decided his poor showing had been due to his carelessness in not measuring his powder charges for consistency when reloading his cartridge cases (Figures 304, 305, 306).

303. This Blackfoot warrior of the 1870s has made extensive use of brass tacks to decorate his Winchester 1873 carbine, quirt handle, bridle, knife sheath, and other gear. Around his neck he wears a necklace made of brass dragon or serpent side plates typically found on Northwest trade guns. (*Courtesy: Wade Lucas*) **304.** A quartet of Ute Indians photographed in 1874. The mounted warrior holds a handful of arrows, his companions (from left) a Winchester Model 1873 rifle, a M1866 Winchester rifle, and another lever-action rifle. (*Courtesy: Smithsonian Office of Anthropology*) **305.** A Cheyenne warrior with an 1866 Winchester carbine. (*Courtesy: Smithsonian Office of Anthropology*) **306.** A Springfield Model 1873 rifle manufactured in 1874 and altered into a classic example of an easily concealed Indian "blanket gun" with a crude iron pistol grip. (*Courtesy: Gary L. Delscamp*)

307. A pair of army Indian scouts holding a Spencer carbine (at left) and a Model 1866 Winchester rifle. The frontier army at various times made very effective use of Indian scouts although the government sometimes repaid them for their services with injustice. **308.** Reservation Indian policemen were often equipped by the federal government. These uniformed Dakota police are armed with Model 1875 Remington revolvers and a Winchester Model 1873 carbine (at left) and probably a Whitney-Kennedy carbine. (*Courtesy: Denver Public Library*) **309.** Choctaw Indian police with Colt Single Actions with 7 1/2-inch barrel except for the man seated at left who appears to be holding a Remington altered to fire metallic ammunition. (*Courtesy: Archives & Manuscripts Division, Oklahoma Historical Society*)

During the 1860s and 1870s, the number and quality of guns in Indian hands increased, as did the variety. This was a period of intense American Indian resistance to advancing white settlement even though the clash of cultures had begun soon after the arrival of the first Europeans on the North American continent. A wagon train on the Bozeman Trail into Montana in 1863 encountered a band of Indians, one of whom wanted to trade two horses for a traveler's young sister. "They were armed with all sorts of guns, some good breechloaders, some good muzzle loaders, shot guns cut off to 16 inches long, six shooters, pistols, horse pistols, and a lot of old fusees not worth picking up. A great many had bows and arrows and spears." Evidence from various sources indicates that at the Custer fight in June 1876 on the "Greasy Grass" or Little Bighorn, Indian warriors had a substantial number of Winchester repeaters including some Model 1873s and other up-to-date arms. Even while the army and government attempted to restrict the firearms trade or distribution of guns at peace conferences except those intended for hunting purposes (usually percussion revolvers and muzzle-loading rifles), an illicit trade in improved firearms, metallic ammunition, and liquor continued, driven by the potential for high profits (Figures 307, 308, 309).

Postscript

As a finale, it's appropriate to recount an anecdote which illustrates the disappointment a collector of antique guns too frequently encounters. Merrill Lindsay in his book *The New England Gun* described a machinery salesman who during the depression of the 1930s often spent his lunch hours looking in the windows of antique shops as he traveled about New England. In one such shop he spied a pair of Belgian flintlock military pistols, clumsy and of late flintlock period from about the 1830s or 1840s. They seemed out of place in a wooden case lined with crimson plush material and with a brass plaque set into the lid. But the salesman's curiosity got the better of him and upon entering the shop and examining the inscription on the plaque found it read "To Gen. George Washington from his friend Benedict Arnold," both of whom were long dead by the time the pistols were made. *Caveat emptor* (buyer beware) is still a most appropriate motto today just as it was then (Figure 310).

310. The gun trader's lament. *I have more than that in it. (*Courtesy: Norm Flayderman*)

Collections of Antique Firearms

Numerous U.S. museums exhibit antique firearms that were significant in American history. I undoubtedly have overlooked a few; state historical societies not listed here may have firearms in their collections. One must remember that exhibits change and galleries may be closed for renovation, so contact the museum in advance to determine the current status and scope of displays. Some museums may have extensive collections but keep in storage many of their weapons, which sometimes are made available to serious researchers. Also be aware that museums located on a military installation, such as the U.S. Naval Academy Museum at Annapolis, may require photo ID for adults to gain access because of security measures. Many of the museums listed have informative Web sites which can be accessed by entering the name of the museum in a search engine.

Also, some arms collectors' associations offer a substantial number of excellent exhibits by members in addition to the hundreds of dealers' tables of guns for sale at their occasional weekend meetings. Many are open only to members and guests, but among the foremost which are available to the public for an admission fee are the Maryland Antique Arms Collectors Association annual gathering in March in Baltimore and the May yearly meeting in Denver of the Colorado Gun Collectors Association.

Arizona State Museum, Tucson

Atlanta History Center, Atlanta, Georgia

Battle Abbey, Richmond, Virginia (Outstanding exhibit of Confederate made and used arms.)

Buffalo Bill Historical Center, Cody, Wyoming (A complex of museums of western art, firearms, and American Indian culture.)

Chickamauga-Chattanooga National Military Park, Fort Oglethorpe, Georgia (Exhibits Claude Fuller collection of U.S. martial arms.)

Colonial Williamsburg, Williamsburg, Virginia

Colorado Historical Society, Denver (Large collection although much of it in storage.)

Colorado Springs Museum, Colorado Springs

Confederate Memorial Hall, New Orleans, Louisiana

Confederate Museum, Greenville, South Carolina

Connecticut State Library Museum of Connecticut History, Hartford (Extensive Colt collection.)

Daughters of Utah Pioneers Pioneer Memorial Museum, Salt Lake City

J. M. Davis Arms Museum, Claremore, Oklahoma (Extensive and eclectic collection of guns in varying states of preservation.)

Fort Ticonderoga, Ticonderoga, New York (Fine exhibit of colonial arms.)

Frazier Historical Arms Museum, Louisville, Kentucky

Gene Autry Museum of Western Heritage, Los Angeles, California (Exhibits many historic guns once owned by individuals well known in western history such as outlaw Frank James and others.)

Harpers Ferry National Historic Park, Harpers Ferry, West Virginia (Exhibits arms made at the site and those associated with John Brown's raid.)

Jim Gatchell Museum, Buffalo, Wyoming (Closed in winter.)

Gettysburg National Military Park, Gettysburg, Pennsylvania (National Park Service museum exhibits an outstanding collection of Civil War era martial arms, both Union and Confederate.)

Huntington Museum of Art, Huntington, West Virginia

Kansas Museum of History, Topeka

Kentucky Military History Museum, Frankfort

Metropolitan Museum of Art, New York, New York

Missouri Historical Society, St. Louis

Missouri State Museum, Jefferson City

Montana Historical Society, Helena

Museum of Nebraska History, Lincoln (Large collection in storage, only a portion on exhibit.)

Museum of the Fur Trade, Chadron, Nebraska (Exhibits an outstanding array of artifacts related to the fur trade including Northwest trade guns and others. A "must see" for anyone interested in that aspect of American history. Closed in winter.)

National Cowboy Hall of Fame and Western Heritage Center, Oklahoma City, Oklahoma

National Firearms Museum, Fairfax, Virginia (Extensive National Rifle Association museum of both antique and modern arms.)

National Museum of American History, Smithsonian Institution, Washington, D.C. (Closed for renovation; expected to reopen in summer 2008.)

Nevada State Museum, Carson City

Old Sturbridge Village, Sturbridge, Massachusetts

Palace of the Governors, Santa Fe, New Mexico

Panhandle-Plains Historical Museum, Canyon, Texas

Philadelphia Museum of Art, Philadelphia, Pennsylvania (Collection includes some outstanding examples of ornate pre-flintlock firearms.)

Remington Arms Co. Museum, Ilion, New York (Displays examples of most models of guns made in the firm's 189-year history. Open only in summer.)

Rock Island Arsenal Museum, Rock Island, Illinois (Collection includes many arms surrendered by American Indians.)

Roswell Museum and Art Center, Roswell, New Mexico (Extensive exhibit of firearms as well as American Indian artifacts.)

C. M. Russell Museum, Great Falls, Montana (Features Browning family collection as well as varied examples of eighteenth- and nineteenth-century firearms including a number of British double rifles.)

Saunders Memorial Museum, Berryville, Arkansas (Closed in winter.)

Smith & Wesson Museum, Springfield, Massachusetts (Changing exhibits at S&W Shooting Sports Center, which sometime include examples of pre-1900 S&W arms.)

Springfield Armory National Historic Site, Springfield, Massachusetts (Major exhibit of U.S. military arms.)

Tennessee State Museum, Nashville

Texas A&M University Museum, College Station

Texas Ranger Hall of Fame and Museum, Waco

Union Pacific Historical Museum, Omaha, Nebraska (Check for current exhibit before visiting; significant collection in storage.)

U.S. Cavalry Museum, Ft. Riley, Kansas (Modest exhibit of guns but also displays other cavalry gear.)

U.S. Naval Academy Museum, Annapolis, Maryland

Valley Forge National Historic Park, Valley Forge, Pennsylvania

Virginia War Museum, Newport News

Wadsworth Atheneum, Hartford, Connecticut (Fine collection of Colts but check availability first since some may be on loan elsewhere.)

West Point Museum, West Point, New York

White House of the Confederacy, Richmond, Virginia

Woolaroc Museum, Bartlesville, Oklahoma (Features famed Phillips collection of Colts including numerous Patersons.)

Wyoming State Museum, Cheyenne

Chronology

(1492–1900)

1492–93	Christopher Columbus makes two voyages to the New World, bringing the first guns, probably in the form of crude hand cannon.
1598	Spanish explorer Luis de Velasco and his party armed with 19 matchlock and 19 wheel lock guns establish the town of Santa Fe, New Mexico.
1607	England establishes colony at Jamestown, Virginia, its first in the New World.
1609	Heavy rain renders the matchlock guns of a Dutch party under Henry Hudson useless as they are threatened by Indians. French explorer Samuel de Champlain kills or mortally wounds three Iroquois Indians with a single shot leading to "strained" relations between the French and Iroquois Confederacy.
1670	Hudson's Bay Company receives charter to trade in North America.
1719	Martin Meylin about this time establishes himself as a gunsmith at Lancaster, Pennsylvania, one of the earliest makers of rifles in the colonies; guns made in this area would evolve into the famed Pennsylvania-Kentucky rifle.
1728	About this time Britain begins production of what becomes known as the Brown Bess musket, used by her army throughout the world for the next 100 years.
1756–63	French and Indian War.
1763	France adopts a model .69 caliber musket which in modified form becomes the pattern for the first official U.S. muskets made at Springfield and Harpers Ferry national armories (see 1794, below).
1769	William Watts of England builds a tall wooden shot tower from which molten pellets of lead fall into water, making it possible to produce spherical shot in commercial quantities.
1775–81	American Revolutionary War.
1777	Continental Congress directs that government owned firearms be marked to reduce theft.
1794	Government authorizes establishment of national armories at Springfield, Massachusetts, and Harpers Ferry, (West) Virginia. Musket production begins at Springfield in 1795 and five years later at Harpers Ferry.
1798	Eli Whitney, Sr. receives government contract for muskets and in their manufacture introduces partial interchangeability of parts.
1799	First government contract for U.S. army pistols, 2,000 ordered from Simeon North of Connecticut, patterned after a French model of 1777.
1802	E. I. du Pont pays $6,740 for land along Brandywine River on which to build his first powder mill.
1803	U.S. purchases Louisiana Territory from France for $15 million, doubling size of the country.

1804–06	Lewis and Clark lead Corps of Discovery to Pacific Ocean and return.
1811	John H. Hall patents a breech loading gun design.
1821	Mexico gains independence from Spain. Santa Fe Trail opens as a trade route from Missouri.
1822	Joshua Shaw receives U.S. patent on percussion cap, already in use in England.
1825	Brothers Jake and Sam Hawken, famed makers of Rocky Mountain or plains rifles, form gun making partnership in St. Louis.
1831	Cincinnati is site for what is thought to be first organized American trapshooting contest.
1833	U.S. Army establishes first mounted regiment of dragoons.
1834	U.S. Army begins distribution of its first percussion arms, Hall carbines to the 1st Dragoons.
1836	Battle of the Alamo in San Antonio, Texas.
1837	Ethan Allen patents double action mechanism which he soon applies to production of extensive line of pepperbox revolvers. Production of Colt firearms begins, but ceases in 1842.
1838	U.S. Army makes first purchases of Colt firearms, during war with Seminole Indians in Florida.
1840	Final western fur trade rendezvous held as trade shifts from beaver pelts to bison hides.
1841	U.S. Army adopts what becomes known as the "Mississippi rifle," its first percussion rifle.
1842	U.S. Army adopts its first percussion and its last smoothbore musket.
1844	Samuel F. B. Morse invents the telegraph. Outnumbered Texas Rangers under Captain J. C. Hays and armed with Colt Paterson revolvers defeat Comanche Indians.
1845	U.S. annexes Republic of Texas. Dentist Dr. Edward Maynard patents tape primer for firearms.
1846–48	War with Mexico.
1847	Samuel Colt receives government contract for 1,000 Walker revolvers, reestablishing himself as a firearms maker.
1848	Gold is discovered in California. Christian Sharps patents a breech loading rifle design.
1852	About this time Henry Deringer, Jr. begins to make pocket pistols.
1855	U.S. Army adopts new series of rifled shoulder and hand guns. Rollin White patents rear loading revolver and licenses patent to Horace Smith and Daniel Wesson.
1857	First Smith & Wesson revolver appears, a diminutive .22; Samuel Colt's revolver patents expire opening the market to other makers.
1859	Abolitionist John Brown, Sr. leads raid on Harpers Ferry Armory.
1860	Pony Express service begins between Missouri and California.
1861	Transcontinental telegraph line is completed.
1861–65	American Civil War.
1862	Army receives its first Spencer repeating rifles in December. Henry .44 repeating rifle appears on the market.

1865 Springfield Armory begins conversion of percussion rifle muskets to breechloaders using Allin's "trapdoor" system. John Wilkes Booth assassinates President Lincoln with a pocket pistol made by Henry Deringer, Jr.

1866 First rifle appears bearing the Winchester name.

1867 U.S. purchases Alaska for $7 million, considered a folly by many.

1869 First transcontinental railroad is completed. Rollin White's patent on rear loading revolver cylinder expires.

1870 Smith & Wesson introduces its first large caliber revolver, the .44 Model 3, later known as the *American*.

1871 Commercial hunting for bison hides begins in earnest in winter of 1871–72. Sharps begins production of what it later designates as its Model 1874 rifle, extensively used by professional bison hunters. National Rifle Association (NRA) is incorporated.

1873 Colt begins to market its famed Single Action revolver. Winchester introduces its popular successor to the Model 1866 rifle, the Model 1873. U.S. Army adopts .45 Springfield "trapdoor" rifle and carbine.

1874 First international target match is held at Creedmoor on Long Island.

1876 Custer and the 7th Cavalry defeated by Indians at the Little Bighorn. James "Wild Bill" Hickok is shot to death in Deadwood, Dakota Territory.

1877 Colt markets its first double action revolver.

1878 Colt begins to offer its Single Action revolver in .44-40 caliber, the same cartridge used in the popular Model 1873 Winchester.

1881 A disappointed office seeker mortally wounds President Garfield with a British Bulldog revolver.

1882 William F. Cody organizes first of his wild west shows.

1890 Battle/Massacre of Wounded Knee, last major engagement of the Indian wars.

1892 U.S. Army begins issuing Colt .38 double action revolvers as replacement for the Single Action .45.

1893 First Grand American trapshooting tournament held.

1894 U.S. Army begins to replace "trapdoor" Springfields with new .30 Krag-Jorgensen, a smokeless powder small-caliber repeater.

1898 War with Spain.

Bibliography

Books

Ahearn, Bill. *Flintlock Muskets in the American Revolution and Other Colonial Wars.* Lincoln, R.I., 2005.

Arese, Count Francesco. *A Trip to the Prairies and in the Interior of North America, 1837-1838.* New York, 1975.

Bailey, De Witt. *British Military Flintlock Rifles, 1740-1840.* Lincoln, R.I., 2002.

Baker, T. Lindsay, and Billy R. Harrison. *Adobe Walls: The History and Archeology of the 1874 Trading Post.* College Station, Tex., 1986.

Baumann, Ken. *Arming the Suckers, 1861-1865.* Dayton, Ohio, 1989.

Benson, Susan W., ed. *Barry Benson's Civil War Book: Memoirs of a Confederate Scout and Sharpshooter.* Athens, Ga., 1992.

Bilby, Joseph G. *A Revolution in Arms: A History of the First Repeating Rifles.* Yardley, Pa., 2006.

———. *Civil War Firearms.* Conshohoken, Pa., 1996.

Bronson, Edgar Beecher. *Cowboy Life on the Western Plains.* New York, 1910.

Brown, M. L. *Firearms in Colonial America: The Impact on History and Technology, 1492-1792.* Washington, D.C., 1980.

Chartrand, Rene. *Uniforms and Equipment of the United States Forces in the War of 1812.* Youngstown, N.Y., 1992.

Coates, Earl J., Michael J. McAfee, and Don Troiani. *Don Troiani's Civil War Cavalry and Artillery.* Mechanicsburg, Pa., 2002.

———. *Don Troiani's Civil War Infantry*. Harrisburg, Pa., 2006.

Cronan, Rudolf. *The Army of the American Revolution and Its Organization.* New York, 1923.

Davies, Paul J. *C. S. Armory Richmond.* Ephrata, Pa., 2000.

Dillin, Capt. John G. W. *The Kentucky Rifle.* York, Pa., 1959.

Dykstra, Robert R. *The Cattle Towns.* New York, 1979.

Eberhart, L. D., and R. L. Wilson. *The Deringer in America, Vol. II, The Cartridge Period.* Lincoln, R.I., 1993.

Edwards, William B. *Civil War Guns.* Harrisburg, Pa., 1958.

Farrington, Dusan P. *Arming and Equipping the U.S. Cavalry, 1865-1902.* Lincoln, R.I., 2004.

Flayderman, Norm. *Flayderman's Guide to Antique American Firearms...and Their Values.* Iola, Wis., 2001.

Ford, Alice, ed. *Audobon, By Himself.* Garden City, N.Y., 1969.

Frasca, Dr. Albert J., and Robert H. Hill. *The .45-70 Springfield.* Northridge, Calif., 1980.

———. *The .45-70 Springfield Book II.* Springfield, Ohio, 1997.

Fuller, Claude E., and Richard D. Steuart. *Firearms of the Confederacy.* Huntington, W.V., 1944.

———. *Springfield Shoulder Arms 1795-1865,* New York, 1931.

———. *The Rifled Musket.* Harrisburg, Pa., 1958.

Garavaglia, Louis A., and Charles G. Worman. *Firearms of the American West, 1803-1865.* Albuquerque, N.M., 1984.

———. *Firearms of the American West, 1866-1894.* Albuquerque, N.M., 1985.

Gardiner, Howard C. *In Pursuit of the Golden Dream.* Dale L. Morgan, ed. Stoughton, Mass., 1970.

Gilkerson, William. *Boarders Away II.* Lincoln, R.I., 1993.

Gooding, S. James. *The Canadian Gunsmiths, 1608 to 1900.* West Hill, Ontario, 1962.

Graham, Ron, John A. Kopec, and C. Kenneth Moore. *A Study of the Colt Single Action Army Revolver.* Dallas, Tex., 1978.

Grant, U. S. *Personal Memoirs of U. S. Grant.* New York, 1894.

Green, Carol C. *Chimborazo: The Confederacy's Largest Hospital.* Knoxville, Tenn., 2004.

Guthman, William H. *U.S. Army Weapons, 1784-1791.* American Society of Arms Collectors, 1975.

Hanson, Charles E. Jr. *The Hawken Rifle: Its Place in History.* Chardon, Neb., 1979.

———. *The Plains Rifle.* Harrisburg, Pa., 1960.

Hartzler, Daniel D. *Arms Makers of Maryland.* York, Pa., 1977.

Heitman, Francis B. *Historical Register and Dictionary of the U.S. Army, Vol. 2.* Washington, D.C., 1903.

Henry, Paul. *Ethan Allen and Allen & Wheelock.* Woonsocket, R.I., 2006.

Holland, Barbara. *Gentlemen's Blood.* New York, 2003.

Hosmer, Richard A. *The .58- and .50-Caliber Rifles and Carbines of the Springfield Armory, 1865-1872.* Tustin, Calif., 2006.

Houze, Herbert G. *Samuel Colt: Arms, Art and Invention.* New Haven, Conn., 2006.

———. *The Winchester Model 1876 "Centennial" Rifle.* Lincoln, R.I., 2001.

———. *Winchester Bolt Action Military and Sporting Rifles, 1877 to 1937.* Lincoln, R.I., 1998.

Huddleston, Joe D. *Colonial Riflemen in the American Revolution.* York, Pa., 1978.

Jinks, Roy G. *History of Smith & Wesson.* North Hollywood, Calif., 1977.

Kindig, Joe, Jr. *Thoughts on the Kentucky Rifle in Its Golden Age.* York, Pa., 1960.

Lewis, Col. Berkeley R. *Small Arms and Ammunition in the United States Service.* Washington, D.C., 1956.

Mallory, Franklin B. *The Krag Rifle Story.* Dover, Del., 1979.

Marcot, Roy M. *Civil War Chief of Sharpshooters Hiram Berdan, Military Commander and Firearms Inventor.* Irvine, Calif., 1989.

———. *Spencer Repeating Firearms.* Irvine, Calif., 1983.

McAulay, John D. *Civil War Breechloading Rifles.* Lincoln, R.I., 1987.

———. *Civil War Carbines Volume II.* Lincoln, R.I., 1991.

———. *Civil War Pistols.* Lincoln, R.I., 1992.

———. *Civil War Small Arms of the U.S. Navy and Marine Corps.* Lincoln, R.I., 1999.

———. *Rifles of the U.S. Army, 1861-1906.* Lincoln, R.I., 2003.

———. *U.S. Military Carbines.* Woonsocket, R.I., 2006.

McChristian, Douglas C. *The U.S. Army in the West, 1870-1880: Uniforms, Weapons, and Equipment.* Norman, Okla., 1995.

McCulloch, Ian M., and Timothy J. Todish, editors. *Through So Many Dangers: The Memoirs and Adventures of Robert Kirk, Late of the Royal Highland Regiment.* Fleischmanns, N.Y., 2004.

McDowell, R. Bruce. *Evolution of the Winchester.* Tacoma, Wash., 1985.

Madis, George. *The Winchester Book.* Lancaster, Tex., 1971.

Moller, George D. *American Military Shoulder Arms, Volume I.* Niwot, Colo., 1993.

Moore, C. Kenneth. *Colt Revolvers and the U.S. Navy, 1865-1889.* Jenkintown, Pa., 1987.

Neumann, George C. *Battle Weapons of the American Revolution.* Texarkana, Tex., 1998.

Noe, David, Larry W. Yantz, and James B. Whisker. *Firearms from Europe.* Rochester, N.Y., 1999.

Parsons, John E. *Henry Deringer's Pocket Pistol.* New York, 1952.

———. *The First Winchester.* New York, 1955.

———. *The Metropolitan Museum of Art Catalog of a Loan Exhibition of Percussion Colt Revolvers and Conversions, 1836-1873.* New York, 1942.

———. *The Peacemaker and Its Rivals.* New York, 1950.

Pate, Charles W. *Smith & Wesson American Model in U.S. and Foreign Service.* Woonsocket, R.I., 2006.

Peterson, Harold I., editor. *Encyclopedia of Firearms.* New York, 1967.

———. *Arms and Armor in Colonial America, 1526-1783.* Harrisburg, Pa., 1956.

Riling, Ray. *The Powder Flask Book.* New York, 1953.

Rodenger, Jeffrey L. *NRA: An American Legend.* Ft. Lauderdale, Fla., 2002.

Rollinson, John K. *Pony Trails in Wyoming.* E. A. Brininstool, ed. Caldwell, Idaho, 1941.

Roosevelt, Theodore. *The Wilderness Hunter.* New York, 1926.

Rosa, Joseph G. *The Gunfighter, Man or Myth.* Norman, Okla., 1982.

Russell, Carl P. *Firearms, Traps, and Tools of the Mountain Men.* New York, 1967.

Russell, Osborne. *Journal of a Trapper.* Aubrey L. Haines, ed. Lincoln, Neb., 1955.

Sellers, Frank M. *Sharps Firearms.* Dallas, Tex., 1978.

Sellers, Frank M., and Samuel E. Smith. *American Percussion Revolvers.* Ottawa, Ontario, 1971.

Serven, James E. *Colt Firearms from 1836.* La Habra, Calif., 1954.

Silva, Lee A. *Wyatt Earp, A Biography of the Legend, Volume I: The Cowtown Years.* Santa Ana, Calif., 2002.

Starr, Stephen Z. *The Union Cavalry in the Civil War, Volume I.* Baton Rouge, La., 1979.

Steele, J. R., and William R. Harrison. *The Gunsmith's Manual: A Complete Handbook for the American Gunsmith.* New York, 1883.

Swayze, Nathan L. *'51 Colt Navies.* Yazoo City, Miss., 1967.

Sword, Wiley. *Firepower from Abroad: The Confederate Enfield and the LeMat Revolver.* Lincoln, R.I., 1986.

———. *The Historic Henry Rifle.* Lincoln, R.I., 2002.

Trefethen, James B. *Americans and Their Guns.* Harrisburg, Pa, 1967.

Twain, Mark. *Roughing It.* New York, 1913.

Ware, Donald L. "Remington Army and Navy Revolvers 1861-1888." Unpublished manuscript.

Wilkerson, Don. *Colt's Double-Action Revolver Model of 1878.* Marceline, Mo., 1998.

Williamson, Harold F. *Winchester: The Gun That Won the West.* New York, 1963.

Wilson, R. L. *The Book of Colt Firearms.* Minneapolis, Minn., 1993.

———. *Theodore Roosevelt: Outdoorsman.* Agoura, Calif., 1994.

Wilson, R. L., and L. D. Eberhart. *The Deringer in America, Vol. I, The Persussion Period.* Lincoln, R.I., 1985.

Winant, Lewis. *Firearms Curiosa.* New York, 1955.

Worman, Charles G. *Gunsmoke and Saddle Leather.* Albuquerque, NM, 2005.

Periodicals

American Heritage Magazine

The American Rifleman

Collections of the Kansas State Historical Society

The Gun Report

Kansas History

Man at Arms

Military Collector & Historian

Museum of the Fur Trade Quarterly

The Rampant Colt

Remington Society of America Journal

Government Documents

Military Laws and Rules and Regulations for the Armies of the United States, Washington, D.C., 1806.

Index

Illustrations are in **boldface**.

Acknowledgments

One of the most pleasant aspects of historic firearms research one often encounters is the willingness of individuals to share their knowledge, information concerning items in their collections, and photos. I found this to be true in this as well as earlier writing projects. I've used this analogy before, but like the turtle one finds sitting on a fence post contentedly sunning itself, you know it had help getting up there. The following individuals and organizations have been most helpful to me "getting up there," and to them I offer sincere thanks.

Individuals: Bill Ahearn; Dale C. Anderson; Steve Augutis, N. Flayderman & Co., Inc.; Jerry A. Barrows, James D. Julia, Inc.; Erich Baumann; Joseph Bilby; Bob Butterfield; Michael F. Carrick; Richard T. Colson, Historian, Springfield Armory National Historic Site; Edward F. Cornett; Bill Curtis of the United Kingdom; Leland E. DeFord III; Gary L. Delscamp; John F. Dussling; Wayne T. Elliott; Robert Everhart; Fred Fellows; Norm Flayderman, always ready with assistance and encouragement; Ronald Gabel; Louis A. Garavaglia, earlier co-author and my long time "saddle pard"; S. James Gooding; James D. Gordon, quick to share his collection and wealth of knowledge; Neil and Julia Gutterman, whose Web site was often a source of photos; Charles L. Hill, Jr.; Richard K. Halter; George Hart; John J. Hayes, another friend always willing to share; Ralph Heinz; Paul Henry; Lloyd Jackson; Roy Kinzie; Dr. John J. Kudlik; Wade Lucas; John D. McAulay; Robert McNellis; Kenneth L. McPheeters; Roy Marcot; Greg Martin; Donald Mitchell; Harold R. Mouillesseaux; George C. Neumann; Robert P. Palazzo; Charles W. Pate; Ron Paxton, a photographer with skill and imagination; Herb Peck, Jr., whose passing was a loss to all; Ron Peterson; Jerry Pitstick; Rudi Prusok; Charles Rollins; Larry T. Shelton, Missouri State Capitol Museum; Jeff Sipling; C. W. Slagle; Rod Smith; Donald Snoddy, Union Pacific Railroad Museum; Dr. and Mrs. Thomas Sweeney; Thomas T. Trevor; Steven W. Walenta; Donald L. Ware; Dale E. Watts, Kansas State Historical Society; James Wertenberger; John A. Williams; Mrs. W. H. Wood; and Mark Wright.

Institutions: Arizona Historical Society; Colorado Historical Society; Colorado Springs Pioneers' Museum; Connecticut State Library; Denver Public Library; Fort Davis National Historic Site; Kansas State Historical Society; Idaho State Historical Society; Library of Congress; Los Angeles County Museum of Natural History; Mississippi Department of Archives and History; Montana Historical Society; Museum of the Fur Trade; National Archives; National Cowboy Hall of Fame; National Park Service, Chickamauga and Chattanooga National Military Park; National Park Service, Morristown National Historical Park; San Antonio Conservation Society; Smithsonian Institution; State Historical Society of North Dakota; State Historical Society of Wisconsin; Tombstone Courthouse State Historic Park; United States Military Academy, West Point; University of Oklahoma Library; Washington State Historical Society; and Wells Fargo Bank History Room.

And thanks to Bruce H. Franklin and his staff at Westholme Publishing who efficiently and cheerfully steered this project to completion.